TOMORROW'S PROFESSOR

IEEE Press
445 Hoes Lane, P.O. Box 1331
Piscataway, NJ 08855-1331

Editorial Board
Roger F. Hoyt, *Editor in Chief*

J. B. Anderson	A. H. Haddad	R. S. Muller
P. M. Anderson	R. Herrick	W. D. Reeve
M. Eden	G. F. Hoffnagle	D. J. Wells
M. E. El-Hawary	S. Kartalopoulos	
S. Furui	P. Laplante	

Kenneth Moore, *Director of IEEE Press*
Karen Hawkins, *Senior Editor*
Lisa Dayne, *Assistant Editor*
Linda Matarazzo, *Assistant Editor*
Savoula Amanatidis, *Production Editor*

IEEE Education Society, *Sponsor*
ED-S Liaison to IEEE Press, Robert Herrick

Technical Reviewers
Professor Pierre Belanger, *McGill University*
Dr. Albert Henning, *Redwood Microsystems*
Professor Michael Lightner, *University of Colorado*
Professor Robert Herrick, *Purdue University*

Also from the IEEE Press . . .

The New Engineer's Guide to Career Growth and Professional Awareness
edited by Irving J. Gabelman
1996　Softcover　300 pp　IEEE Order No. PP4119　ISBN 0-7803-1057-8

Managing Your First Years in Industry: The Essential Guide to Career Transition and Success
David J. Wells
1995　Softcover　200 pp　IEEE Order No. PP3707　ISBN 0-7803-1021-7

The Internet for Scientists and Engineers: Online Tools and Resources, 1996 Edition
Brian J. Thomas
Copublished with SPIE
1996　Softcover　520 pp　IEEE Order No. PP5379　ISBN 0-7803-1194-9

The Unofficial IEEE Brainbuster Gamebook: Mental Workouts for the Technically Inclined
compiled by Donald R. Mack
1992　Softcover　144 pp　IEEE Order No. PP3186　ISBN 0-7803-0423-3

TOMORROW'S PROFESSOR

Preparing for Academic Careers in Science and Engineering

Richard M. Reis
Stanford University

IEEE Education Society, *Sponsor*

IEEE PRESS

The Institute of Electrical and Electronics Engineers, Inc., New York

This book may be purchased at a discount from the
publisher when ordered in bulk quantities. Contact:

IEEE PRESS Marketing
Attn: Special Sales
445 Hoes Lane, P.O. Box 1331
Piscataway, NJ 08855-1331
Fax: (732) 981-9334

For more information about IEEE PRESS products,
visit the IEEE Home Page: http://www.ieee.org/

©1997 by the Institute of Electrical and Electronics Engineers, Inc.
345 East 47th Street, New York, NY 10017-2394

*All rights reserved. No part of this book may be reproduced in any form,
nor may it be stored in a retrieval system or transmitted in any form,
without written permission from the publisher.*

Printed in the United States of America
10   9   8   7   6   5

ISBN 0-7803-1136-1
IEEE Order Number: PP5602

Library of Congress Cataloging-in-Publication Data

Reis, Richard M. (date)
   Tomorrow's professor: preparing for academic careers in science
and engineering / Richard M. Reis; IEEE Education Society, sponsor,
   p. cm.
   Includes bibliographical references and index.
   ISBN 0-7803-1136-1
   1. Scientists—Vocational guidance—United States—Handbooks, manuals,
etc.   2. Scientists—Vocational guidance—Canada—Handbooks, manuals,
etc.   3. Engineers—Vocational guidance—United States—Handbooks,
manuals, etc.   4. Engineers—Vocational guidance—Canada—Handbooks,
manuals, etc.   5. College teachers—Vocational guidance—Handbooks,
manuals, etc.   6. College teaching—Vocational guidance—Handbooks,
manuals, etc.   I. IEEE Education Society.
II. Title.
Q149.U5R45   1997   96-37355
507'.1'1—dc21   CIP

To Nancy
who more than any one else brings out the teacher in me

Contents

Preface xiii

Acknowledgments xv

Introduction xvii

PART I Setting the Stage 1

Chapter 1 The Academic Enterprise 3
   Unlike Any Other Institution   4
   Key Characteristics   6
   Governance and Decision Making   13
   Institutional Issues   18
   A New Look at Scholarship   20
   Seven Sample Schools   22
   Vignette #1: A Place for Scholarship in Undergraduate Education   27
   Summary   29
   Appendix: Doctorate-Granting Institutions   30
   References   32

Chapter 2 Science and Engineering in Higher Education 37
   Comparisons Across the Institution   38
   Departments of Science   47
   Departments of Engineering   48
   Interdisciplinary Collaboration   51
   Scholarship Across the Disciplines   53
   Vignette #2: Science at a Metropolitan University   54
   Summary   56
   References   57

Chapter 3 New Challenges for the Professoriate 59
Forces for Change in Teaching and Research 60
Implications for Faculty Scholarship 71
Vignette #3: The Laboratory Without Walls 76
Summary 78
References 79

PART II Preparing for an Academic Career 81

Chapter 4 Your Professional Preparation Strategy 83
The Decision to Pursue an Academic Career 85
Supply and Demand—What is Going on Here? 90
The Three-Pronged Preparation Strategy 95
Vignette #4: A Ph.D. Career in Industry 102
Summary 104
References 104

Chapter 5 Research as a Graduate Student and Postdoc 107
Choosing a Graduate School or Postdoc Institution 108
Choosing a Research Topic 111
Choosing a Dissertation Advisor/Postdoc Supervisor 118
Writing Your Own Research Proposals 123
Carrying Out Your Research—An Example 125
Publishing 126
Attending Conferences and Other Professional Meetings 129
Presentations 131
Supervising Other Researchers 133
Managing Research Projects and Programs 135
Networking 137
Vignette #5: The Research Continuum 137
Summary 140
References 141

Chapter 6 Teaching Experiences Prior to Becoming a Professor 143
Why Teach as a Graduate Student or Postdoc? 144
Types of Teaching Experiences 148
How to Find the Right Teaching Opportunities 155
Preparing for a Successful Experience 156
Your Teaching Portfolio 157
Vignette #6: Teaching as a Postdoc 160
Summary 162
References 163

PART III Finding and Getting the Best Possible Academic Position 165

Chapter 7 Identifying the Possibilities 167
Explore Now, Search Later 168
Deciding What You Want 170
Researching What Is Out There 173
Preparing for the Search 176
Vignette #7: From Industry to Academia 177
Summary 180
References 180

Chapter 8 Applying for Positions 183
Setting the Stage 185
Preparing Your Application Materials 192
The Application Process 206
Positions Outside Academia 214
Vignette #8: Diversity Issues in the Hiring of Science and Engineering Faculty—An Illustration from Astronomy 216
Summary 217
References 218

Chapter 9 Getting the Results You Want 221
Your Negotiating Approach 222
General Principles for Responding to Academic Job Offers 224
Dual-Career Couples 232
What to Do if You Do Not Get the Offer You Want 233
Vignette #9: The Dual-Career Job Search 237
Summary 239
References 240

PART IV Looking Ahead to Your First Years on the Job – Advice from the Field 241

Changing Gears 243

Chapter 10 Insights on Time Management 245
Setting the Stage 245
Vignette #10: Establish Your Absence 246
Vignette #11: Set Long-Term Goals 248
Vignette #12: Keep Something on the Burner 250
Vignette #13: How to Help New Faculty Find the Time— One Department Chair's Approach 252

In Addition: Sources of Faculty Stress, Faculty
Efficiency, the Urgency Addiction, and Achieving
Balance in Our Lives   253
Conclusions   259
References   260

Chapter 11   Insights on Teaching and Learning   261

Setting the Stage   261
Vignette #14: Five Elements of Effective Teaching   266
Vignette #15: Developing Engaged and Responsive Learners   268
Vignette #16: Team Teaching in an Interdisciplinary Program   271
Vignette #17: The Upside-Down Curriculum   272
In Addition: Characteristics of Successful Teachers, Course
Planning, Teaching, and Learning with Technology, and
Developing a Teaching Portfolio   275
Conclusions   286
References   286

Chapter 12   Insights on Research   289

Setting the Stage   289
Vignette #18: Keeping Your Research Alive   294
Vignette #19: A High-Leverage Approach to Industry–
University Collaboration   296
Vignette #20: Multidisciplinary Research and the
Untenured Professor   298
Vignette #21: Cross-University Collaborations   300
In Addition: Writing Research Papers   302
Conclusions   305
References   307

Chapter 13   Insights on Professional Responsibility   309

Setting the Stage   309
Vignette #22: Service to Your Department and Your Profession   313
Vignette #23: Consulting and Other Industry Relationships   315
Vignette #24: Teaching and Learning Standards   318
Vignette #25: Professional Responsibility and Academic Duty   320
In Addition: Appropriating the Ideas of Others,
Conflict of Interest, and Freedom of Information?   322
Conclusions   325
References   325

Chapter 14   Insights on Tenure   327

Setting the Stage   327
Vignette #26: Leveraging Wherever Possible   333

Vignette #27: Understanding the Priorities   335
Vignette #28: A Second Chance at Tenure   337
Vignette #29: Taking Another Direction   340
Vignette #30: Lessons Learned   342
In Addition: The Ten Commandments of Tenure Success,
   Tenure as a Political Process, and Getting Help Along the Way   344
Conclusions   350
References   350

Chapter 15   Insights on Academia: Needed Changes   351

Helping Graduate Students and Postdocs Prepare
   for Academic Careers   351
Helping Graduate Students and Postdocs Find
   Academic Positions   354
Helping Beginning Faculty Succeed   353
Conclusions   361
References   362

PART V   Appendixes   363

Appendix A   Possible Items for Inclusion in a Teaching Portfolio   363
Appendix B   Statement of Personal Philosophy Regarding
             Teaching and Learning   367
Appendix C   Professional Associations for Academic Job
             Seekers in Science and Engineering   369
Appendix D   Questions to Ask Before Accepting a
             Faculty Position   373
Appendix E   Sample Offer Letters   381
Appendix F   Elements Found in Most Successful Proposals   387
Appendix G   Common Shortcomings of Grant Proposals   391

Index   393

About the Author   415

Preface

This book is intended primarily for graduate students and postdocs interested in academic careers in science and engineering. It should also be of interest to college juniors or seniors considering graduate school in one of these fields. In addition, I hope professional scientists and engineers in government and industry who are contemplating a return to academia as professors will profit from the material. The book should also be of benefit to beginning faculty, and to all faculty and administrators in a position to encourage and support those interested in becoming professors.

Schools of science and engineering produce a number of "products" of value to society. The first is graduates at the bachelor's, master's, and, in certain cases, doctoral level. The second is courses that can be taken in one form or another by industry employees. The third is all forms of scholarship including basic research, the integration and application of knowledge, and the development of new courses and methods of instruction. The key to all three of these products is a fourth product, professors, whom we want to be well prepared, highly motivated, and strongly supported.

There are approximately 1500 four-year institutions of higher education in the United States and Canada. Virtually all new faculty hires at these institutions, particularly at the assistant professorship level, have doctorate degrees. Of these 1500 institutions, approximately 250, or 17%, offer doctorates in one or more fields of science or engineering. These schools also employ approximately 55% of the total number of professors at four-year institutions. Thus, while the "producers" of Ph.D.s are also the "buyers" of Ph.D.s, the remaining 1250 schools also hire a significant number of Ph.D.s as professors. Of these schools, approximately 700 grant both bachelor's and master's degrees, while approximately 550 are liberal arts schools primarily offering four-year degrees.

This book represents a new way to help individuals prepare for, find, and succeed at careers as science or engineering professors. It derives from a course I teach at Stanford University. It also builds on my background as an engineer in industry and as a director of a nonprofit scientific and educational society. It further profits from my experiences as a college professor, career counselor, associate dean, and executive director of two Stanford University research centers with extensive relationships among graduate students, faculty, government, and industry.

I have taught at a variety of schools in the United States and Canada, including community colleges, institutions offering bachelor's and master's degrees, and those schools with a strong emphasis on research and the granting of doctorate degrees. Yet, as I began writing this book, it became obvious to me that I needed further information about both the schools students attended for their doctorates, as well as the other nondoctorate-granting institutions where many wished to go to pursue an academic career.

As a consequence, I identified as sources of more in-depth information seven sample schools in the United States and Canada covering the spectrum of institutions of interest to most future science and engineering professors. These schools are representative of the four major categories of four-year institutions defined by the Carnegie Foundation for the Advancement of Teaching. The schools and their classifications are: Bucknell University in Lewisburg, PA (Private Baccalaureate), Memorial University of Newfoundland in St. John's, Nfld. (Public Doctorate), the University of Michigan in Ann Arbor, MI (Public Research), the Rochester Institute of Technology in Rochester, NY (Private Master's), San Jose State University in San Jose, CA (Public Master's), Stanford University in Stanford, CA (Private Research), and the University of New Orleans in New Orleans, LA (Public Doctorate).

During a three-year period, I also talked at length with over 70 faculty, graduate students, and postdocs at some 20 additional institutions in all fields of science and engineering. Their comments and insights have proved invaluable. Indeed, quotes from most of these individuals appear in the pages that follow. Thirty of them are also the subjects of the vignettes appearing throughout the book.

Acknowledgments

I want to thank the following individuals for sharing their experiences and insights with me: Jim Adams, Emily Allen, Guy Blaylock, Diann Brei, Ralf Brinkmann, Amir Bukhari, Patricia Burchart, Mark Cutkosky, Elizabeth Drotleff, Renate Fruchter, Mahmoud Haddara, Eloise Hamann, John Hennessy, Charles Holloway, Dan Huttenlocher, Ben Knapp, Elizabeth Komives, Nety Krishna, Charles Kruger, Michael Kutilek, Tava Lennon, Al Levin, Martin Ligare, Michael Lightner, Susan Lord, Thomas Magnanti, Michele Marincovich, Joanne Martin, Robert McGinn, Susan Montgomery, Jim Patell, Jim Plummer, Bezhad Razavi, Michael Reed, Kirk Schulz, Noel Schulz, Alan Schwettman, George Springer, Pam Stacks, Robert Sutton, James Sweeney, Kelly Johansen-Trottier, Kody Varahrayan, Sally Veregge, Roger Verhelst, Sam Wood, and Candice Yano.

I would like a special note of appreciation to go to the students in the Stanford Future Professors of Manufacturing program who have contributed so much through their participation in my ongoing Proseminar in Manufacturing Education. They are: Dina Birrell, James Bradley, Kyle Cattani, Eliav Dehan, Wendell Gilland, Andrew Hargadon, Neil Kane, David Kasmer, Christopher Kitts, Jeff Kundrach, Constantinos Maglaras, Christopher Marselli, Mark Martin, David Owens, Sanjay Rajagopalan, Keith Rollag, Glen Schmidt, Steven Spear, and Jan Van Mieghem.

I owe a particular debt to the individuals behind the 30 vignettes: Thalia Anagnos, Ernest Boyer, Alison Bridger, Keith Buffinton, Mary Anne Carroll, Geoffrey Clayton, Jo Anne Freeman, Lance Glasser, Mark Hopkins, Paul Humke, Rollie Jenison, Donald Kennedy, Ruthann Kibler, Hau Lee, Lew Lefton, Paul Losleben, Brian Love, Nancy Love, Nino Masnari, Kim Needy, Greg Petsko, Shon Pully, Martin Ramirez, Joseph Reichenberger, Eve Riskin, Ulrike Salzner, Gerald Selter, Cheryl Shavers, Sheri Sheppard, Susan Smith-Baish, Susan Taylor, and Norm Whitley.

A number of individuals took the time to read portions of the manuscript covering their area of specialty or interest, and their comments were particularly helpful: Rafael Betancourt, Ron Bracewell, Claudio Coelho, Russ Hall, Stacy

Holander, Paul Hurd, Seymore Keller, Christopher Kitts, Michael Kutilek, Fritz Prinz, Evan Reis, Jim Schneider, Carolyn Tajnai, Robert Tschirgi, Decker Walker, and Steve Yencho. In addition, the following individuals read the entire manuscript, and in the process, offered in-depth comments and suggestions: Pierre Belanger, Yasser Haddara, Albert Henning, Michael Lightner, Vince Mooney, Ron Reis, Rick Vinci, and especially Robert Herrick, the IEEE Education Society Liaison to the Press.

I want to particularly acknowledge the contribution of Albert Henning. It was he, more than anyone else, who emphasized my need to go beyond the boundaries of Research I and II universities, to think deeply about the broader role of scholarship at colleges and universities, and to examine carefully the reality behind much of today's higher education rhetoric. His patient and thoughtful analysis of all my drafts has resulted in a far better final product than would have been the case without his contribution.

I want to thank Deanna Reis for her very helpful editing, and Elizabeth Drotleff for her editorial help and constructive input on Chapters 10–14.

Harrianne Mills provided very helpful suggestions with respect to graphic approaches. Kiersten Lammerding is mainly responsible for the layout, graphic design, and many of the illustrations in the book.

Vince Mooney was my research assistant who, in the process of gathering and synthesizing information, had the opportunity to talk with over 40 faculty at 20 different schools. As a result, he gained considerable insight into the workings of academia, and I gained valuable information, a broader network of faculty colleagues, and a good friend.

I also wish to acknowledge the financial support for this project from Stanford University's Center for Integrated Systems (Robert Dutton), Electrical Engineering Department (Joseph Goodman), and School of Engineering (Jim Gibbons and Jim Plummer).

Finally, let me point out that I see this book as the beginning of a process, not an end. While much of the material is time-independent, I expect that periodic updates of certain statistics and trends will be desirable. Also, material that could not be included in the book because of space limitations, such as additional vignettes or an elaboration of a process or procedure, should also be of interest to some readers. Furthermore, it is likely we could all profit from sharing information, opinions, ideas, and stories. To promote the above, I have created a World Wide Web Home Page (**http://cis.stanford.edu/structure/reis.html**) and an electronic mail address (**Reis@stanford.edu**). Through the Home Page, you can access updated information and comments from readers, while the e-mail address can be used for correspondence. I very much look forward to our exchanges.

Richard M. Reis
Stanford University

Introduction

We begin by setting the stage in Part I for the more specific work to follow. Chapter 1, "The Academic Enterprise," is a look at the unique characteristics of higher education. Graduate students and postdocs of course, have been part of such enterprises for some time. Yet, most lack an accurate understanding of how colleges or universities function, and the important ways in which they differ from other organizations in society.

The place of science and engineering in academia is examined in Chapter 2, "Science and Engineering in Higher Education." Here, we look at the similarities, and more importantly, the differences among various science and engineering departments and disciplines. We also discuss the prospects for cross-disciplinary collaboration among these various fields.

Part I concludes with Chapter 3, "New Challenges for the Professoriate." It begins by examining the significant forces currently impacting higher education. These include the prospects of decreasing government funding, the changing relationship between industry and academia, the increasing use of communication and computational tools, the rising costs of doing research, and the greater focus on interdisciplinary programs. We then discuss the implications the above factors have for faculty scholarship and the preparation of tomorrow's professors of science and engineering.

Part II, "Preparing for an Academic Career," begins with Chapter 4, "Your Professional Preparation Strategy." Here, we explore the decision to pursue an academic career, particularly in light of the current situation with respect to supply and demand. We then outline a three-pronged preparation strategy to prepare you for an academic career while maintaining your options for careers in government and industry.

Chapter 5, "Research as a Graduate Student and Postdoc," looks at how to apply the above strategy to the many research activities that you need to complete prior to becoming a professor. Some activities, such as choosing a research topic and identifying an advisor, are required of all Ph.D. students and postdocs, while others, such as writing proposals and supervising other researchers, although not specifically required, will nevertheless put you ahead of most of your competition in looking for academic, government, and industry positions.

Part II concludes with Chapter 6, "Teaching Experiences Prior to Becoming a Professor." The chapter begins with a discussion of the benefits of acquiring such experiences, followed by a look at some innovative ways to go beyond teaching assistantships to developing and presenting lectures, conducting laboratory sessions, and even teaching full courses at your own or another institution. The chapter concludes with a discussion of the teaching portfolio, and how it can be used to capture the successes of your teaching experiences for presentation to potential employers.

How to find, and then get, the best possible academic position is the subject of Part III. As the supply and demand for new assistant professorships in science and engineering shift, anxiety among graduate students and postdocs about obtaining an academic appointment increases. However, as will be seen in Chapter 4, the situation is more complex, and in many instances more positive, than recent headlines would suggest. At the very least, if you follow the strategy discussed in Chapters 4–6, you are apt to be in a much more competitive position than many of your colleagues who have not done so.

It is important to keep in mind that finding "a" job is not the issue; finding the "best" possible academic position, the right one for you and for the hiring department and school, is the real goal. Detailed suggestions on how to do all of the above are the subjects of Chapter 7–9.

Chapter 7, "Identifying the Possibilities," explains that in seeking an academic position, it is essential for you to explore before you search. You need to compare what is available (types of institutions, positions, and locations) with what you need and want (capabilities, interests, and values). Only then will you be in a position to search–apply for specific jobs.

Chapter 8, "Applying for Positions," is a detailed discussion of the job search process. Here, we examine how new faculty positions are established, what departments look for in new faculty, how to find out what is available, and the time frame for academic openings in your field. We then discuss the preparation of your application materials, including cover letters, curriculum vitae, and letters of recommendation. Conferences, campus visits, and the all-important academic job talk are looked at next. The chapter concludes with an examination of jobs outside academia, and how you can accept one of them while keeping your options open for a future academic position.

Chapter 9, "Getting the Results You Want," begins with a look at the negotiation process by examining in some detail the principles you need to use in responding to academic job offers. We then explore some of the special problems faced by dual-career couples, in particular those in which both members are seeking faculty positions. The chapter concludes by discussing what to do if you did not receive an academic job offer, or received one that is unacceptable to you.

Your shift from a graduate student or postdoc to beginning professor is going to be exciting and dramatic. At such a time, the most valuable thing you could probably do would be to ask a half dozen professors at the institution to which you are going what they feel it takes to succeed as a beginning professor. If you pick the right assistant, associate, and full professors, the advice you receive could be invaluable. In Part IV, "Looking Ahead to Your First Years on the Job—Advice from the Field," we do the next best thing by capturing insights for success in five key areas from science and engineering professors across North America. The five areas are: time management (Chapter 10), teaching and learning (Chapter 11), research (Chapter 12), professional responsibility (Chapter 13), and tenure (Chapter 14).

Each chapter begins with an introduction, followed by four or five vignettes on faculty who provide insights on the theme under discussion. These vignettes are followed by a detailed, "In Addition" section describing other sources you can turn to for further information and understanding. The chapters conclude with a section summarizing the main ideas from both the vignettes and the readings.

While you can earnestly follow the suggestions described in Chapters 1–14, unless academia does its share to support you, your success could be limited. Changes are required to help graduate students and postdocs obtain meaningful teaching experience, participate directly in the research development process, obtain the best possible academic position, and then succeed in their chosen careers as faculty. The book concludes with Chapter 15, "Insights on Academia: Needed Changes." It suggests ways in which administrators and senior faculty can provide an environment that will enable tomorrow's professors to prepare for, find, and succeed at academic careers in science and engineering.

SETTING THE STAGE

CHAPTER 1

The Academic Enterprise

George P. Shultz, former U.S. secretary of state, of labor, and of treasury, was also a senior officer in the Bechtel Corporation, and former dean of the Business School at the University of Chicago. He is currently on the faculty at the Stanford University Graduate School of Business. Shultz was asked recently to compare the three types of organizations in which he had spent so much time; industry, government and academia. He replied, "When I worked in industry I had to be careful if I asked someone to do something because there was a very good chance they would do it. When I worked in government I didn't have that problem. But at the university I very quickly came to understand that it was... inappropriate to ask"[1].

Tongue-in-cheek as his comment may be, Shultz is hinting at something important about how colleges and universities differ from industry, government, labor unions, churches, hospitals, and virtually every other institution in society.

Clark Kerr, president emeritus of the University of California, supports Shultz's point in a more formal way by noting that [2]:

[American colleges and universities] have mostly been comparatively privileged entities of tolerant societies exercising great self-restraint toward them. And their principal participants—the faculties—have had more leeway to conduct their lives according to their individual wishes than most other members of the modern labor force—they have not viewed themselves, or been viewed by others, as "employees." It has been a world of comparative institutional autonomy and comparative individual academic freedom.

As a possible future professor, it is important for you to understand the unique features of an institution in which you may spend the rest of your professional life. We begin the development of such an understanding in this chapter, first with a brief look at the evolution of higher education in North America. This historical discussion is followed by an examination of the key characteristics of academia, including governance and decision making. Some of the critical issues currently facing all colleges and universities are examined next. A new concept of scholarship originally proposed by Ernest Boyer, former president of the Carnegie Foundation for the Advancement of Teaching, is then introduced. This scholarship concept forms the basis for important discussions in the chapters to follow. We then introduce seven sample schools representing the types of four-year institutions to which most science and engineering Ph.D.s and postdocs will go as new professors. The chapter concludes with a vignette describing Ernest Boyer's views on the role of scholarship in undergraduate education.

UNLIKE ANY OTHER INSTITUTION

With all the downsizing and restructuring taking place in higher education, you might think colleges and universities are looking more, not less, like other institutions. Hahnemann University in Philadelphia, PA is a case in point. The Hahnemann administration recently threatened to fire any faculty member, tenured or not, who is not able to attract research grants providing between 50 and 100% of his or her salary. As Leonard L. Roos, dean of the Hahnemann School of Medicine, put it, "If IBM expects that of its employees, why can't we expect it of the academic community? It's a big business" [3].

Another industry-like characteristic, increasing demands for accountability and productivity, has resulted in mandated minimum college and university teaching loads in some states. Hawaii and Florida, for example, now require 12 hours of classroom instruction per week or the equivalent for faculty in four-year institutions [4].

On the other hand, industry has reduced its number of management levels, put more decision making in the hands of those who actually do the "value-added" work, sought consensus across functions, and so on. Could it be that private enterprise is taking on some of the characteristics long associated with colleges and universities? Perhaps, but fundamental differences remain in the culture, governance, mission, methods of generating income, employment security, and accountability between academia and other organizations with which we are familiar.

Historical Perspective

Before looking more closely at these differences, let us consider a little history. Higher education in the United States and Canada began during the 17th century as an outgrowth of both the medieval European universities and the British universities of Oxford and Cambridge. In these so-called colonial colleges, teaching was central. It was viewed, "... as a vocation—a sacred calling—an act of dedication honored as fully as the ministry" [5, Ch.1, p.4]. It was during this time that the self-governing nature of universities developed, as well as the idea that universities were "communities of scholars" [6, p.3]. It was also during this period that the notion of "Town and Gown" developed as a way of "separating" scholars from the local lay population [6, pp.22–23].

The number of institutions and students remained small until the passage in the United States in 1862 of the Morrill Act establishing land grant colleges and universities. Through this act, every state was granted 30,000 acres of land for each senator and representative it had in Congress. The land was then to be sold and the proceeds invested to create and maintain institutions that were to emphasize agriculture and mining (A&M) as a way to produce better educated farmers and engineers. The universities of Arizona, California, Illinois, Texas, and Washington are just a few such institutions formed during this period. The late 19th century was a time when colleges were to provide "useful studies," and when "going to college" was viewed as a way of "getting ahead." As one undergraduate put it in 1871, "A degree from Harvard is worth money in Chicago" [6, p.29]. By the end of the century, 59 separate land grant colleges had been established in 44 states under the Morrill Act [7]. Many of you have attended, are now attending, or will eventually teach at such institutions.

A second significant advance occurred in the 1890s with the establishment of research-oriented private universities such as Johns Hopkins, Chicago, Cornell, and Stanford [8]. A further growth period occurred after World War I with the passage of additional legislation and the involvement of state universities in large-scale applied research.

However, the Golden Age of higher education was clearly the one during the three decades following World War II. The 1950s and 1960s was a period of unprecedented expansion, both in the size of existing institutions and the number of new institutions. In the United States, it was a result of federally funded research, an outgrowth of experiences at the Massachusetts Institute of Technology's Lincoln Laboratory (radar) and the University of Chicago (atomic bomb) during the war. Expansion was also due to the GI Bill and subsequent equal opportunity funding initiatives, and the requirements of a labor force trained in emerging engineering fields, particularly electronics and computers [2, p.22].

Historian John Thelin put it this way [9]:

> By 1965 one could speak of an "academic revolution" in which American society had come to rely on and accept the expertise of colleges and universities, indicative of an "information society" whose foundation was a "knowledge industry." Student enrollments had grown, both in actual numbers and as a proportion of total population, such that higher education had been transformed from an elite to mass access.

During the period right after World War II to the early 1970s, the number of college and university professors and students approximately tripled. The number of institutions also grew, as did the number of graduate programs [10, p.229]. The most rapid growth in faculty occurred in the 1960s.

Most of these faculty will soon retire, a fact clearly relevant to those of you considering academic careers. You should be aware, however, that anticipated increase in demand expected from such retirements will be at least partly offset by advances in teaching productivity through instructional television and computers, the increasing use of part-time faculty, particularly at the community college level, and the downsizing or even elimination of some departments. We will discuss the supply and demand topic in greater detail in Chapter 4, "Your Professional Preparation Strategy."

KEY CHARACTERISTICS

There are currently about 3600 (1600 public, 2000 private) accredited institutions of higher education in the United States, up from approximately 3400 in 1987, the last time a survey of such institutions was conducted by the Carnegie Foundation for the Advancement of Teaching [11, p.A17]. These institutions enroll approximately 14.5 million students and award slightly over two million degrees, a quarter of which are in science and engineering [12, pp.2/6–2/10]. While there is considerable variation among fields, Ernest L. Boyer, the Foundation's late president, points out that overall: "There is now more higher education than ever in history... and predictions of decline are simply not supported by the facts" [14].

A similar growth pattern exists in Canada. Current full and part-time enrollment in higher education exceeds 870,000, up from 630,000 in 1980–1981 [13, p.29]. Canada currently has 80 institutions offering bachelor's, master's, and doctorate degrees [13, pp.64–67].

The Carnegie Classification

The Carnegie Foundation for the Advancement of Teaching groups accredited U.S. institutions into 11 categories based largely on their mission. The categories are: Research universities I&II (Res. I&II), Doctoral universities I&II (Doc. I&II), Master's (comprehensive) universities and colleges I&II (MA I&II), Baccalaureate (liberal arts) colleges I&II (BA I&II), Associate of Arts colleges (AA), Professional schools (Prof.), and specialized institutions (Spec.). (Note: A new term, Metropolitan University, not formally part of the Carnegie classification, has recently come into use among a number of Master's institutions located in urban areas.) Table 1-1 describes the basis for these categories. Institutions are classified according to the highest level of degree they award, the number of degrees conferred by discipline, and, in some cases, the amount of federal research support they receive, and the selectivity of their admissions. Table 1.2 lists the number of schools by Carnegie classification. Figure 1-1 shows the proportion of institutions by category. Figure 1-2 summarizes the above informa-

Table 1.1 The 1994 Carnegie Classification of Definitions for Four-Year Institutions

	CLASS I	CLASS II
Research Universities	These institutions offer a full range of baccalaureate programs, are committed to graduate education through the doctorate, and give high priority to research. They award 50 or more doctorate degrees each year. In addition, they receive annually $40 million or more in federal support.	These institutions offer a full range of baccalaureate programs, are committed to graduate education through the doctorate, and give high priority to research. They award 50 or more doctorate degrees each year. In addition, they receive annually between $15.5 million and $40 million in federal support.
Doctoral Universities	These institutions offer a full range of baccalaureate programs and are committed to graduate education through the doctorate. They award at least 40 doctorate degrees annually in five or more disciplines.	These institutions offer a full range of baccalaureate programs and are committed to graduate education through the doctorate. They award annually at least 10 doctorate degrees—in three or more disciplines—or 20 or more doctorate degrees in one or more disciplines.
Master's (Comprehensive) Colleges and Universities	These institutions offer a full range of baccalaureate programs and are committed to graduate education through the master's degree. They award 40 or more master's degrees annually in three or more disciplines.	These institutions offer a full range of baccalaureate programs and are committed to graduate education through the master's degree. They award 20 or more master's degrees annually in three or more disciplines.
Baccalaureate (Liberal Arts) Colleges	These institutions are primarily undergraduate colleges with major emphasis on baccalaureate degree programs. They award 40% or more of their baccalaureate degrees in liberal arts fields and are restrictive in admissions.	These institutions are primarily undergraduate colleges with major emphasis on baccalaureate degree programs. They award fewer than 40% of their baccalaureate degrees in liberal arts fields or are less restrictive in admissions.

Source: The Carnegie Foundation for the Advancement of Teaching, *A Classification of Institutions of Higher Education*, 1994 ed. Reprinted with permission.

Table 1.2 Number of Colleges and Universities by Carnegie Classification, 1994	
INSTITUTIONS	TOTAL
Doctorate-granting	236
Research I	88
Research II	37
Doctorate I	52
Doctorate II	59
Master's-granting	532
MA I	439
MA II	93
Baccalaureate-granting	633
BA I	163
BA II	470
Associate of Arts colleges	1,480
Professional schools and specialized institutions	690
Tribal colleges	29
Total	3,600

Source: Carnegie Foundation for the Advancement of Teaching, *A Classification of Institutions of Higher Education*, 1994 ed. Reprinted with permission.

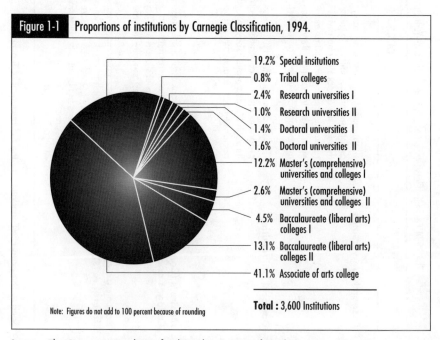

Figure 1-1 Proportions of institutions by Carnegie Classification, 1994.

- 19.2% Special insitutions
- 0.8% Tribal colleges
- 2.4% Research universities I
- 1.0% Research universities II
- 1.4% Doctoral universities I
- 1.6% Doctoral universities II
- 12.2% Master's (comprehensive) universities and colleges I
- 2.6% Master's (comprehensive) universities and colleges II
- 4.5% Baccalaureate (liberal arts) colleges I
- 13.1% Baccalaureate (liberal arts) colleges II
- 41.1% Associate of arts college

Total: 3,600 Institutions

Note: Figures do not add to 100 percent because of rounding

Source: The Carnegie Foundation for the Advancement of Teaching

tion, and also includes data on the enrollment of students and degrees granted. (Note: The small differences in institutional totals between Table 1.2 and Figure 1-2 are due to differences in reference dates.)

Canada does not use the Carnegie classification, although it is not that difficult to "assign" Canadian schools to the Carnegie categories. We will refer to the Carnegie classification often throughout this book. It provides a convenient way to examine the characteristics of colleges and universities of interest to you as a future science or engineering professor. While science and engineering teaching takes place at almost all types of colleges and universities in the classification, we will concentrate on those 1500 U.S. and Canadian schools offering four or more years of higher education, i.e., Res. I&II, Doc. I&II, MA I&II, and BA I&II. Virtually all of these schools require a doctorate degree of their newly hired faculty.

Where New Faculty Come From—Where New Faculty Go

The Appendix lists all 236 U.S. doctorate-granting institutions. These schools award approximately 41,000 doctorates per year, with the top 35 schools awarding 16,874, or 41.5% of the total. It is from these schools that essentially all U.S.-educated science and engineering faculty will come, but it is certainly not where all of them will go. Of all the faculty at four-year institutions, approximately 55% are at Research and Doctoral schools and 45% are at Master's and Baccalaureate schools [14].

An interesting example of where faculty come from and where they go can be found in my own academic neighborhood. Stanford University and the University of California, Berkeley (UCB) are both Research I institutions. San Jose State University is a Master's I institution. All are within 50 miles of each other. Stanford has 221 full-time engineering faculty of which 109, or 49%, received their Ph.D.s from either Stanford or UCB. Yet, at San Jose State University with a total of 97 engineering faculty, 36, or 37%, are also from Stanford or UCB.

Of course, some of you now attending Research universities as graduate students and postdocs attended other types of schools as undergraduates. You have been exposed to Master's and Baccalaureate schools, but probably at a time when you were not yet considering an academic career. The key point in all of the above is this:

> **Graduate students and postdocs preparing for academic careers must consider not only the 250 or so schools from which new faculty come, and to which, of course, a number return, but the other 1250 or so schools to which almost half will go as new professors.**

Teaching and Research Emphasis

The relative importance of teaching and research varies by type of institution. As will be discussed later, there is increasing talk about putting more emphasis on all forms of scholarship, including teaching at research universities. However,

Figure 1-2 U.S. higer education in 1993: students, institutions, and degrees.

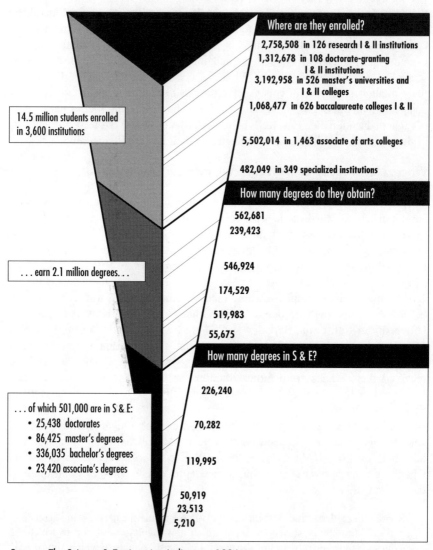

Source: The Science & Engineering Indicators, 1996

most of this talk comes from university presidents and deans, not from faculty. As Boyer points out, "Almost all colleges pay lip service to the trilogy of teaching, research, and service but when it comes to making judgments about professional performance, the three are rarely assigned equal merit" [5, Ch.2, p.15]. Whether or not the institutional incentives and faculty reward systems can be restructured to bring about this change is a matter of much debate. We will look more closely at such efforts at various points throughout this book.

While some Research universities, in at least some departments, do an excellent job of teaching undergraduates, generally speaking, more emphasis on teaching occurs in Master's and Baccalaureate schools. Evidence for this difference can be seen in how faculty from each type of school responded to questions about the importance of teaching and research in the awarding of tenure at their institution.

Table 1.3 shows the percentage of faculty who replied "very important" in responding to a series of questions on this subject. These results indicate that research publications and grants are perceived as much more important in tenure decisions at Research and Doctorate-granting institutions, with the reverse being true with respect to teaching at Master's and Baccalaureate (liberal arts) schools.

There is a tendency among some academics to view the Carnegie Classification as a hierarchy topped by selective liberal arts colleges and major research universities. Boyer discounts this view, pointing out that the classification is not an attempt to build a pyramid in terms of quality. "It doesn't talk about

Table 1.3 Impact of Teaching and Research on Tenure Decisions at Various Types of Colleges and Unversities

QUESTIONS	Percentage of faculty who answered "very important" to the following questions, by type of institution:			
	RES. I&II	DOC. I&II	MASTER'S I&II	L.A. I&II
How important is the number of publications for granting tenure in your department?	56	55	30	8
How important are research grants received by the scholar for granting tenure in your department?	40	35	19	9
How important are student evaluations of courses taught in granting tenure in your department?	10	9	37	45
How important are observations of teaching by colleagues and/or administrators for granting tenure in your department?	4	6	20	29
How important are recommendations from current or former students for granting tenure in your department?	3	6	13	30

Source: E. L. Boyer, *Scholarship Reconsidered: Priorities of the Professoriate*. Princeton, NJ: The Carnegie Foundation for the Advancement of Teaching, 1990, Appendix A—National Survey of Faculty, 1989. Reprinted with permission.

quality, or hierarchy in terms of good or bad," he states. "It is not a measure of creativity or innovation. It talks about the level of complexity of program. It doesn't do more, and it shouldn't do more. It's a beginning point, not an end point" [11, p.A17].

Boyer's comments not withstanding, faculty at a number of schools are feeling the pressure to help their institution "move up" in the Carnegie categories. From 1987 to 1994, there was a total of 433 category changes. Sixteen Research II institutions shifted, all to the Research I category. Eighteen Doctoral institutions shifted, one to Research I, 16 to Research II, and one "down" to Doctoral II. The shifts among MA I&II and BA I&II are not as easy to interpret since what is "up" and "down" is less clear in these categories [11, p.A20].

During the last seven years, the University of Alabama at Birmingham moved "up" three steps from the Doctoral II category to the Research I category. Commented Kenneth J. Rooren, executive vice president at Birmingham, "We've been striving to get to the top...We didn't develop a strategy to become a Research I institution, but we did develop a strategy to gain excellence and breadth" [11, p.A17].

Recently, the University of California, Irvine, announced plans to "vault" into the ranks of the top 30 research universities by expanding its presence in biomedicine, the neurosciences, and related fields [15].

A more disparaging view of this situation was expressed by a faculty member at a public institution in New York, who noted that her university had gone from "state supported, to state assisted," and this had put tremendous pressures on what was basically a teaching university to raise indirect cost support through research grants.

The teaching–research balance can also vary quite a bit among institutions of the same type, and even within a specific institution. A professor of biology at a large Master's I institution commented that her department has always had a strong teaching emphasis because the senior professors who were hired in the 1960s, when teaching was the dominant activity, continued to hire like-minded faculty in the 1970s and 1980s. She noted, however, that the Chemistry Department chairman was a "U.C. wannabe," U.C. being the University of California, and that his department had a much stronger research emphasis in its retention, tenure, and promotion process.

The above examples raise a number of important questions. What is actually meant by excellence and breadth? If more schools move "up" than "down," does this mean that total research funding has increased? What does this imply about future success if research support from the government decreases? Is there a measurable trend toward better teaching, and if so, how does this relate to tenure and promotion? We will look more closely at these questions in later chapters. The point to be made here is that, as a future professor, it is important for you to find out about the future goals, as well as the funding history, of an institution or department before accepting a position for which there might be quite different expectations and responsibilities two–five years down the road.

GOVERNANCE AND DECISION MAKING

Organizational Structure

Although there is some variation from one institution to the other, most colleges and universities have similar governance structures. As an example, consider the University of Arizona (Figure 1-3). The president, who reports to the Arizona Board of Regents, is the chief executive officer, chief spokesman, and chief fundraiser for the university. The senior vice president for academic affairs and provost is the chief academic officer. The senior vice president for business affairs is the chief fiscal and operations officer for the university. Both report to the president. The president and the senior vice president for academic affairs and provost are both tenured faculty members. Four vice presidents with responsibilities appropriate to their titles report to the senior vice president for academic affairs and provost. They are the vice president for academic services and undergraduate education, the vice president for institutional planning, the vice president for research and graduate studies, and the vice president for student affairs. None of these four must be tenured faculty members. The deans of the 11 colleges, all of whom are tenured professors, report to the senior vice president for academic affairs and provost. The department chairs report to the various deans, and the faculty report to the department chairs [16].

Figure 1-3 looks like the typical top-down organization chart you would expect to find in most private sector companies. Such is not the case. In most Research and Doctoral institutions, there is very little top-down governance and control. The real power lies with the faculty, and decision making is from the bottom up.

Gerhard Casper, president of Stanford University, explains the politics of academia this way:

> Many people think the universities are hierarchical because they have a president with a fancy title, and three or four levels of professors, but they are not hierarchical. Power comes from the bottom up. The most important decisions are those concerning admissions, curriculum and faculty appointments, and these are areas where the university president has almost no power. In most circumstances, I'm the man with the pail and broom [17].

There is more top-down control at some master's and baccalaureate institutions, yet even here, faculty usually have considerable decision-making authority. Generally speaking, department chairs, deans, provosts, and presidents have the power to block some decisions made by the faculty, for example, those concerning retention, tenure, and promotion. In certain cases, they can also reallocate resources such as space and faculty billets. They can also force cuts in administrative and support services. However, for the most part, decisions about curriculum, grading, and research policies are made by faculty committees.

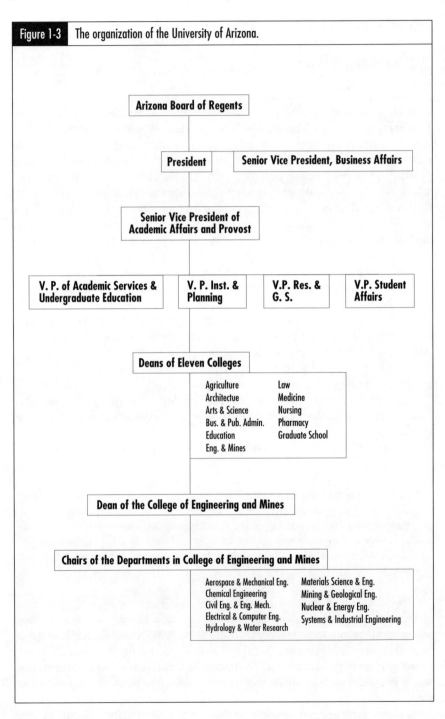

Figure 1-3 The organization of the University of Arizona.

Some of these committees operate at the department, or even subdepartment level. Others operate at the school or college level, and still others, such as the faculty senates, at the institution level. Some are technically advisory to the chairs, deans, and provost, and can thus be overruled by these people, but most are true decision-making bodies.

The Multiuniversity

One of the biggest changes in higher education governance over the last 30 years has been the move from single-campus governance to large, complex, and heterogeneous multicampus systems [18]. Such systems involve two or more campuses with a single state-wide governing board. Three-fourths of all students in public higher education in the United States attend a multicampus university [19, pp.358].

There are pluses and minuses to such organizations. On the one hand, state-wide governing boards can often provide a broader, longer term perspective on what is important to students across the state. They may be in a better position to decide what to eliminate or preserve among the programs of the various campuses. They should also be able to play a role in fostering intercampus programs. On the other hand, they can pose a threat to campus autonomy, and can make faculty-shared governance more difficult. How these conflicting interests will play out over the next few years is unclear. As Ami Zusman, coordinator for academic affairs, Office of the President, University of California System, notes [19, pp.358–359]:

> During the 1990s, systemwide leadership (boards, administrators, and faculties) will be particularly important in responding to an environment of continuing budget constraints, demands for greater institutional accountability, questions of student access, and other changing conditions... Whether systemwide leadership will be able to exercise initiative is uncertain, however in light of both campus concerns for autonomy and state demands for more direct control.

The Institution of Tenure

Nothing distinguishes academia from other organizations in society more clearly than the institution of tenure. The 1940 Statement of Principles on Academic Freedom and Tenure, drafted by the Association of American Colleges, sets out the reasons for academic tenure this way [20]:

> Tenure is a means to certain ends; specifically: (1) freedom of teaching and research and of extramural activities and (2) a sufficient degree of economic security to make the profession attractive to men and women of ability. Freedom and economic security, hence, tenure, are indispensable to the success of an institution in fulfilling its obligations to its students and to society.

This statement was drafted against the background of the Great Depression. Today, you can hear arguments that tenure is no longer needed, and protection against unwarranted firings is now accomplished through legislation applying to employees of all institutions. Still others have argued that, in the parlance of modern business and social language, tenure has enabled dysfunctional behavior by allowing some tenure faculty to be shielded from accountability for their decisions and actions. Nevertheless, tenure is an institution "that professors around the world take to be the prime guarantor of their freedom to seek and deliver truth" [21]. Do not look for it to disappear soon. As Philip Altbach, professor of higher education at Boston College, puts it, "The Professoriate sees tenure as one of its most important perquisites and has defended it vigorously. Administrators and policy makers have recognized the centrality of tenure to the self-concept of the profession" [10, Ch.10, p.237].

This being the case, tenure is something you will need to pay a great deal of attention to in your first years as a new professor. We will look at tenure, and how to improve your chances of getting it, in Part IV, "Looking Ahead to Your First Years on the Job—Advice from the Field."

Power and Money

In a given college or university, power lies less with the consumers of resources than with the providers of resources. This reality explains why the largest departments are not always the most powerful, i.e., have the greatest prestige, are the most listened to by other departments, and have the most faculty on university-wide committees.

Income enters academia from many sources: tuition, funding from state legislators in the case of public institutions, the federal government for both public and private institutions, interest on endowment, and donations from alumni and friends. How resources flow into a college or university affects where the institution puts its emphasis. As Gerald Salancik of Carnegie-Mellon University points out, "Prestigious private universities are more likely than prestigious public universities to have outstanding law and business schools, and prestigious public universities are more likely to have outstanding engineering and agricultural schools" [22, p.66]. As we noted earlier in this chapter, these differences reflect the historical basis of these institutions. According to Salancik [2, p.68]:

> Private schools, lacking the subsidies of a state appropriation process, operate very differently than public universities. They survive by accumulating endowment, which depends on their graduates' wealth and willingness to part with it as they age. Not surprisingly, private schools pay a lot more attention to their undergraduates' experiences. They want their students to leave with warm feelings and fond memories.

It has been found that high-level administrators, such as directors of development, were more highly compensated in private colleges and universities, where presumably their results were more valued, than in public colleges and universities. The reverse was true for athletic directors [23].

Nevertheless, tuition and endowment are almost never enough to balance the books. For example, at Stanford University, tuition, which now runs close to $25,000 per year, still covers only half the University's annual operating budget. As a consequence, doctorate granting institutions and, increasingly, other four-year institutions are relying on research grants to help support the enterprise.

Grants from government and industry are, of course, designed to pay for faculty and graduate student salaries, equipment, and travel associated with doing research. However, this use is not the only reason such grants are sought. Universities have an agreement with the U.S. federal government by which the latter will fund both the direct and indirect costs of the research it chooses to support. These costs include the obvious, such as portions of faculty salaries, student research assistantships, laboratory fees, and travel to conferences to present papers on the research. They also involve an overhead amount for indirect costs, those costs that cannot be identified directly with a research activity, but which nevertheless contribute to its operation. These include the costs of libraries to buy and house scientific journals, heating and lighting buildings that researchers use, and even a portion of the costs of roads and other university maintenance. Each university negotiates its indirect cost rate with the government. For private universities, this can run as high as 50–60% of the direct costs of a research grant.

The value of overhead funds is that they are *discretionary,* and can be used by campus administrators to fund a variety of projects not directly related to a specific research grant. As Jeffrey Pfeffer, professor of organizational behavior at the Stanford Graduate School of Business, puts it [24]:

> The most precious resource in any organization is an incremental resource, not already spoken for, that can then be used to solve the organization's current problems —problems that are more difficult to address using the current resources because of the conflict involved in reallocation.

Not surprisingly, Salancik and others have found that a department's power is proportional to its contribution to the overhead pool. Indeed, they determined that the major predictors of the differences between the power of departments was the grant monies they each provided [22, p.68].

There is now some indication that the U.S. government may no longer fund the full cost of doing sponsored research. With full-cost recovery, the university did not really care what research faculty "chose" to do since the university always had its costs covered. If full cost recovery is no longer the case, what impact will it have on the research a university "chooses" to support? The

Canadian government has also changed the way it funds provincial education, and this change is having a significant impact on its colleges and universities. We will discuss these and other factors in greater detail in Chapter 3, "New Challenges for the Professoriate." We will also look much more closely at the importance of such grants to beginning professors in Chapter 5, "Research as a Graduate Student and Postdoc."

INSTITUTIONAL ISSUES

It is unlikely that you are hearing the term Golden Age on today's college and university campuses. Higher education is currently facing unprecedented pressures and challenges. Here is just a brief list of some of the recent institutional issues making headlines around the country [25]:

- Demands for multiculturalism in the curriculum

- Scrutiny of racially based undergraduate admissions quotas

- Attempts to control violence and hate crimes on campus

- Graduate student teaching assistants going on strike

- Increased prominence of university–industry partnerships and technology transfer activities

- Occasional allegations of scientific misconduct/fraud

- The alleged mismanagement of indirect cost funds

- Dramatic consequences from massive budget cuts brought about by a national economic downturn, and often accompanied by explicit "downsizing" mandates from state legislatures

You can add to the above list anxiety about federal funding, increased global competition, and the current difficulties of graduate students and postdocs in finding industry and academic positions. We see these consequences in downsizing, outsourcing, and greater demands for accountability at all levels of higher education. Everywhere there is a move to consolidate and focus. As one university president noted, "My university can do anything, but it can't do everything."

A computer science professor at a large Master's institution put it more personally when he said:

I've been asked to do four things: improve the quality of my teaching, teach more students, and do it all in less time, and for less cost. I know how to do any three, if I don't have to worry about the fourth. But I can't figure out how to do all four at the same time.

In addition to the retrenchment and reallocation issues noted above, colleges and universities are facing a number of other challenges. Student body compositions and expectations are changing. In general, students are more conservative than they were 20 years ago, and they, and their parents, are demanding more relevance in the curriculum and greater attention to the quality of undergraduate teaching. These demands are particularly pertinent when college costs rise at the same time students have difficulty in enrolling in required courses.

As the demands for improvement in undergraduate education increase, there is increasing tension over the balance between teaching and research. As mentioned above, tenure and promotion, even at four-year colleges, still give greater weight to research and publications than to teaching, particularly undergraduate teaching. There are some trends in the other direction, based in part on a call for an expanded definition of scholarship discussed below. However, given the financial rewards and institutional prestige that come with a greater research emphasis, it is likely that tensions around the appropriate incentive and reward systems in higher education will only increase.

Of particular interest to you as a future professor of science and engineering are the significant changes taking place in the nature of science and engineering research at colleges and universities. These include growing university–industry collaborations, the commercialization of some science and engineering research, greater shifts toward directed and applied research, and the general move toward "big science" and "big engineering" projects involving millions and even billions of dollars [19, p.352]. These programs can provide a number of benefits, but also create a number of potential problems. We will discuss this important topic in more detail in Chapter 2, "Science and Engineering in Higher Education," and Chapter 3, "New Challenges for the Professoriate."

Ironically, even at a time of turmoil, higher education in North America remains the envy of the world. Nowhere else do as many people have as much access to as high a quality of education as they do in the United States and Canada. Furthermore, students from all over the world, particularly at the graduate level, continue to apply to U.S. and Canadian institutions in significant numbers [12, p.2/17], [13, p.6].

The number of students applying to graduate schools in the United States and Canada raises the issue of supply and demand for Ph.D.s in science and engineering and how universities should respond to it. As we will see in Chapter 4, "Your Personal Preparation Strategy," there is currently an oversupply of Ph.D.s in a number of science and engineering fields relative to full-time openings for such Ph.D.s in industry and academia. Should academia just let market forces take care of the problem? Should they continue their admissions policies, but warn students of the possible nonavailability of some jobs? Should they instead admit fewer students, eliminate or reduce foreign students, or downsize their research programs? The answers are not clear. More importantly, as a prospective science or engineering professor, what does all this mean to you? Realism and market forces, combined with your passion for a particular option, are likely to determine what you do. In Chapter 4, we will outline a strategy to help you prepare for whatever choice you make.

A NEW LOOK AT SCHOLARSHIP

Undoubtedly, you have already heard much about the teaching and research balance debate at your current institution. One of the most important tasks as you move toward your first academic position is to determine what kind of ratio best suits your interests and capabilities. In doing so, you may want to identify institutions, particularly in the Master's and Baccalaureate categories, where other forms of scholarship are valued in addition to traditional research. In at least some institutions, there is a move to place more value on a larger range of activities in the retention, tenure, and promotion process. Ernest Boyer has done the most to articulate the need for such a move. As he noted in the Carnegie Foundation for the Advancement of Teaching 1990 special report, *Scholarship Reconsidered: Priorities of the Professoriate* [5, Ch.2, p.16]:

> We believe the time has come to move beyond the tired old "teaching versus research" debate and give the familiar and honorable term "scholarship" a broader, more capacious meaning, one that brings legitimacy to the full scope of academic work...Specifically, we conclude that the work of the professoriate might be thought of as having four separate, yet overlapping, functions. These are: the scholarship of *discovery*; the scholarship of *integration*; the scholarship of *application*; and the scholarship of *teaching*.

The scholarship of discovery is what most of us think of as traditional research. The scholarship of integration is work that makes connections across disciplines, but is interdisciplinary, not just multidisciplinary. It is an attempt to put specialties in a large context, and is in part a response to the demands of industry which is increasingly dealing with problems not bounded by specific disciplines. The application of knowledge involves service tied directly to a faculty member's special field of knowledge, and includes such things as medical service, serving clients in psychotherapy, working with industry on the design of a new microprocessor, and testing a new software application in the local school system. The scholarship of teaching, in the words of Boyer, "both educates and entices future scholars...It means not only transmitting knowledge, but transforming and extending it as well" [5, Ch.2, p.16].

The relationship among the various forms of scholarship is circular, not linear, as shown in Figure 1-4. Each form benefits from, and contributes to, the other.

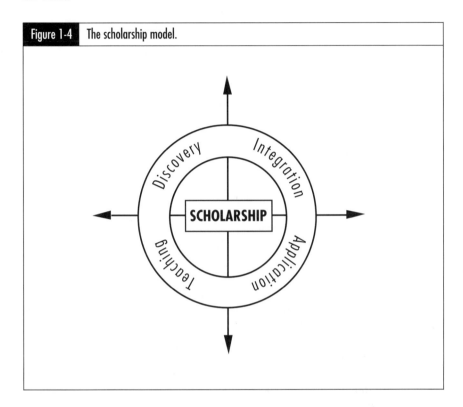

Figure 1-4 The scholarship model.

Boyer notes that the scholarship of teaching was emphasized in the colonial college, the scholarship of application in the land grant institutions of the last century, and that the scholarship of discovery, and to some extent integration, has emerged in the post-World War II university. He believes that with the diversity found in today's institutions of higher education, it should be possible to more fully promote and reward all forms of scholarship.

This perspective is all very well, of course, but the real test is how this plays out in the retention, promotion, and tenure process. In 1990, Boyer made the following observation [5, Ch.2, p.28]:

> Today at most four-year institutions, the requirements of tenure and promotion continue to focus heavily on research and on articles published in journals, especially those that are referred. Good teaching is expected, but it is often inadequately assessed. And the category of "service," while given token recognition by most colleges, is consistently underrated too.

Today, however, particularly with better tools available for the assessment of the various forms of scholarship, there is evidence of a movement, albeit a modest one, toward the acceptance of this broader view of scholarship [26],[27]. A number of presidents of metropolitan universities (primarily Master's I and II universities located in urban environments) have signed a declaration stating, among other things, that the "creation, interpretation, dissemination, and application of knowledge are the fundamental functions of our universities," and that faculty research must [28]:

> seek and exploit opportunities for linking basic investigations with practical application, and for creating synergistic interdisciplinary and multidisciplinary scholarly partnerships for attacking complex metropolitan problems, while meeting the highest scholarly standards of the academic community.

Not surprisingly, at Baccalaureate and Master's schools, the scholarship of teaching and application is given greater weight in tenure decisions than at Research and Doctoral schools. In a recent survey, 45% of the faculty at liberal arts colleges and 37% of the faculty at Master's colleges rated student evaluations of courses taught as "very important" for granting tenure in their department. This result compared with 10 and 19%, respectively, for faculty at Research and Doctorate-granting institutions [5, Ch.2, p.30].

While certainly not common, one can point to examples where the scholarship of integration, application, or teaching made a difference with respect to retention, promotion, and tenure for faculty at Research and Doctorate-granting institutions. Nino Masnari in the Electrical Engineering Department at North Carolina State University (scholarship of integration), David Kelly in the Mechanical Engineering Department at Stanford University (scholarship of application), and Susan Montgomery in the Chemical Engineering Department at the University of Michigan (scholarship of teaching) are three specific examples. Whether or not these examples will remain the exception, or whether they foretell of at least a modest trend in the broader scholarship direction, is something we will discuss at greater length in the chapters to come.

SEVEN SAMPLE SCHOOLS

As noted in the introduction, over 70 faculty, graduate students, and postdocs from a range of schools across North America have provided material for this book. In addition to acquiring this broad input, I have examined, as sources of more in-depth information, seven institutions representative of the four-year colleges and universities in the Carnegie classification. The seven schools, which are located across North America, are Bucknell University (BA I), Memorial

University of Newfoundland (Doc. I), the University of Michigan (Res. I), Rochester Institute of Technology (MA I), San Jose State University (MA I), Stanford University (Res. I), and the University of New Orleans (Doc. II). Table 1.4 looks at the institutions by date of founding, type (public or private), location, and enrollment. There is at least one school from each of the four major Carnegie classifications. One school has approximately 4000 students, four have from 10,000 to 18,000 students, and two from 30,000 to 37,000 students. All offer at least bachelor's degrees in engineering and in one or more of the natural sciences. Below is a brief description of each school.

Table 1.4 Seven sample schools

Bucknell University	Lewisburg, PA, BA I, Private Est. 1895, Enrollment: 3,696
Memorial University of Newfoundland	St. John's, Nfld., Doc. I, Public Est. 1925, Enrollment: 17,500
Rochester Institute of Technology (R.I.T.)	Rochester, NY, MA. I, Private Est. 1912, Enrollment: 13,004
San Jose State University (S.J.S.U.)	San Jose, CA, MA. I, Public Est. 1857, Enrollment: 26,000
Stanford University	Stanford, CA, Res. I, Private Est.: 1891, Enrollment: 13,549
University of Michigan	Ann Arbor, MI, Res. I, Public Est. 1817, Enrollment: 36,288
University of New Orleans	New Orleans, LA, Doc. II, Public Est. 1956, Enrollment: 16,000

Bucknell University (BA I)

Bucknell University is a private, 100-year-old baccalaureate institution of about 3600 students. It is located in Lewisburg, PA, 60 miles from Harrisburg. According to the university catalog, Bucknell is [29]:

> ...a highly selective, primarily undergraduate institution offering a broad curriculum of studies in the humanities, social sciences, and natural sciences, as well as professional studies in engineering, education, and management. Bucknell benefits from its focus on the liberal arts and the professions, its modest size, its location, and the

large number of qualified applicants attracted by the competitive environment of the private colleges along the East Coast. The University's primary responsibility is to provide wide educational opportunities within a collegiate setting to a controlled number of talented men and women.

According to Martin Ramirez, Assistant Professor of Biology at Bucknell:

> I like working in a small liberal arts college with strong science and engineering programs. But one problem faced by many faculty who have professional spouses is that of finding career opportunities for both of them in such a small, isolated location. I know some faculty who live quite a distance from Lewisburg so that their spouses can work in Harrisburg, or even Rochester.

Memorial University of Newfoundland (Doc. I)

Memorial University of Newfoundland (MUN), in St. John's, NFLD, is a public institution founded in 1925 as a memorial to those Newfoundlanders who died in World War I. Newfoundland was a British colony until 1949, and MUN continues to be influenced by British traditions. It is, however, a modern Canadian university, the largest in the Maritime provinces, with extensive undergraduate and graduate programs in the natural sciences and in engineering and applied sciences.

A particularly impressive MUN feature is found in the Ocean Engineering Research Center (OERC) which serves as the focus of ocean engineering research and teaching in Canada. MUN is the only university in Canada to offer the bachelor's degree in naval architecture and the master's and Ph.D. in ocean engineering. The Center's facilities include a 60-meter wave/tow tank and the international class facilities of the Institute for Marine Dynamics of the National Research Council of Canada [30].

Naval Architecture Professor, Mahmood Haddara, describes MUN this way:

> MUN has one of the best naval architecture departments in the world, and so it seemed natural for me to want to come here. I particularly wanted to study ship motion, and with all the work going on off-shore these days (platforms), this is the perfect place for me to teach and do research.

The University of Michigan, Ann Arbor (Res. I)

The University of Michigan, Ann Arbor, is a public institution of some 36,000 students founded in 1817. It offers undergraduate programs in the arts and sciences, architecture, business administration, education, engineering, fine arts, natural resources, nursing, professional studies, and military training. It has 12 undergraduate and 18 graduate schools. The College of Engineering, founded in 1853, is one of the oldest in the United States. It established the nation's first programs in electrical engineering (1895), chemical engineering (1898), aeronautical engineering (1914), nuclear engineering (1953), and computer engineering (1965) [31].

Diann Brei, Assistant Professor of Mechanical Engineering, talks about her time at the University of Michigan this way:

> To me, it's like building a house. The first year, I made plans and acquired the materials. The second year, I built the foundation. The third year, I put up the frame, and people began to see that it was going to be a very good house. Between the third year and tenure, I plan to put up the walls, finish the house, and decorate. Hopefully, people will then say that for this town (my research area), this is one of the best houses!

Rochester Institute of Technology (MA I)

The Rochester Institute of Technology (R.I.T.) is a private institution of about 13,000 students founded in 1829. It is located in Rochester, NY. According to the school's bulletin [32]:

> R.I.T. offers programs in science, computer science, allied health, engineering, business, hotel management, graphic arts and photography, as well as the liberal arts and includes the National Technical Institute for the Deaf. Most programs include a cooperative education component which provides full-time work experience to complement classroom studies.

Explains Mark Hopkins, R.I.T. professor of electrical engineering:

> A particularly interesting feature of R.I.T. is its close relationship with surrounding industry. This relationship allowed me to spend eight years as a professor at R.I.T. while also working as a senior scientist at the nearby Xerox research facility. There is no question in my mind that this benefited both me and my students a great deal.

San Jose State University (MA I)

San Jose State University is the oldest (1857) school of what is now a 20-campus California State University system, the largest senior higher education system in the United States. (It is not to be confused with the nine-campus University of California system, associated with the University of California, Berkeley.) San Jose State University enrolls about 26,000 regular students and another 30,000 in its extended education program, the latter reaching out to the community (Silicon Valley) with a variety of professional certificate programs and other educational experiences. The University offers baccalaureate and master's degrees and professional credentials in more than 150 disciplines, including the professions, business, social work, engineering, science, technology, education, social science, the arts, and the humanities. It is located in San Jose, CA, about 40 miles south of San Francisco [33].

According to Sally Veregee, professor of biology:

> My department tends to hire older people. By that, I mean assistant professors in their middle, to even late 30s. We look for good postdoc or industrial experience first. We've found it makes a real difference.

Stanford University (Res. I)

Stanford University is a private institution established in 1891 by Leland and Jane Stanford in memory of their son, Leland, Jr., who died of typhoid fever in Florence, Italy, in 1884. It enrolls approximately 6500 undergraduates, and a similar number of graduate students, of which approximately 2300 are from other countries. It offers a full range of bachelor's, master's, doctorate, and professional degrees in over 70 fields across seven schools of business, earth sciences, education, engineering, humanities and sciences, law, and medicine. It is adjacent to Palo Alto, CA, about 25 miles north of San Jose [34].

George Springer, chairman of the Aeronautics and Astronautics Department, offers this comment:

> From my faculty, I expect one–two conference presentations per year, three–four journal papers every two years and at least one seminal paper prior to tenure.

University of New Orleans (Doc. II)

The University of New Orleans was established in 1956 by the Louisiana Legislature to bring public-supported higher education to the state's largest urban community. According to the university catalog [35]:

> The University offers extensive learning experiences and academic training at both the undergraduate and graduate levels to nearly 16,000 students in more than one hundred degree programs. Over thirty-five thousand degrees have been offered since the first graduating class of 115 in 1962, nearly one-quarter of which are at the masters or doctoral level. Programs of study are offered through six academic colleges: Business Administration, Education, Engineering, Liberal Arts, Sciences and Urban and Public Affairs.

Norm Whitley, professor of mechanical engineering at U.N.O., referred to his work this way:

> I have been particularly interested not only in the teaching of ethics, but in the ethics of teaching. I've been working with my colleagues on developing a code of ethics that I think will have a real impact on how U.N.O. contributes to this important issue.

Vignette # 1 A Place for Scholarship in Undergraduate Education

Our first vignette takes a closer look at some of the views of Ernest Boyer whose work on a new scholarship paradigm is having an important impact on higher education. The vignette is based on his writings and on a speech produced by the National Technological University and given in March 1995 to the National Science Foundation sponsored Engineering Faculty Forum at the University of Maryland in College Park, MD.

Ernest L. Boyer, late President of the Carnegie
Foundation for the Advancement of Teaching

> The most urgent obligation higher learning now confronts is to reaffirm the undergraduate experience in its full breadth and rediscover the essentialness of teaching. This means making it possible for all students to become scholars through the discovery, the integration and application of knowledge—and through the transmission of knowledge, which will keep the flame of scholarship alive [36].

So says Ernest L. Boyer, former chancellor of the State University of New York, late president of the Carnegie Foundation for the Advancement of teaching, and author of the landmark publication, *Scholarship Reconsidered: Priorities for the Professoriate*. Boyer passed away in December 1995.

In a March 14, 1995 speech, Boyer suggested that his four-part notion of scholarship may be a way to define the purpose of an undergraduate education [36].

> Suppose that we tell every undergraduate that he or she is entering the world of scholarship, and that in so doing we bring the language of the professoriate and the language of the freshman together right from the start by letting the student know what it means to be a scholar. We tell them that it means someone who does discovery, someone who learns to integrate, someone who learns how to apply their knowledge and someone who knows how to share what they've learned.

Through such an approach, undergraduates could, for example, begin doing research projects with senior professors. In Boyer's words, "The student could learn what discovery is all about by starting a conversation with faculty when they are a freshman, not when they are working on their Ph.D."

Boyer believed that the integration of knowledge can be placed in the context of liberal learning or general education, and not just treated as "something you get out of the way." The scholarship of application can be a part of every undergraduate's experience through some type of field work or service. It is important that such service come from a specific area or specialty, for as Boyer said, "Don't confuse doing good with doing scholarship, important as both may be." In other words, "To be considered scholarship, service activities must be tied directly to one's special field of knowledge and relate to, and flow directly out of, this professional activity"[5, p.22].

Sharing, or the scholarship of teaching, would mean that students would have to regularly present their ideas in some form of open discourse. After all, commented Boyer [36]:

> One hundred years ago to get a degree you didn't turn in green stamps to the registrar. You stood up and had a declamation and that's when you earned your degree. You didn't get your degree by stacking up units, but by showing that you were educated enough to write a paper, stand up and present it, and defend it in an open discourse of sharing knowledge. That was the test of an educated person.

Boyer understood that you may not be able to do that today, but he suggested that students might be required to take a senior seminar in groups of 20 or so in which they develop a paper and present it orally while others critique it. It would be the final culmination of their sharing of knowledge, or teaching. As he put it [36]:

> It would be a fascinating way to rethink what the undergraduate experience is about. You can have everything in it that we have today, but by doing it this way you would give it a greater purpose. You'd have a shared culture so that a senior professor and an incoming freshman would have a common language about discovering knowledge, integrating it, applying it, and sharing it.

Boyer, whose ideas on the broader notion of scholarship will be referred to throughout this book, had a long and distinguished career in education. He came to the Carnegie Foundation in 1979 after serving as U.S. Commissioner of Education (under President Carter). In his seven years as chancellor of the State University of New York, he oversaw the operations of 64 campuses and 350,000 students. In 1990, he was named Educator of the Year by *U.S. News and World Report*, and in 1994, he received the Charles Frankel Prize in the Humanities, a Presidential citation. Boyer had been named by three U.S. presidents—Nixon, Ford, and Carter—to national commissions, and former Secretary of State George Shultz appointed him to chair the State Department's National Overseas Schools Advisory Council. Boyer received his Ph.D. from the University of Southern California, and was a postdoctoral fellow in medical audiology at the University of Iowa Hospital. Called by many "an evangelist of education" for his dedication to spreading ideas about teaching and learning, Ernest Boyer will be long remembered and sorely missed [26].

SUMMARY

We began this chapter by pointing out that colleges and universities differ in important ways from other institutions in society. We first looked briefly at the history of higher education in North America, and then examined in greater detail some key characteristics of today's colleges and universities. We introduced the Carnegie classification of higher education that divides most four-year institutions into four categories: Research I&II, Doctorate Granting I&II, Master's I&II, and Liberal Arts I&II. This classification formed the basis for a comparison of schools along a number of dimensions having to do with mission, type of students, and research and teaching emphasis.

An important difference between higher education and the rest of society is in its governance and decision making. We looked at the "bottom-up" approach to decision making and the powerful role faculty play in academia. We examined some of the sources of power in higher education, in particular the role of discretionary funding that results from research contracts and grants. Concerns about the balance between teaching and research has led some schools to consider new forms of scholarship such as those of integration, application, and teaching, in addition to the more research-oriented scholarship of discovery.

We then examined some of the critical challenges facing higher education in general. These challenges include significant restructuring brought about by massive budget cuts, calls for greater relevance in the curriculum, demands for increasing productivity, and the implications of greater university–industry collaborations.

Finally, we introduced seven sample institutions representative of the various four-year colleges and universities in the Carnegie classification. Focused discussions with faculty at these institutions will provide a way of comparing and contrasting the range of schools of interest to future science and engineering professors. We concluded this chapter with our first vignette highlighting Ernest Boyer's views on the role of scholarship in undergraduate education.

In Chapter 2, we will look more closely at the place of science and engineering in academia, and the similarities and differences within, and among, science and engineering departments. Although you will most likely accept a position in a single department, you will certainly have to, and hopefully want to, work with colleagues in other science and/or engineering disciplines, as well as some in completely different fields. You will be both competing and cooperating with these colleagues, and understanding how they think and operate will be essential to your future success.

Appendix Doctorate-Granting Institutions

Research Universities I

Arizona State University	Carnegie Mellon University	Cornell University
Boston University	Case Western Reserve University	Duke University
Brown University	Colorado State University	Emory University
California Institute of Technology	Columbia University	Florida State University

Appendix Doctorate-Granting Institutions (*Continued*)

Georgetown University
Georgia Institute of Technology
Harvard University
Howard University
Indiana University at Bloomington
Iowa State University
Johns Hopkins University
Louisiana State University and A&M College
Massachusetts Institute of Technology
Michigan State University
New Mexico State University
main campus
New York University
North Carolina State University
Northwestern University
Ohio State University
main campus
Oregon State University
Pennsylvania State University
main campus
Princeton University
Purdue University
main campus
Rockefeller University*
Rutgers University at New Brunswick
Stanford University
State University of New York at Buffalo
State University of New York at Stony Brook
Temple University
Texas A&M University
main campus
Tufts University
University of Alabama in Birmingham
University of Arizona

University of California at Berkeley
University of California at Davis
University of California at Irvine
University of California at Los Angeles
University of California at San Diego
University of California at San Francisco*
University of California at Santa Barbara
University of Chicago
University of Cincinnati
main campus
University of Colorado at Boulder
University of Connecticut
University of Florida
University of Georgia
University of Hawaii at Manoa
University of Illinois at Chicago
University of Illinois at Urbana-Champaign
University of Iowa
University of Kansas
main campus
University of Kentucky
University of Maryland at College Park
University of Massachusetts at Amherst
University of Miami
University of Michigan at Ann Arbor
University of Minnesota-Twin Cities
University of Missouri at Columbia
University of Nebraska at Lincoln
University of New Mexico
main campus
University of North Carolina at Chapel Hill
University of Pennsylvania

University of Pittsburgh
Pittsburgh campus
University of Rochester
University of Southern California
University of Tennessee at Knoxville
University of Texas at Austin
University of Utah
University of Virginia
University of Washington
University of Wisconsin at Madison
Utah State University
Vanderbilt University
Virginia Commonwealth University
Virginia Polytechnic Institute and State University
Washington University
Wayne State University
West Virginia University
Yale University
Yeshiva University

Research Universities II

Auburn University
Brandeis University
Brigham Young University
Clemson University
George Washington University
Kansas State University
Kent State University
main campus
Lehigh University
Mississippi State University
Northeastern University
Ohio University
main campus
Oklahoma State University
main campus

Rensselaer Polytechnic Institute
Rice University
Saint Louis University
Southern Illinois University at Carbondale
State University of New York at Albany
Syracuse University
main campus
Texas Tech University
Tulane University
University of Arkansas
main campus
University of California at Riverside
University of California at Santa Cruz
University of Delaware
University of Houston
University of Idaho
University of Mississippi
University of Notre Dame
University of Oklahoma at Norman
University of Oregon
University of Rhode Island
University of South Carolina at Columbia
University of South Florida
University of Vermont
University of Wisconsin at Milwaukee
University of Wyoming
Washington State University

Doctoral Universities I
Adelphi University
American University
Andrews University
Ball State University
Boston College
Bowling Green State University

Catholic University of America
City University of New York Graduate School and University Center
Claremont Graduate School
Clark Atlanta University
College of William and Mary
Drexel University
East Texas State University
Florida Institute of Technology
Fordham University
Georgia State University
Hofstra University
Illinois Institute of Technology
Illinois State University
Indiana University of Pennsylvania
Loyola University of Chicago
Marquette University
Miami University
New School for Social Research
Northern Arizona University
Northern Illinois University
Nova University
Old Dominion University
Polytechnic University
Saint John's University (N.Y.)
Southern Methodist University
State University of New York at Binghamton
Teachers College of Columbia University
Texas Woman's University
Union Institute
United States International University
University of Akron
main campus
University of Alabama
University of Denver
University of Louisville
University of Maryland

at Baltimore
University of Memphis
University of Missouri at Kansas City
University of Missouri at Rolla
University of North Carolina at Greensboro
University of North Texas
University of Northern Colorado
University of Southern Mississippi
University of Texas at Arlington
University of Texas at Dallas
University of Toledo
Western Michigan University

Doctoral Universities II
Baylor University
Biola University
Clark University
Clarkson University
Cleveland State University
Colorado School of Mines
Dartmouth College
De Paul University
Duquesne University
Florida Atlantic University
Florida International University
George Mason University
Hahnemann University
Idaho State University
Indiana State University
Indiana University-Purdue University at Indianapolis
Loma Linda University
Louisiana Tech University
Michigan Technological University
Middle Tennessee State University
Montana State University

| Appendix | Doctorate-Granting Institutions (*Continued*) |

New Jersey Institute of Technology	University of Alabama in Huntsville	University of the Pacific
North Dakota State University main campus	University of Alaska at Fairbanks	University of Puerto Rico Rio Piedras campus
	University of Central Florida	
Pace University New York campus	University of Colorado at Denver	University of San Diego
	University of Detroit Mercy	University of San Francisco
Pepperdine University	University of La Verne	University of South Dakota
Portland State University	University of Maine	University of Southwestern Louisiana
Rutgers University at Newark	University of Maryland	University of Tulsa
San Diego State University	Baltimore County	Wake Forest University
Seattle University	University of Massachusetts	Wichita State University
Seton Hall University	at Lowell	Worcester Polytechnic Institute
State University of New York College of Environmental Science and Forestry	University of Missouri at Saint Louis University of Montana	Wright State University main campus
Stevens Institute of Technology	University of Nevada at Reno	
Tennessee State University	University of New Hampshire	
Texas Christian University	University of New Orleans	* Indicates that an institution meets
Texas Southern University	University of North Dakota main campus	the criteria for more than one Carnegie Category.

Source: The Carnegie Foundation for the Advancement of Teaching, "*A Classification of Institutions of Higher Learning,*" 1994 edition. Reprinted with permission.

REFERENCES

[1] Presentation at the Global Manufacturing Associates Forum, Stanford University, Stanford, CA, June 14, 1995.

[2] C. Kerr, "American society turns more assertive: A new century approaches for higher education in the United States," in P. G. Altbach, R.O. Berdahl, and P. J. Gumport, Eds., *Higher Education in American Society*. Amherst, NY: Prometheus Books, 1994, foreword. Copyright ©1994. Reprinted by permission of the publisher.

[3] K. Mangan, "Hahnemann U. angers faculty with threat to fire those who don't attract grant money," *The Chronicle of Higher Education*, vol. XLI, no. 6, p. A20, Oct. 5, 1994.

[4] A. Levin and J. Nidiffer, "Faculty productivity: A re-examination of current

attitudes and actions," unpublished paper, Institute of Educational Management, Harvard Graduate School of Education, 1993.
[5] E. L. Boyer, *Scholarship Reconsidered—Priorities of the Professoriate.* Princeton, NJ: The Carnegie Foundation for the Advancement of Teaching, 1990.
[6] J. R. Thelin, "Campus and commonwealth: A historical interpretation," in P. G. Altbach, R. O. Berdahl, and P. J. Gumport, Eds., *Higher Education in American Society.* Amherst, NY: Prometheus Books, 1994, ch. 1.
[7] L. Grayson, *The Making of an Engineer: An Illustrated History of Engineering Education in the United States and Canada.* New York: Wiley, 1993, ch. 3, p. 42.
[8] L. Veysey, *The Emergence of the American University.* Chicago, IL: University of Chicago Press, 1965, ch. 1, p. 23.
[9] J. R. Thelin, "Campus and commonwealth: A historical interpretation," in P. G. Altbach, R. O. Berdahl, and P. J. Gumport, Eds., *Higher Education in American Society.* Amherst, NY: Prometheus Books, 1994, ch.1, p. 33. Reprinted by permission of the publisher. Note: The following author's note accompanied this passage: C. Jencks and D. Riesman, *The Academic Revolution* (Garden City, NY: Doubleday Anchor, 1968); C. Kerr, *The Uses of the University* (Cambridge, MA: Harvard University Press, 1964); R.M. Rosenzweig with B. Turlington, *The Research Universities and Their Patrons* (Berkeley, CA: University of California Press, 1982); R. Geiger, *To Advance Knowledge: The Growth of American Research Universities* (New York and Oxford: Oxford University Press, 1986).
[10] P. G. Altbach, "Problems and possibilities: The American academic profession," in P. G. Altbach, R. O. Berdahl, and P. J. Gumport, Eds., *Higher Education in American Society.* Amherst, NY: Prometheus Books, 1994.
[11] J. Evangelauf, "The new Carnegie classification," *The Chronicle of Higher Education*, vol. XL, no. 32, Apr. 6, 1994.
[12] National Science Board, *Science and Engineering Indicators—1996.* Washington, DC: U.S. Government Printing Office, 1996 (NSB 96-21), ch. 2.
[13] *Education Quarterly Review* (catalogue 81-003), Ottawa, Ont., Canada: Statistics Canada, 1994.
[14] The *Chronicle of Higher Education—Almanac Issue*, vol. XLII, no. 1, p. 22, Sept. 1, 1995.
[15] W. Cellis, 3rd, "The big stars on campus are now research labs," *The New York Times*, vol. CXLIV, no. 49,900, p. 18, Dec. 4, 1994.
[16] *University of Arizona Bulletin.* Tucson, AZ: University of Arizona, 1994–1995, p. 6.
[17] W. H. Honan, "New pressures on the university," *Education Life, New York Times Supplement*, vol. CXLV, no. 40936, p. 17, Jan. 9, 1994.

[18] C. Kerr and M. Gade, *The Guardians: Boards of Trustees of American Colleges and Universities*. Washington, DC: Association of Governing Boards of Universities and Colleges, 1989, p. 16.

[19] A. Zusman, "Current and emerging issues facing higher education in the United States," in P.G. Altbach, R.O. Berdahl, and P. J. Gumport, Eds., *Higher Education in American Society*. Amherst, NY: Prometheus Books,1994, ch. 15.

[20] E. Arden, Ed., "Academic freedom and tenure," *ACADEME: Bulletin of the American Association of University Professors*, vol. 79, no. 2, p. 111, Mar.–Apr. 1993.

[21] M. Edmundson, "Bennington means business," *The New York Times Magazine*, p. 45, Oct. 23, 1994.

[22] G. R. Salancik, "Power and politics in academic departments," in *The Compleat Academic: A Practical Guide for the Beginning Social Scientist*. New York: Random House, 1987, ch. 3.

[23] J. Pfeffer and A. Davis-Blake, "Understanding organizational wage structures: A resource dependence approach," *Academy of Management Journal*, vol. 30, 1987, pp. 437–455.

[24] J. Pfeffer, *Managing with Power: Politics and Influence in Organizations*. Boston, MA: Harvard University Press, 1994, ch. 5, p. 94.

[25] From P. G. Altbach, R. O. Berdahl, and P. J. Gumport, Eds., *Higher Education in American Society*. Amherst, NY: Prometheus Books, 1994, introduction, he publisher.

[26] C. Leatherman, "The legacy of Ernest Boyer, 'evangelist of education,' " *The Chronicle of Higher Education*, vol. XLII, no. 17, p. A18, Jan. 5, 1996.

[27] E. Boyer, "Assessing scholarship," *ASEE Prism*, vol. 4, no. 7, pp. 23–24, Mar. 1995.

[28] "Declaration of metropolitan universities," unpublished document, president's office, San Jose State University, San Jose, CA, Mar. 1996.

[29] *Bucknell 1994–95 Catalog*. Lewisburg, PA: Bucknell University, 1994, p. ii.

[30] *Faculty of Engineering and Applied Science Calendar—1994–1995*. St. John's, Nfld., Canada: Memorial University of Newfoundland, 1994, p. 5.

[31] *University of Michigan Bulletin*. Ann Arbor, MI: University of Michigan, vol. 24, no. 15, 1994–1995, p. 9.

[32] *Rochester Institute of Technology Bulletin*. Rochester, NY: Rochester Institute of Technology, vol. 9, no. 2, July 12, 1994, p. 1.

[33] *San Jose State University 1992–93 Catalog*. San Jose, CA: San Jose State University, 1994, p. 11.

[34] *Stanford University Courses and Degrees, 1993–94*. Stanford, CA: Stanford University, 1993, p. 6.

[35] *General Graduate Catalog.* New Orleans, LA: University of New Orleans, vol. XXXVI, no. 1, July 1994, p. 7.

[36] Speech given to the National Science Foundation, Engineering Faculty Forum, National Technological University, University of Maryland, College Park, Mar. 14, 1995.

CHAPTER 2

Science and Engineering in Higher Education

> Disciplines and departments are ranked into hierarchies, with the traditional academic specialties in the arts and sciences along with medicine and, to some extent law, at the top. The "hard" sciences tend to have more prestige than the social sciences or humanities. Other applied fields, such as education and agriculture, are considerably lower on the scale. These hierarchies are very much part of the realities and perceptions of the academic profession.
>
> *Philip Altbach, Professor of Higher Education,*
> *Boston College* [1]

Clark Kerr, president emeritus of the University of California, once joked that universities consisted of hundreds of individual faculty united only by their common desire to find a parking place. Faculty do indeed act more independently than other types of employees, as was pointed out in Chapter 1. Nevertheless, how they think, and what they actually do, depends to a large extent on the specific discipline to which they belong. Power and influence, financial compensation, types of students, ease of publication, expenditures for research and development, number of like-minded colleagues, and even agreement on what constitutes quality work in a given field can vary considerably across departments within a college or university.

These factors are examined in Chapter 2, with particular attention to their impact on science and engineering. Similarities and differences among science, engineering, and other disciplines, such as the humanities and social sciences, are examined first. We then look in more detail at departments within science and within engineering. This examination is followed by a discussion of the prospects for cross-disciplinary collaboration among the various fields. We then return to the model of scholarship introduced in Chapter 1 with a look at differences in its various forms across disciplines. The chapter concludes with a vignette on the issues faced by a dean of science at a major master's-granting institution.

COMPARISONS ACROSS THE INSTITUTION

Faculty assign different levels of importance to their discipline, their department, and their college or university. In a recent survey, 77% of the faculty respondents said their academic discipline was very important to them, while 53% said the same thing about their department, and only 40% felt this way about their college or university [2, Appendix A]. While faculty identify closely with their discipline, an understanding of other disciplines is also important. As a prospective faculty member, you need to consider the following:

- Those outside your department and discipline will be your institutional colleagues. You will share the same employer and higher level administration, many of the same resources, a number of the same problems, and at the undergraduate level at least, many of the same students.

- In many cases, interesting cross-disciplinary scholarship opportunities will exist with colleagues in other departments and disciplines.

- You will compete with colleagues outside your department and discipline for resources, influence, and attention.

- At times, you will find it is easier to learn from, confide in, and be mentored by colleagues in other parts of your college and university.

For these, as well as other reasons, you will want to become knowledgeable about the similarities and differences existing across the college or university where you become a professor. In Chapter 4, "Your Professional Preparation Strategy," we suggest ways to begin acquiring this understanding by "practicing" at the institution you now attend. In this chapter, we set the stage for this examination by looking at the differences with respect to degree of development, power and influence, type of graduate students, number of postdocs, number of faculty, financial compensation, ease of publication, and expenditures for research and development.

Degree of Discipline Development

Disciplines and fields differ in their degree of development. This differential is particularly evident across the natural and social sciences. "Hard" sciences such as physics and chemistry are regarded as more developed than the "soft" sciences such as the political and social sciences [3]. In this context, "more developed" means those disciplines having more evolved paradigms or shared theoretical structures, and which in general share a greater level of consensus about methods, what constitutes quality research, and course prerequisites [4, p.106].

Sociologist Steven Cole uses six measures to determine the degree of development of a scientific field. They are: (1) development of theory, (2) degree of quantification of ideas, (3) degree of cognitive consensus, (4) level of theory predictability, (5) rate at which work becomes obsolete, and (6) rate of growth of knowledge [4, p.107].

According to this scheme, physics, chemistry, and biochemistry are relatively developed fields, geology, botany, and zoology are less developed, whereas economics, sociology, anthropology, and political science are the least developed [4, p.106].

Cole does make a distinction between knowledge at the research frontier and knowledge at the core. Physics has greater agreement at the core, where there is a relatively small number of theories or exemplars, than does sociology, but both have considerable disagreement at the frontiers of knowledge where knowledge is broader and more diverse [4, p.135]. Nevertheless, the overall *perceptions* among faculty as to the degree of development of their fields is pretty much as stated above [4, p.135]. Philosopher Lawrence Laudan puts it this way [4, p.108]:

> To anyone working in the humanities or social sciences, where debate and disagreement between rival factions are pandemic, the natural sciences present a tranquil scene indeed. For the most part, natural scientists working in any field or subfield tend to be in agreement about most of the assertions of their discipline. They will typically agree about many of the central phenomena to be explained and about the broad range of quantitative and experimental techniques appropriate for establishing "factual claims." Beyond this agreement about what is to be explained, there is usually agreement at the deeper level of explanatory and theoretical entities. Chemists, for instance, talk quite happily about atomic structure and subatomic particles. Geologists, at least for now, treat in a matter-of-fact fashion claims about the existence of massive subterranean plates whose motion is thought to produce most of the observable (i.e. surface) tectonic activity—claims that, three decades ago, would have been treated as hopelessly speculative. Biologists agree about the general structure of DNA and about many of the general mechanisms of evolution, even though few can be directly observed.

Where does engineering fit into this picture? The likely answer is, somewhere in between the "hard" and the "soft" sciences. Engineering disciplines that "derive" from the more developed natural sciences, such as chemical engineering (chemistry), electrical engineering and mechanical engineering (physics), and civil engineering (geology and physics), share some of the developmental characteristics of these disciplines. Fields such as industrial engineering, management engineering, and operations research share more of the characteristics associated with business, economics, and sociology. As a colleague in industrial engineering noted, "We often beat up on each other in low paradigm fields such as organizational behavior. In such fields, there is always a subgroup of people who think what you do is garbage, and you just have to learn to live with it."

Politics and Influence

The discipline differences discussed above can have a very real impact on academic politics, as sociologists Beyer and Lodahl noted in their study of the governance in British and American universities [5]:

> ...the higher predictability of greater paradigm development tends to increase consensus over means and goals...This serves to reduce conflicts within departments, and may reduce the potential for conflict and misunderstanding with the administration. Second, faculty members who have more consensus can form stronger and more effective coalitions than those in fields rife with internal conflicts.

Jeffrey Pfeffer, professor of organizational behavior in the Stanford University Graduate School of Business, has studied politics and influence in organizations extensively. He found that more paradigmatically developed academic disciplines such as physics and chemistry had department heads who tended to stay in their jobs for longer periods of time. According to Pfeffer [6, Ch. 3, p.52]:

> When there is consensus in the department about research methods, curriculum content and other such issues, it matters less who heads the department...This unity has obvious advantages for dealing with other units. There is more stability, and the leader knows that his or her position is relatively secure.

Pfeffer also notes that departments in more developed fields tend to have longer chains of courses, that is, one course serving as a prerequisite for another. He sees such chains as a reflection of the relatively high agreement on the core concepts in the field and how these concepts and skills are allocated to specific courses [6, p.152].

Types of Graduate Students

Another area where there are significant differences among disciplines is in the nationality and gender, race, and ethnicity of graduate students. These differences bear directly on the future faculty population in various fields since it is from this pool that the vast majority of new faculty will come. Table 2.1 compares the number of foreign students with the total number of students in various fields who earned doctorates in the United States.

Table 2.1 Earned Doctorate Degrees by Citizenship, 1993

	Foreign students	% of total
Total, all degrees	9,923	26
S&E	8,087	33
Natural sciences*	3,191	31
Math and CS	866	44
Social and behavioral	1,247	18
Engineering	2,783	51

* Natural sciences include all physical, environmental, biological, and agricultural sciences. Social and behavioral sciences include psychology, sociology, and other social sciences.

Source: National Science Board, *Science and Engineering Indicators—1996*, Appendix A, pp. 58–59.

The high percentages of foreign students, 44% in mathematics and computer science and 51% in engineering, reflect the relatively great paradigm development of these fields. Also, to a large extent, they are "culturally and politically neutral." Add to this neutrality their relative practicality as seen by many countries throughout the world, and it is not surprising that most foreign students are in these fields. A similar situation exists in Canadian universities [7].

Given the worldwide pool from which to draw, it is also not surprising that foreign students are often among the best in their fields. About 50% of the foreign students in U.S. and Canadian universities seek academic positions in North America after graduation or a period as a postdoc. In so doing, they add to the cultural mix and diversity that enrich academia. They also contribute to the current large supply of students seeking postdoc and academic positions. Also, most of these foreign students and postdocs did not attend U.S. or Canadian schools as undergraduates, so they often do not share the same understanding about college life as their North American counterparts.

Table 2.2 looks at doctorate degrees by sex and field in the United States, and Table 2.3 does the same for race/ethnicity.

Women received almost half of all social and behavioral sciences, and almost one third of the natural science degrees at the doctoral level. These numbers represent a doubling of the female participation rates over the last 15 years. However, women still received relatively few engineering or mathematics/computer science degrees at the doctoral level, 9 and 20%, respectively.

Table 2.2 Earned Doctorate Degrees by Sex and Field, 1993

	Female students	% of total
Total, all degrees	15,108	38
S&E	7,652	30
Natural sciences	3,221	31
Math and CS	401	20
Social and beh. sci.	3,509	49
Engineering	521	9

Source: National Science Board, *Science and Engineering Indicators—1996*, Appendix A, pp. 56–57.

The number of doctorates obtained by underrepresented minorities has increased over the last 15 years in all fields of science and engineering, especially in the social and natural sciences. However, this growth is from a small base. These populations still represent only 4% of all natural science and 2% of engineering and computer science doctorate degrees [8]. Similar patterns exist in Canadian universities [7].

One way to help increase the number of minority graduate students in science and engineering is to have more faculty role models who can mentor such students. This mentoring can be a source of considerable pleasure and satisfaction. As with all mentoring, it can also take a great deal of time. Furthermore, it is not always a good idea for women and minority faculty to be seen as only mentoring women and minority students. In addition to pressures to serve as mentors, there is often the pressure to serve on faculty committees. As one woman colleague noted: "Every committee seems to feel they need to have an X (where X equals your group, i.e., Hispanics, Blacks, women, etc.). The fewer the X's around, the more likely it is that you will be contacted" [9]. To help

Table 2.3 Earned Doctorate Degrees by Race/Ethnicity, and Field, 1993*

	White	Asian	Black	Hispanic	Native American
Total, all degrees	23,993	2,009	1,275	972	119
S&E	13,535	1,602	452	536	41
Natural sci.	5,943	684	135	228	17
Math and CS	886	156	14	23	2
Social and beh. sci.	4,684	237	253	220	20
Engineering	2,022	525	50	65	2

Source: National Science Board, *Science and Engineering Indicators—1996*, Appendix A, pp. 58–59.
*U.S. citizens and permanent residents only.

with this problem, administrators and mentors of women and minority faculty must take the lead in providing support, and in some cases, offsetting time, for new faculty. We will look more closely at how to balance these pressures in a later chapter.

Number of Postdocs

Another element of interest to tomorrow's professors is the number of postdoctoral appointments in various fields. As can be seen from Table 2.4, there are far more appointments relative to earned doctorate degrees in the natural sciences than in mathematics and computer sciences, the social sciences, and engineering.

Table 2.4 Postdoctoral Appointments and Earned Doctorate Degrees in Various Fields, 1991

Field	Number Postdoctorates	Number Earned Doctorates
Natural sciences	19,153	10,141
Math and computer sciences	324	1,837
Social sciences	967	6,653
Engineering	1,953	5,042
Total	22,397	23,673

Source: National Science Board, *Science and Engineering Indicators—1993*, pp. 292.

While it is fairly common for an engineering Ph.D. to go directly into a tenure-track faculty position, such is not the case in the natural sciences. This difference has significant implications for the preparation and job search strategy of future science and engineering professors, and will be examined in detail in Part III, "Finding and Getting the Best Possible Academic Position."

Number of Faculty

Across all institutions of higher education in the United States, the natural sciences has the largest number of full-time faculty, 101,681 out of a total faculty of 526,222. By contrast, engineering has only 24,680 full-time faculty. Eighty percent of the natural sciences faculty are male, and the comparable percentage in engineering is 94.2 [10].

There is some evidence that the male/female ratio is beginning to shift. While current data are not available by discipline, a recent study of all full-time faculty shows that women make up almost 41% of faculty in the first seven years of their academic careers. This number compares with 28% at the senior faculty level. Newly hired women outnumber newly hired men at liberal arts colleges, although only a third of the new hires at research and doctoral institutions are women [11].

Financial Compensation

Another dimension of obvious interest to prospective as well as current faculty is financial compensation. Table 2.5 gives the average salaries of full-time science and engineering faculty at four-year public and private U.S. academic institutions.

These figures are for full-time faculty members on nine- or ten-month contracts. Most faculty receive additional compensation during the summer for teaching, research, consulting, or employment in government or industry. For

Table 2.5 Average Faculty Salaries in Selected Fields at Public and Private Four-Year Institutions, 1995–1996

	New assistant professor	All ranks
Engineering		
Public	47,081	60,640
Private	48,458	65,244
Physics		
Public	37,452	53,996
Private	36,007	55,273
Life sciences		
Public	36,120	49,451
Private	33,323	46,894
Mathematics		
Public	36,330	47,860
Private	34,782	47,531
Social sciences		
Public	33,193	46,047
Private	32,677	47,783

The figures are based on reports covering 100,862 faculty members at 329 public four-year institutions and 53,459 faculty members at 531 private four-year colleges and universities. The figures cover full-time faculty members on nine- or ten-month contracts.

From: *The Chronicle of Higher Education.* vol. XLII, no. 28, p. A18, Mar. 22, 1996. Source: College and University Personnel Association. Reprinted with permission.

our purposes, the absolute values are less important than are the relative rankings. The position of engineering at the top is due in large measure to competitive pressures from employment opportunities in government and industry.

Ease of Publication

Another interesting difference among disciplines is the ease or difficulty of publishing scholarly papers. Journals in the natural sciences have significantly lower rejection rates than those in the social sciences [4, Ch. 3, p.114]. This difference is often taken as evidence of higher levels of consensus in the natural sciences, but other factors such as the space available in journals, the number of subdisciplines, and what are called field-specific norms can also have an impact. As sociologist Steven Cole points out [4, Ch. 3, p.114]:

> Physics journals prefer to make "Type I" errors of accepting unimportant work rather than "Type II" errors of rejecting potentially important work. This policy often leads to the publication of trivial articles with little or no theoretical significance, deficits which are frequently cited by referees in social science fields in rejecting articles. Other fields, such as sociology in the United States, follow a norm of rejecing an article unless it represents a significant contribution to knowledge. Sociologists prefer to make Type II errors.

Another factor that affects, if not the ease of publication, then at least the number of publications is collaboration with other investigators. Today, single-author publications are rare. Of the ten most cited articles of 1993, none was by a single author. When it comes to multiple authorship, nothing beats high-energy physics. Carlo Rubbia and Simon van de Meer were awarded the 1984 Nobel Prize in physics. The results of the experiments that led to this prize were reported in two articles in the journal *Physics Letters* published in 1983. The articles were published under the names of 59 and 138 joint authors, respectively! [12].

Different publication rates may also correlate closely to differences in the perceived rate of advancement in the field. In a recent survey 80% of the faculty in the biological sciences, 59% in the physical sciences, 55% in engineering, yet only 38% in the social sciences and 32% in the humanities strongly agreed with the statement, "exciting developments are now taking place in my field" [2, Appendix A, Table A-32].

Expenditures for R&D

Perhaps nowhere are the differences among the disciplines more evident than in the sums of money spent on research and development. Table 2.6 shows the expenditures for academic R&D by field in the United States.

Table 2.6 Expenditures for Academic R&D, by Field, 1993

Field	Millions of current dollars
Natural sciences	
Physical sciences	2,124
Astronomy	252
Chemistry	736
Physics	928
Other	209
Mathematical sciences	272
Computer sciences	597
Environmental sciences	1,318
Life sciences	10,828
Agricultural sciences	1,558
Biological sciences	3,536
Medical sciences	5,285
Other	446
Total	15,139
Engineering	
Aeronautics & astronautics	206
Chemical	269
Civil	367
Electrical/electronics	696
Materials	301
Mechanical	480
Other	830
Total	3,151
Social sciences	895
Grand total	19,185

Source: National Science Board, *Science and Engineering Indicators—1996*, Appendix A, p. 173.

The figures are from both federal and nonfederal sources. Given our earlier discussion, it is not surprising that engineering and the physical sciences receive so much more support than the social sciences. Most interesting, however, is how the life sciences dominate the picture, consuming $10.828B or 56% of the total. While more than half of this life sciences figure goes for medical research and development, $3.536B, or 18% of the total science and engineering R&D budget, still goes to the biological sciences.

We will discuss the impact of changing academic R&D funding on the preparation of science and engineering professors in greater detail in the next chapter, "New Challenges for the Professoriate."

DEPARTMENTS OF SCIENCE

There are approximately 1500 colleges and universities in the United States and Canada that offer at least a bachelor's degree in one or more of the natural sciences. The most common fields are biology, chemistry, geology, mathematics, and physics. Even schools offering no degrees in these fields will usually offer courses in them as part of a general education requirement. A number of schools also offer degrees in astronomy and geophysics, while a smaller number do so in meteorology, statistics, and other natural sciences.

Since biology, chemistry, mathematics, and physics are taught at most four-year schools, at least the potential pool of openings for science professors is quite broad. This breadth is clearly not the case for fields such as geophysics and meteorology, but, of course, there are also far fewer doctoral graduates in these fields. The ratio of academic openings to available doctoral graduates may even be higher than in high-volume fields like biology and chemistry. However, if the number of such schools is small, then so is the range of opportunities.

Earlier in this chapter, we looked at the relationship among the sciences in terms of a developmental hierarchy. You can see this hierarchy in the course requirements for different science majors. Table 2.7 shows the relationship between mathematics and science course requirements for various mathematics and science bachelor's degrees at Stanford University. Mathematics majors are not actually required to take any science courses to receive their degree. Physics

Table 2.7 Math and Science Requirements for Bachelor's Degrees in the Following Fields at Stanford University

	Bachelor's degree fields			
	Mathematics	Physics	Chemistry	Biology
Require courses in:				
Mathematics	X	X	X	X
Physics		X	X	X
Chemistry			X	X
Biology				X

Source: Stanford University, Courses and Degrees, 1993–1994.

majors must take physics and mathematics. Chemistry majors must take chemistry, physics, and mathematics, and biology majors must take all of the above in addition to biology. Of course, majors in each of these fields often take other science courses. Yet, the actual requirements tell you something important about the hierarchy in science, and also the number of faculty needed in various fields. Mathematics courses of one kind or another are required of every bachelor's degree graduate, no matter what his or her major. This requirement is less so for physics, and even less for chemistry and biology. In almost all schools, mathematics majors are a very tiny percentage of the total student body. But the number of mathematics faculty at these schools can be quite large due to the demand for "service" courses for other majors.

Supply depends not only on the number of doctorates awarded in the various sciences each year, but also on the percentage of such degree holders who seek academic positions. This percentage can vary quite a bit among the different science disciplines. We will look much more carefully at this variation, as well as the whole demand/supply situation, in Chapter 4, "Your Professional Preparation Strategy," and Part III, "Finding and Getting the Best Possible Academic Position."

As a future science professor, it is important for you to have some idea of what goes on in science departments other than the one to which you will be appointed since, as we will see, cross-disciplinary interactions are becoming more common. A first step is to take a look at what is taught in other fields. The easiest way to obtain this information is to peruse course descriptions in college catalogs or on the Internet. Also, take the time to wander around the buildings, classrooms, offices, and laboratories of other science departments. It can actually be quite interesting to compare the activities, layouts, overheard conversations, displays, and even laboratory smells with those of your own department. The differences will be quite revealing, and can tell you much about the professional activities of your future colleagues.

DEPARTMENTS OF ENGINEERING

There are approximately 425 colleges and universities in the United States and Canada that offer at least a bachelor's degree in one or more fields of engineering or engineering technology. Represented across these 425 schools are over 261 different engineering degree programs ranging from aerospace engineering, to manufacturing systems engineering, to nuclear engineering. Some of these programs, such as computational and neural systems engineering (California Institute of Technology) and fire protection engineering (University of Maryland, Worcester Polytechnic Institute), are unique to one or two schools. Others, such as electrical engineering, civil engineering, and mechanical engineering, are found in almost all the schools.

The ten most common graduate engineering departments, in decreasing order, are as follows [13]:

- Electrical engineering/electrical and computer engineering
- Mechanical engineering
- Civil engineering/civil and environmental engineering
- Chemical engineering
- Computer science/computer systems engineering
- Industrial engineering/engineering management
- Materials sciences and engineering
- Nuclear engineering
- Aeronautics/astronautics engineering
- Biomedical engineering.

There are many more electrical engineering departments than there are industrial engineering departments, and as you would expect, there are fewer faculty openings in any given year in industrial engineering than in electrical engineering. However, there are also fewer Ph.D. graduates in industrial engineering than in electrical engineering, so the ratio of the number of openings to the number of graduates could be the same or even higher in industrial engineering. Of course, it could also be lower. The situation is complicated further by the fact that approximately 90% of industrial engineering Ph.D.s seek academic positions after graduation, whereas approximately 30% do so in electrical engineering. As with meteorology and physics, the number of schools from which you will have opportunities is much smaller in industrial than in electrical engineering. Since virtually any school that offers engineering has an electrical engineering department, your potential pool of schools covers all regions of North America and all types of institutions (Research, Doctoral, Master's, and Baccalaureate).

As with future science professors, future engineering professors should have some idea of what takes place in other engineering departments. The comment made earlier with respect to college catalogs, the Internet, and looking around different science departments applies to engineering as well. Engineering fields are connected to each other. There is much overlap among them, particularly in fields derived from similar scientific disciplines. Table 2.8 shows the relationships among 10 engineering fields and 28 disciplines that support these fields. Fluid mechanics, for example, is fundamental to aeronautics and astronautics, chemical engineering, civil engineering, mechanical engineering, and petroleum engineering. The controls discipline is fundamental to aeronautics, electrical engineering, and mechanical engineering. The study of thermodynamics is critical to aeronautics and astronautics, chemical engineering, civil engineering,

Table 2.8 Typical Disciplines in Engineering Departments

DISCIPLINES	A&A	Ch E	Ci E	C Sc	C Sy	EE	In E	MS	ME	PE
Artificial Intelligence				■						
Chem. Separation & Process Control		■								■
Circuits						■				
Contract Tort & Environmental Law			■							
Controls	■					■			■	
Crystallography								■		
Databases				■						
Design	■	■	■						■	
Economic and Financial Analysis							■			
Electromagnetics						■				
Electronics (Materials)						■		■		
Fluid Mechanics	■	■	■						■	■
Heat, Mass, and Energy Transport		■						■	■	
Information Systems & Processing				■	■					
Management							■			
Manufacturing & Production							■		■	
Materials	■		■					■	■	
Operations Research							■			
Organizations			■				■			
Planning			■				■			
Risk Analysis							■			
Software Design				■						
Software Theory				■						
Structural Mechanics	■		■						■	
Systems	■	■	■	■	■	■	■		■	
Telecommunications					■	■				
Thermodynamics		■							■	■

Adapted with permission from David Fryberg, Civil Engineering Department, Stanford University.

electrical engineering, materials science, and mechanical and petroleum engineering. This distribution of disciplines across departments can form the basis for faculty cross-disciplinary collaborations, a subject we will turn to next.

INTERDISCIPLINARY COLLABORATION

> The best institutions of the future are those that can reorganize themselves to address scientific and educational questions in an interdisciplinary way. The institutions that will have difficulty are those that keep the same rigid structure that prevents pollination among disciplines.
>
> *Mark C. Rogers, Vice Chancellor for Health Affairs*
> *Duke University* [14, p.1]

Programs to promote interdisciplinary interaction are evident in many university-based research centers that in one way or another attempt to bring collaborators together around common problems or themes. A number of these centers are supported by a combination of outside funds from industry and government. There are currently over 6000 university-related and other not-for-profit centers devoted to research in the physical and life sciences and engineering in the United States and Canada [15, Ch. 3, p.37]. While there is some evidence that a shake-out is occurring in a number of such centers, many will survive and prosper over the coming years.

It is simply no longer true that all problems must be solved in a disciplinary context, and this change is one of the reasons for the prevalence of such centers. True, the disciplinary context does provide a way of focusing thinking and resources, and over the years has resulted in significant advances in all areas of science and engineering. Yet, increasingly, many of the problems faced by society are "systems level" problems whose solutions, if they exist at all, require expertise and perspectives from more than one discipline. Furthermore, working with colleagues in other disciplines can produce fresh insights into one's own discipline-based research.

Another reason for the development of such centers is that industry, under the right circumstances, will support them. As the shrinking federal research dollar is spread over a greater number of institutions, higher education has turned to industry for additional support. This support still provides less than 10% of the total R&D funding at colleges and universities, and is not likely to grow significantly in the near future. However, these funds are usually discretionary. They often support seed projects that form the basis for follow-on government support.

Interdisciplinary centers also expose graduate students to the thinking of other scientists and engineers outside their immediate discipline. Working with a range of individuals who have differing perspectives and skills is excellent training for the interdisciplinary opportunities that await these students as new professors.

Nevertheless, the challenges these interdisciplinary collaborations present are formidable. First, there is the problem of the participants not having a common language or of assigning different meanings to the same words or terms. In the Stanford Integrated Manufacturing Association, with which I am associated, the word "manufacturing" has a very different meaning to a professor of operations research than it does to a professor of mechanical design. To the former, it may mean organizing the work flow in a system with various equipment constraints or bottlenecks; to the latter, it may mean redesigning the equipment to function more effectively.

Then there are the challenges of interacting with industry that result from different cultures, expectations, and reward systems, as well as the conflicts around the tradeoffs between short- and long-term goals. It is not always easy for faculty to work on research problems leading to publications and Ph.D. dissertations, while at the same time meeting industry's desire for shorter term payoffs.

There are also potential conflict of interest issues. As Ami Zusman, coordinator for Academic Affairs, Office of the President, University of California system, points out [16]:

> Industrial support may hinder the flow of research information because industrial sponsors often require researchers to delay release of potentially marketable results. It may also alter research priorities; a 1985 Harvard study found that 30 percent of a national sample of researchers said that they choose research topics based on how marketable the results might be.

We will look at all these issues in greater detail in Chapter 12, "Insights on Research." For now, it is worth keeping in mind the admonition of Robert Tschirgi, former professor of neurosciences at the University of California, San Diego, who notes that [17]:

> Interdisciplinary collaborations occur when the practitioners reach a development in their field that clearly requires input from other fields. It is not something that can be imposed by the administration, or by pious conviction that it is simply a "good thing."

SCHOLARSHIP ACROSS THE DISCIPLINES

In Chapter 1, we introduced the broader concept of scholarship developed by the Carnegie Foundation for the Advancement of Teaching. We also discussed how support for its various forms (discovery, integration, application, and dissemination) differs among academic institutions. Are there also support differences among disciplines within an institution? This question is difficult to answer because no specific surveys have been conducted on the subject. We can glean some insight into the matter by looking at how faculty respond to questions about the importance of teaching and research, and from informal observations by colleagues at various institutions.

There is not much evidence to suggest differences in activity among disciplines with respect to the scholarship of teaching. Measures such as the importance of student evaluations of courses taught and the observations of one's teaching by colleagues and/or administrators in the granting of tenure do not vary much across disciplines [2, Ch. 2, p.30]. There are, however, some differences in terms of teaching or research interests across disciplines. In a recent survey of full-time faculty at all institutions of higher learning, 53% of education faculty, 51% of business faculty, 44% of physical sciences faculty, 33% of biological sciences faculty, and 27% of engineering faculty said their interests lie primarily in teaching as compared to research [2, Table A-26].

What about the scholarship of integration that seeks connections across the disciplines by placing specialties in a larger context? There is some indication that faculty support such efforts. When asked to respond to the statement, "Multidisciplinary work is soft and should not be considered scholarship," 8% strongly agreed or agreed with reservations, 17% were neutral, while a striking 75% disagreed with reservations or strongly disagreed [2, p.19].

The survey did show some significant differences among fields, with 60% of the faculty in the social sciences, 53% in the biological sciences, 42% in the physical sciences, and 39% in engineering strongly disagreeing with the statement. These differences are probably due to the types of problems these disciplines seek to solve. Greg Petsko, Director of the Rosentiel Medical Sciences Center at Brandeis University, points out that biologists work primarily on systems level problems that lend themselves to contributions from a number of subdisciplines. The same can be said for the social sciences. Yet, there is also some movement in this direction in engineering, and even in the physical sciences, as investigators in these fields respond to industrial and societal problems not restricted to disciplinary boundaries. We will examine this development more closely in the chapters to come.

Of course, there is a difference between voicing support for the scholarship of integration and valuing it in the retention, promotion, and tenure process. Steve Benowitz, writing in *The Scientist,* notes that: "most universities remain bound by traditional departmental structures for administrative and curricular purposes, including peer review, tenure and promotion" [14, p.4]. He goes on to point out that: "many academic administrators' advice, until young faculty have established a track record within a discipline, is that publications should be in that discipline and not outside or shared by too many colleagues [14, p.4].

Finally, what about the scholarship of application? In general, fields such as engineering, business, education, and agriculture are more open to this type of work than are the natural sciences. Within academia, engineering is usually viewed as an applied field, in part because of its close association with industry. How does this view impact tenure and promotion decisions? While the research emphasis in academic engineering remains focused on the discovery of new knowledge, there has been some movement toward acceptance of a more applied scholarship. In recent years, the School of Engineering at Stanford University has awarded tenure or promotion to a number of faculty for what is essentially applied research based in large measure on responses to industry problems in such areas as microprocessor design, compiler development, hydrology, rapid prototyping, and supply chain management. For the most part, the evaluation of these contributions remains in the hands of faculty, but even here, there has been an increase in the acceptance of letters of support from researchers in government and industry.

Science has its applied side as well. It would be a mistake to put all of engineering on the technology side of the ledger and all of science on the theoretical side. Robert McGinn, professor of industrial engineering and engineering management at Stanford University and acting director of its Science, Technology and Society Program, points out that engineering existed long before science, and that the relationship between the two has evolved into "an intimate association of mutually beneficial interdependence" [15, Ch. 3, p.19]. Even a single individual's activities often defy simple classification as "science" or "technology": for example, a molecular biologist creating an organism with desired commercial properties may at times function as an engineer, at times as a scientist, and at times as both simultaneously. This dual role might also be true for an electrical engineer designing a microprocessor or a low-power battery. In this sense, a technical activity can have a dual, scientific–technological character. Modern science and technology are not only interdependent, they overlap [15, p.27].

Vignette #2 | **Science at a Metropolitan University**

Master's institutions, many of which are also called Metropolitan Universities, face unique challenges with respect to science and engineering education. The following vignette looks at some of these challenges from the perspective of a dean of science at San Jose State University in San Jose, CA.

Gerald Selter
San Jose State University

What is it like to be the dean of science at a large, public, metropolitan university undergoing major changes in mission, funding, student composition, and industrial relations? "Well it's not easy, but then again, it's certainly not boring," says Gerry Selter, dean of the College of Science at San Jose State University, a Master's I institution of over 26,000 students located in San Jose, CA.

Selter has no trouble listing a dozen or so challenges that he and many other deans are currently facing. These include: responding to state funding cutbacks and the subsequent need to find additional sources of support; figuring out how to recruit and support new faculty in an area known for its high cost of living; determining the appropriate research/teaching/service mix for faculty retention, tenure, and promotion; assessing the impact of advanced technology on faculty productivity; dealing with faculty accountability pressures from state legislatures, satisfying demands for a more interdisciplinary curriculum; promoting industrial interactions at a campus in the heart of Silicon Valley; and deciding on the proper relationship between the College of Science and an extension education program that enrolls an additional 30,000 students.

Unrelated as some of these issues might first appear, they are often connected in interesting ways. For example, cutbacks in state funding have led to prohibitions on faculty raises beyond the cost of living. This, in turn, has resulted in some pressure to promote faculty as the only means of providing real salary increases. "We have to be very careful here," says Selter. "It dilutes the promise of scholarship if we promote for the wrong reasons." With Selter's support, these pressures have led to a reexamination of the requirements for promotion and tenure, which in turn has led to a more comprehensive understanding of what it takes be successful as a science professor at this particular university.

"In the College of Science we have specifically moved to recognize different forms of scholarship as the criteria for RTP (retention, tenure, and promotion)," notes Selter. He continues:

> Five years ago, we were striving to be like the University of California. That has changed, and it was a big breakthrough for us to recognize and acknowledge that we are a metropolitan university and that teaching is really important. We believe that there are many ways to contribute, but high-quality teaching is absolutely essential. Other forms of scholarship can, and do impact teaching, so we want to provide for a diversity of contributions. One model does not have to fit all. You can make your contribution through research, applications, and service to the University, but these must be in addition to a significant teaching contribution.

Another compelling issue is faculty productivity. If the state of California mandates productivity increases, but provides little financial support for doing so, what do you do? Well, for one thing, you figure out how to get the college and the rest of the university more involved in raising funds from industry and through other sources such as extension education. Selter's strategy is to get faculty involved in fund raising by supporting them through time buy-outs to write large college-wide grants. His first attempt was successful. He gave a faculty member full release time for a semester to write a proposal to the National Science Foundation entitled, "Collaborative for Excellence in Teacher Preparation." According to Selter:

> This is a five million dollar, five year project encompassing San Francisco State University, four community colleges, and numerous high schools. The goal is to significantly increase the number and quality of underrepresented students

who become science and mathematics teachers, and to infuse multimedia technology into K–12 classroom instruction. This is exactly the kind of project (to effect systemic reform) that characterizes us as a metropolitan university.

These are just a few of the many challenges that Selter and deans like him face every day. And yet, Selter, who twice received his department's Professor of the Year Award, seems up to the task. As one biology professor puts it, "He is very open with the faculty and consequently we know where we stand on all sorts of issues of importance to us." A chemistry professor echoes this view by noting, "He works extremely hard and always puts the needs of the college above his own. He has a great deal of respect from the faculty for his efforts on our behalf." Another professor in the college commented that Selter is still new on the job, and that with an even "newer" president on board, the jury is still out. "Yet," she says, "I have confidence that he has the ability to represent us well in what are going to be some difficult times ahead."

Selter has experienced a meteoric rise from professor in the Chemistry Department to acting department chairman, to associate dean, to interim dean, to dean, all in just two years. "I didn't seek any of these positions," he says, "but at this point I have to admit that I do find it quite challenging."

SUMMARY

We began this chapter by examining the similarities and differences among disciplines across the academic institution. We saw that there is a discipline hierarchy based on perceived development of the field. This hierarchy helps to determine such factors as status and prestige, politics and influence, financial compensation, types of students, ease of publication, and expenditures for research and development. We then took an overall look at science departments and engineering departments, the subjects taught in each discipline, and the possible relationships among disciplines within science and engineering. These relationships can form a basis for cross-disciplinary collaboration. The trend toward such collaboration was discussed next, with a look at both the prospects and problems associated with such interactions. We then returned to the model of scholarship introduced in Chapter 1 with a look at differences in such scholarship across disciplines. The chapter concluded with a vignette on the issues faced by Gerry Selter, dean of the College of Science at San Jose State University, a large master's-granting institution in San Jose, CA.

This is an unsettling time for academic science and engineering. Significant forces are at work, both internal and external, that will almost certainly transform the way science and engineering is carried out at colleges and universities. In the next chapter, we conclude our setting of the academic stage by examining some of these forces and their implications for tomorrow's professors of science and engineering.

REFERENCES

[1] P. Altbach, "Problems and possibilities: The American academic profession," in P. G. Altbach, R. O. Berdahl, and P. J. Gumport, Eds., *Higher Education in American Society*. Amherst, NY: Prometheus Books, 1994, ch. 10, pp. 232–233. Copyright ©1994. Reprinted with permission of the publisher.

[2] E. Boyer, *Scholarship Reconsidered: Priorities of the Professoriate*. Princeton, NJ: The Carnegie Foundation for the Advancement of Teaching, 1990.

[3] T. Kuhn, *The Structure of Scientific Revolutions*, 2nd ed. Chicago, IL: University of Chicago Press, 1970, pp. 200–210.

[4] S. Cole, *Making Science: Between Nature and Society*. Cambridge, MA: Harvard University Press, 1992, ch. 3.

[5] J. Beyer and T. Lodahl, " A comparative study of patterns of influence in United States and English universities," *Administrative Science Quarterly*, vol. 21, pp. 104–129, 1976.

[6] J. Pfeffer, *Managing with Power: Politics and Influence in Organizations*. Boston, MA: Harvard Business School Press, 1994.

[7] *Education Quarterly Review* (catalogue 81-003). Ottawa, Ont., Canada: Statistics Canada, 1994, p. 4.

[8] National Science Board, *Science and Engineering Indicators—1996*. Washington, DC: U.S. Government Printing Office, 1996 (NSB 6-21), ch. 2, p. 2–20.

[9] S. Taylor and J. Martin, "The presented-minded professor," Stanford Graduate School of Business, Research Paper Series #717, Dec. 1983.

[10] *The Chronicle of Higher Education—Almanac Issue*, vol. XLII, no. 1, p. 22, Sept. 1, 1995.

[11] D. K. Magner, "The new generation," *The Chronicle of Higher Education*, vol. XLII, no. 21, pp. 17–18, Feb. 2, 1996.

[12] *Physics Letters*, 122B, no. 1, p. 103, Feb. 24, 1983.

[13] *Directory of Engineering Graduate Studies and Research, 1993*. Washington, DC: American Society of Engineering Education, 1993, pp. 526–533.

[14] S. Benowitz, "Wave of the future: Interdisciplinary collaborations," *The Scientist*, vol. 9, no. 13, June 26, 1995.

[15] R. E. McGinn, *Science, Technology and Society*. Englewood Cliffs, NJ: Prentice-Hall, 1991.

[16] A. Zusman, "Current and emerging issues facing higher education in the United States," in P. G. Altbach, R. O. Berdahl, and P. J. Gumport, Eds., *Higher Education in American Society*. Amherst, NY: Prometheus Books,

1994, ch. 15, pp. 352–353. Copyright ©1994. Reprinted with permission of the publisher. Note: The following author's note accompanied this passage: National Science Foundation, *Selected Data on Academic Science/ Engineering R&D Expenditures FY 1989* (Washington, DC: NSF, 1990); R. L. Geiger, "Research universities in a new era: From the 1980s to the 1990s," in *Higher Learning in America, 1980–2000*, A. Levine, Ed. (Baltimore, MD: Johns Hopkins University Press, 1993, pp. 67–85); J. L. Nicklin, "University deals with drug companies raise concerns over autonomy, secrecy," *Chronicle of Higher Education*, pp. A25–26, Mar. 24, 1993.

[17] R. D. Tschirgi, private correspondence with the author, Dec. 8, 1995.

CHAPTER 3

New Challenges for the Professoriate

> Higher education is not in danger. But we would be wise to ask whether the particularly quaint way in which universities now do their work will survive the transformation of information technology. It may, but I don't think so. I expect to see major changes—changes not only in execution of the mission of universities but in our perception of the mission itself.
>
> *William A. Wulf*
> *University of Virginia in Charlottesville* [1, p.47].

Changes are underway in academia that will have an important impact on tomorrow's professors of science and engineering [2, pp.125–131]. Some of these changes, such as the tremendous increase in the use of communication tools, for example, the Internet, and analysis tools, for example, computer modeling, have already pervaded the research establishment, and are moving into teaching and curriculum development. The direction, duration, and impact of other changes, such as the level of government funding, supply of graduate students and postdocs, and number of Ph.D. programs, are not as easy to discern. In addition, mixed signals are appearing in a number of areas. Do government agencies want more or less university interaction with industry? Is a shift underway toward more basic research support from federal sources? Is another shift imminent in the way industry funds academic research? What impact will the new education delivery systems, including such things as "private teaching companies," have on the teaching activities of typical faculty, the structure of academic departments, and even the locations where students go to receive an education? While universities are not going to return to the world from which they are emerging, it is also not entirely clear just where they are heading.

Transitions are unsettling, and under such circumstances, says John Hopcroft, dean of engineering at Cornell University, "The best thing you can do is help prepare your faculty for change." To succeed in this environment, faculty must become much more versatile and broad-based than ever before. At the same time, they must also develop a depth of knowledge in an area where they can make a unique contribution.

We begin this chapter with a look at some of the factors likely to bring about significant changes in teaching and research. We then discuss the implications such changes have for the broader view of faculty scholarship. We conclude with a vignette describing a cross-university collaboration using state-of-the-art communication tools. Along with Chapters 1 and 2, this completes our setting of the stage for the work to follow on preparing for an academic career in science and engineering.

FORCES FOR CHANGE IN TEACHING AND RESEARCH

Not all the forces affecting higher education are well understood at this time, and some important new ones may still emerge. Furthermore, some forces are pulling in opposite directions, making more than one outcome possible. Nevertheless, as illustrated in Figure 3-1, the following factors seem likely to have a significant impact on the future of teaching and research in science and engineering:

- The increasing use of communications tools

- The increasing use of computational tools

- The increasing focus on interdisciplinary programs

- The prospects of decreased government funding

- The increasing costs of doing research

- The changing role of industry in academic research.

These factors form the basis for a set of preparation activities to be discussed in Part II, "Preparing for an Academic Career." At a minimum, you will be expected to be familiar with the issues they raise. If you also have some direct

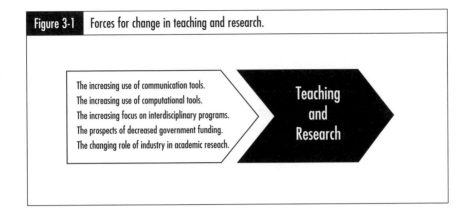

Figure 3-1 Forces for change in teaching and research.

experience with at least a few of them, your appeal as a prospective faculty member will be strengthened considerably.

As we will see, many of the above factors are helping to facilitate a move toward a scholarship that includes not only the generation of new knowledge, but also its integration, application, and dissemination. Let us then examine each of these factors in turn, followed by a look at their implications for current and future science and engineering professors.

The Increasing Use of Communications Tools

> While we may not end up with a "calculus teacher for the nation", there is no question that the new telecommunications technologies are going to challenge traditional teachers and traditional teaching in fundamental ways.
>
> *Thomas Magnanti*
> *Massachusetts Institute of Technology*

There is still a great deal of hype with respect to the communications revolution, and some of it is not even new. You have only to read articles about how the expansion of radio in the 1920s would reduce the need for more universities, or the predicted effect of television and videophones on education in the early 1950s to realize that political, economic, and social considerations have as much to do with whether things really change as does the actual technology. Yet, certain inventions, for example, the telephone, had a utility so high and a cost so low that their value was quickly realized. We are now at a point with the new communications technologies where their impact on the teaching and research communities could be quite profound. Digital libraries, multimedia, CD-ROM, the Internet, and videoconferencing are increasing in popularity, not only because they make it possible to do old things in new ways, for example, video broadcasts of traditional "talk and chalk" presentations, but they also make it possible to do new things in new ways, for example, the electronic sharing of class notes among students.

How many times, in how many courses, and at how many locations do we need to have faculty lecturing on the third law of thermodynamics? Could not these lectures be modularized in one set of outstanding presentations, available on demand to students in such courses as physics, chemistry, and mechanical engineering? If so, what impact would this change have on departmental course offerings, funding allocations, and even the demand for faculty?

The National Science Foundation sponsored Synthesis Coalition is an example of a program making extensive use of communications tools in undergraduate engineering education. The Coalition is a union of eight diverse institutions whose goal is to design, implement, and assess new approaches to undergraduate teaching and learning. These approaches include an emphasis on multidisciplinary synthesis, teamwork, communication, hands-on and laboratory experiences, open problem formulation and solving, and the use of "best practices" from industry [4].

According to Jeff Aldrich, Synthesis information resource coordinator [4]:

> The Synthesis Coalition's strategy in pursuing these goals is distinctive in its focus. We promote the innovative use of information technologies, to attract and to inspire teachers and learners with approaches that go far beyond current teaching and learning strategies. And we concentrate on the extension and integration of engineering knowledge—across disciplines, among institutions, and within society as a whole—throughout the individual's career.

With respect to research, communications advances provide new and different possibilities for collaboration among investigators next door and around the world. As Stephen Stiger, professor of statistics at the University of Chicago, puts it [5]:

> It seems plausible that [the] expanding electronic network will eventually lead to a weakening of our sense of institutional identity and a fundamental change in the intellectual competition that organizes our enterprise. Individual faculty may be in closer contact with collaborating colleagues at other universities (or with graduate students working under their direction in other countries) than with faculty and students in slightly different specialties down the hall. The importance of the geographic unit may be eclipsed by intellectual disciplinary units that are international in scope.

Communication tools allow for more than just the sharing of information via e-mail or on the World Wide Web. They make possible real-time collaboration, including the remote sharing of data, the operation of equipment, and the carrying out of experiments. Hypermedia—hyper links to documents, video, and sound—dramatically increase the information bandwidth, and are beginning to make long-distance collaboration a reality. Through such "teleresearch," investigators are not only able to see each other via video displays on their desks, but also to interact with shared graphics and other information on their screens. As discussed further in the vignette at the end of this chapter, such collaborations have the potential of breaking what sociologists refer to as the "eight-meter rule," which is based on the observation that the most significant interactions take place among people who are in close physical proximity to each other.

The two prototyping experiments referred to below are to be carried out at geographically distributed sites via the Internet. So, too, is work on problems associated with the AIDS pandemic that involve researchers cooperating and sharing data from sites throughout the world [2, p.126].

The Increasing Use of Computational Tools

> Can you imagine that they used to have libraries where the books didn't even talk to each other?
>
> *Marvin Minsky,*
> *Massachusetts Institute of Technology* [1, p.47]

The communications revolution is built on the even more fundamental computer revolution, and the latter, like the former, has the potential to significantly affect teaching and research. As with telecommunications, it has taken over 30 years for the impact of the computer to be truly felt in teaching. In fact, it was not until the marriage of telecommunications and computing and the development of such things as relational databases and low-cost portable systems, linked through ever-expanding networks, that the real revolution began. The technology is right and the motivation exists in part because of the promise of increased productivity, but more importantly, because of the ability to allow professors to do exciting new things, while at the same time establishing connections among their various types of scholarship.

Two likely influences on how faculty will teach and do research in the near future are the "new" books and libraries made possible by the computer. William Wulf, professor of engineering and applied sciences at the University of Virginia, describes the changes in store for books this way, [1, pp.48–49]:

> They [electronic books] will let you "see the data" behind a graph by clicking on it. They will contain multidimensional links so that you can navigate through the information in ways that suit your purpose rather than the author's. They won't contain references to sources, but the source material itself; for example, the critique of a play will include its script and performance. They will have tools that will let you manipulate equations, trying them on your own data or modifying them to test scientific hypotheses. They will let you annotate and augment documents for use by later readers, thus making a book a "living document."

It remains to be seen where such books will have their greatest utility. In most cases, there will still be the need for someone (the author?) to organize and synthesize information. However, the flexibility this new technology provides is certain to affect scholarship at the undergraduate as well as graduate level.

The quote at the beginning of this section refers to the fact that libraries are moving from collections of books to facilitators of retrieval and dissemination. Libraries will soon have software agents collecting, organizing, relating, and summarizing on behalf of their users. As a result, says Wulf, "They will 'spontaneously' become deeper, richer, and more useful" [1, p.49].

New communications and computer tools have the potential to increase productivity while improving the quality of the learning experience. Yet, faculty need to have the right incentives to invest in such efforts.

A professor at Wayne State University in Detroit used to teach an engineering course with a maximum enrollment of 25 students. This number was the largest he felt he could interact with effectively. Along came the new technologies that made possible distant learners at their workstations at the Ford Motor company, two-way voice communication, instant feedback on "pop quizzes" via electronic keyboards, and excellent graphics that were prepared on his computer in advance and then scanned into the coursework database. The professor found that with the time savings resulting from not having to put drawings on the board, and the ability to have instant two-way communication with any student in the class, he could interact with 125 students just as effectively as with the original 25.

As you can imagine, this increased productivity was terrific from the point of view of the administration. However, for the professor, all was not so great. He was putting in a significant amount of extra time preparing the material that led to the completed graphics, and an even greater amount of time interacting via e-mail with students after class. He probably doubled the time he invested in the course over his previous approach, and yet he was teaching five times as many students even more effectively. The administration, on the other hand, had not yet turned this into a win–win situation by "rewarding" him with a more reasonable teaching commitment.

That computers have made it possible to solve research problems at speeds never before imagined is common knowledge. What is perhaps less well known is their ability to attack problems in entirely new ways. For example, the advent of what are called massively parallel computers makes possible the systematic analysis of many parts of a single problem. This analytic ability enables investigators to study phenomena involving large amounts of data, e.g., enormous assemblies of particulates such as atoms, molecules, cells, and organisms [2, p.127]. These computers are making significant contributions in such diverse areas as image and speech recognition, fluid dynamics, computational chemistry, neural modeling, semiconductor device physics, and weather forecasting.

What is even more remarkable is that computers have made possible the solution of otherwise intractable problems. An example is the use of computers to keep track, decipher, and order information from the worldwide Human Genome project which, when completed, will have generated some 10 trillion bits of information [2, p.127].

These kinds of efforts led a recent symposium sponsored by the American Association for the Advancement of Science to recommend that computing be recognized as "the third branch of science along with theory and experimentation" [6].

The computer's impact has been particularly profound in the area of computational prototyping, which involves the use of high-speed computers to create virtual models of complex real-world systems or artifacts. Paul Losleben, senior research scientist in the Electrical Engineering Department at Stanford University, is a coinvestigator on a program called "Computational Prototyping for 21st Century Semiconductor Structures." An important aspect of this effort includes plans to use computational prototyping as a partial substitute for costly laboratory experiments [7]. (See the vignette at the end of this chapter for further details.)

John Seely Brown, director of the Xerox Palo Alto Research Center in Palo Alto, CA, talks about "the shift from mass and energy to computers and communications." Nowhere can this shift be seen more clearly than in the Department of Energy's recent decision to award Intel Corporation of Santa Clara, CA, a $45 million contract to build the world's fastest computer, "a machine powerful enough to simulate nuclear weapons explosions and potentially halt the need for underground testing" [8].

Other investigators at Carnegie-Mellon University, Stanford University, and the University of California, Berkeley, are researching ways to develop both virtual and physical prototype environments in which users will be able to design, test, and debug a product before it is built. Once the virtual prototype is "built" and "tested" in the computer, the computer will transform the description of the virtual prototype to an actual plan for manufacturing the physical prototype [9].

It is worth noting that not all cross-university research programs are limited to Research I universities. Consider North Carolina's Center for Advanced Electronic Materials Processing. It coordinates research activities among three Research I universities (Duke University, North Carolina State University, and the University of North Carolina at Chapel Hill), and two Master's I institutions (University of North Carolina at Charlotte and North Carolina Agricultural and Technical State University) [10].

The Increasing Focus on Interdisciplinary Programs

As noted in Chapter 2, discipline-based research provides the core of our knowledge about the universe. Such research has led to many of the fundamental breakthroughs in science and engineering. Discipline-centered research also provides the structure around which academia is organized, and is the established way of assessing faculty research quality and productivity. Such research is in no danger of disappearing.

However, increased attention is also being paid to basic and applied research activities requiring the contributions of two or more disciplines and yielding new areas of inquiry and application. Terms such as biophysics, biochemistry, bioengineering, and now biotechnology highlight intersections between biology and other science and engineering disciplines. Astrophysics, geophysics, physical chemistry, and electromechanics are just a few more illustrations of the development of cross-disciplinary fields.

At Stanford University alone, there are over 22 interdisciplinary research centers. Examples include the Center for Computer Research in Music and Acoustics, the Center for Integrated Facilities Engineering, the Center for the Study of Language and Information, and the Western Region Hazardous Substance Research Center, as well as the two centers with which I am associated, the Center for Integrated Systems and the Stanford Integrated Manufacturing Association.

Industry has long understood that the "real world" is not divided along disciplinary lines. As Venkatesh Narayanamurti, dean of engineering at the University of California, Santa Barbara, notes: "While academe has only in the last decade or so begun to embrace the interdisciplinary approach, industry has promoted such collaborations as part of its ethos" [11]. "Systems level" problems ranging from the development of computers and telecommunications networks to automobiles, airplanes, and spacecraft all require the contributions of many basic and applied fields.

Many of the basic science and engineering challenges lie at the intersection of disciplinary boundaries. One example is the emerging field of "smart, or intelligent, materials and structures." Here, investigators with backgrounds in chemistry, materials science, biology, mathematics, computers, and engineering cooperate in developing human-made artifacts that, like their creators, sense and respond to their environment by learning, adapting, and repairing themselves [2, p.129].

In the life sciences, the widespread use of new molecular techniques has blurred the distinctions among many disciplines. In the words of Doug Randall, a plant biochemist from the University of Missouri, "Geneticists use molecular biology and chemistry techniques, chemists use molecular biology and genetics, and plant pathologists need to know molecular biology to study disease resistance" [11].

In some cases, disciplinary combinations can yield whole new fields of inquiry. According to Paul DeHart Hurd, professor emeritus of science education at Stanford University [2, p.130]:

> The contexts of science today, particularly problems that are nonlinear or which have crossdisciplinary character, do not respond to the long standing modes of scientific inquiry. A new science of 'chaos' is being developed to deal with problems that are now seen as disorder, such as weather, tides, turbulence, living populations

of animals, and the behavior of nerve impulses. These are among the messy problems into which chaos theory portends more insight than we have now.

Interdisciplinary research has its share of problems, not the least of which is how to provide rewards (tenure and promotion) and awards (funding) when both universities and funding agencies are still locked into a disciplinary structure and way of thinking. As we will see in the next section, however, there is some evidence that this way of thinking is beginning to change.

As noted earlier in our discussion of the National Science Foundation sponsored Synthesis Coalition, interdisciplinary programs are finding their way into the curriculum, particularly at the undergraduate level. The ability to integrate studies in several technical areas, as well as across economic, social, and political boundaries, is increasingly important. Problems in environmental planning, communications, manufacturing and transportation systems, and urban planning are just a few of the possibilities that come to mind. The University of Michigan, for example, has a number of undergraduate interdisciplinary programs that are both college-wide and university-wide.

Another example of cross-university education collaboration takes place among the schools in the National Coalition for Manufacturing Leadership. The participating schools are: Arizona State University, Auburn University, California State Polytechnic University at San Luis Obispo, California State Polytechnic University at Pomona, Colorado State University, Georgia Institute of Technology, Cornell University, Massachusetts Institute of Technology, Northwestern University, Purdue University, Pennsylvania State University, the University of Michigan, the University of New Mexico, the University of California at San Diego, and Wayne State University. All of the schools have collaborating engineering and management programs, internships for undergraduate and graduate students, and industry participation in their project courses. The schools work together to share graduate curriculum materials, as well as course syllabi and class notes. In some cases, students enrolled in one school take courses via distant learning at other schools in the coalition [12].

The new communications tools create intriguing opportunities to increase the connection and overlap between interdisciplinary research and other forms of interdisciplinary scholarship, including teaching and curriculum development. We will look more closely at these opportunities in the next section.

The Prospects of Decreasing Government Funding

> Congress has placed everything in science "on the table." This includes not just the federal programs that support university research, but also fundamental assumptions about the appropriate relationships between government and academic science.
>
> *Alan G. Kraut*
> *The Chronicle of Higher Education* [13]

As we saw in Chapter 2, federal spending for nonmilitary science grew fairly steadily in the 1980s and early 1990s [14, p.1]. However, when inflation and other increases in the cost of doing research, such as university overhead, are taken into consideration, the rises are less significant (but at least they have been increases). The future is of great concern, however. As the following quotations illustrate, prospects for overall growth in most, if not all, areas are slim indeed:

> Federal support for civilian science is destined to weaken and shrink no matter what its budget, declining to as much as a third if inflation is taken into account. Such cuts portent wide changes in American science and American life.
>
> William J. Broad, science writer
> The New York Times [14, p.2]

> State governments are tightening their budgets, with some public universities experiencing absolute decreases of 20–25% in state funding. This has reduced the ability of state universities to hire scientific and engineering faculty and to fund graduate students. Many state legislators view graduate education as a budget item that must compete with social requirements whose call on the tax dollar is at least as persuasive.
>
> Committee on Science, Engineering and Public Policy [15]

The situation is no better in Canada. There, the federal government has instituted deep cuts in the transfer payments it makes to the provinces. In the past, these payments have been earmarked in specific percentages for such things as transportation, health, and education at all levels. The federal government has loosened up on these restrictions, and the provinces are now free to allocate the funds wherever they choose. This unrestricted allocation means that higher education must compete not only with primary and secondary education, but with the other service sectors as well. As James Sharp, professor of civil engineering at Memorial University of Newfoundland in St. John's, Nfld put it, "We are no longer at the top of even the education priority, so who knows were we are in the whole scheme of things."

The Increasing Costs of Doing Research

The prospects of significant decreases in research funding come at a time when the cost of doing research continues to increase. Some of this increase is due to greater research complexity, resulting in more expensive equipment and other fixed infrastructure costs. A laser costing $15,000 15 years ago now requires $150,000. This type of increase is also true for computers. While the cost per computation has gone way down, the computation needs and costs to support research have gone up greatly.

I believe another reason for the escalating costs is the need to support increasing numbers of graduate students and postdocs (variable costs). Prior to

1990, when faculty wrote research funding proposals, they would look first at the desired research, then decide how many students were required to execute the tasks. The reverse is now occurring in many areas, where the number of graduate students who need to be supported is driving the "need" to do a certain level of research. Some indication of this "need" can be seen in high-energy physics, where both the very expensive facilities and the huge number of graduate students and postdocs are leading to the "necessity" to do even more high-energy physics research. You cannot support a multimillion dollar facility on a contract that has only a few graduate students—the research results would not justify the investment. Hopefully, there is a positive correlation between the number of graduate students and postdocs doing research and the quality and quantity of research output, but this is not always the case.

Leon M. Lederman, president of the Board of Directors of the American Association for the Advancement of Science, sums up the cost of research this way [16, p.5]:

> In all areas of research the last decade's 'easy' problems have been solved, and the cost of creating new understanding of nature has increased considerably. Finally, new regulatory requirements have added to overhead costs and reduced the funds available for the direct costs of research.

The Changing Role of Industry in Academic Research

As noted in Chapter 1, industry's involvement in academic research, particularly in engineering, goes back to the middle of the last century. It was not until the post-WW II period, however, that such support began to grow significantly. While never close to the level provided by government, it nevertheless has played a crucial role in the establishment of certain research programs, particularly those that provide start-up, or seed project grants not funded by federal agencies.

The 1980s saw a significant expansion of these programs, particularly in the form of new university–industrial consortia and in the direct funding of research laboratories.

There are now over 1500 industry–university consortia in the United States and Canada, ranging in size from one or two faculty and the same number of companies, to much larger operations with dozens of companies and affiliated faculty, and annual membership fees into the hundreds of thousands of dollars. They are often interdisciplinary, bringing together investigators from a number of academic departments and industry around common, overarching themes such as telecommunications, manufacturing, polymers, molecular and electronic nanostructures, and human–computer intelligence.

For example, the University of California at Davis uses "organized research units" to form interdisciplinary centers that seek industry and government funding. One such unit for the study of polymers brings together researchers from agriculture, textiles, chemistry, and chemical engineering [11].

Other examples include stand-alone consortia, such as the Semiconductor Research Corporation (SRC) in Raleigh, NC, which uses industrial membership fees to support programs in semiconductor research at universities across the United States. The application process to the SRC is similar to that for government agencies such as the National Institutes of Health, the National Science Foundation, and the Advanced Research Projects Agency.

Initially, much of industry's interest was motivated by the desire to support research of potential benefit to society as a whole (philanthropy), and to have access to promising graduate students and postdocs who might consider working for such supporters.

With the retrenching of the late 1980s and early 1990s, there has been a change in motivation. A number of companies have pulled back from such support, as their corporate laboratories have been cut back to focus on shorter term deliverables. As the recent report, *Reshaping the Graduate Education of Scientists and Engineers* [15], points out:

> The last decade has seen both a rise in international economic competition and cutbacks in basic research at large industrial laboratories. Industry is said to be hiring fewer scientists and engineers and shifting emphasis toward core businesses; industrial grants to universities, an important source of research funds, are said to be reduced and increasingly directed toward incremental, low-risk programs.

Support for such centers may be on the increase again, although with objectives that are still mostly focused on short-term deliverables. The situation varies among consortia, but in all cases, important issues having to do with intellectual property, conflict of interest, and the very role and purpose of university research are raised. As Susan Zolla-Pazner, professor of pathology at New York University Medical Center, notes with respect to biotechnology [17]:

> This interaction among scientists, universities and industry is not new. But the decrease in government support for research, combined with the explosion of new biotechnology products, has intensified the relationship. It is now more productive—and more complicated.

Clearly, industry–university relations are undergoing a transition just as are those between the government and the university. Both will continue, albeit with changing goals and objectives. We will look at the implications this change has for science and engineering faculty in the next section. We conclude here by noting that, overall, faculty seem to want more, not less involvement with industry. With the very real prospect of decreasing government funding, it is not surprising that a recent survey of engineering professors found that some 79% wanted either somewhat more or much more involvement with industry, and only 6% wanted less involvement [18].

IMPLICATIONS FOR FACULTY SCHOLARSHIP

How will the above factors impact faculty scholarship? This question is not so easy to answer, partly because, as noted earlier, many of the forces for change are pulling in opposite directions. However, answers are important, not only because of what they say about your future life as a professor, but because of what they imply about your preparation for an academic career. As a faculty member, you will need to find an appropriate balance, as shown in Figure 3-2, among: (1) cooperation and competition, (2) basic and applied research, and (3) high-risk and low-risk behaviors. Let us look at each of these in turn with an eye toward their implications for faculty scholarship.

Figure 3-2 Finding the appropriate balance among competing behaviors and activities.

Balance Between Cooperation and Competition

Over the next few years, some faculty will become more pessimistic, while others will become more optimistic, depending to a large extent on their versatility and flexibility. Decreasing prospects for external funding, coupled with the increasing costs of doing research (including the time required to develop proposals and grants) have certainly discouraged a number of professors. As a physics faculty member at the California Institute of Technology put it: "Every time you write a proposal for a renewal of your grant, you are playing Russian roulette with people's lives. You soon find that your chief responsibility is no longer to do science at all; it is to feed your graduate students' children" [16, p.5].

One consequence of funding restrictions is that many faculty will need to explore multiple sources of support for specific projects and programs. This approach has the advantage of allowing an investigator to remain untied to the needs and wishes of a particular funding entity. Of course, multiple sources usually mean multiple reports, multiple interactions, and multiple efforts that go into avoiding, or at least reducing, misunderstandings. Nevertheless, John H. Gibbons, director of the Office of Science and Technology Policy, feels the multiple source approach is still the best strategy [19]:

> You can get over-dependent on any one source of funds, just like a farmer can get over dependent on one kind of seed and run into a bad situation by being overspecialized. So it's a healthy thing for the nation for universities and industry to have a closer association than they have in recent decades.

There are indications that as the research funding pie gets smaller, many faculty become less willing to share their research results with others, are more critical reviewers of others' submissions to referred journals, are more disparaging of competing research proposals, and are less willing to collaborate on research projects. As Leon M. Lederman of the American Association for the Advancement of Science comments [16, p.11]:

> ...scientists are also increasingly viewing their fellows as competitors, rather than colleagues, leading to an increasingly corrosive atmosphere. The manifestations of this attitude range from a reluctance to share new results with other scientists, to public bickering about relative priorities in funding different fields.

A biology professor at the University of Minnesota put it more directly when he said, "I feel pressure not to share information about exciting data...for fear that others, with larger labs, will be able to use this information to write a similar proposal" [16, p.10].

Yet, the opposite is also occurring in some areas. Decreasing funding prospects, increasing costs of doing research, increasing availability of communication and collaboration tools, and the increasing emphasis on interdisciplinary research have led to greater collaboration, not only among members of the same department, but across departments and even universities. As we will see in the vignette at the end of this chapter, such collaborations can include a variety of schools, not just major research universities.

Cooperation works particularly well in interdisciplinary areas, where each faculty member or team makes a unique contribution to the overall effort. It is also effective with expensive facilities that can be shared with each user conducting his/her own experiments.

An example of the former is a group of Stanford professors, one in robotics, one in composite materials, and one in cost accounting, who pooled their talents to develop a method of automatically laying up composite materials using robots in such a way that the costs of such procedures can be accurately assessed in advance.

An example of the latter is the Kitt Peak National Observatory near Tucson, AZ, where astronomers from all over the world share time on the many very expensive telescopes at the observatory complex. Geographically distributed research enterprises are occurring across science and engineering, and are likely to be a significant mode of operation in the decade to come.

As noted earlier, interdisciplinary research and teaching are not easy. Competition for scarce resources, departmental territoriality, and the difficulty in obtaining proper recognition are just a few of the many barriers to such efforts. However, encouraging signs exist. Industry, and even some federal agencies, have begun funding interdisciplinary activities. According to Robert Shelton, vice chancellor of the University of California, Davis [11]:

> More and more federal funding agencies, such as NSF, the Department of Energy, as well as private industries insist upon it [interdisciplinary research]. Many of them want a clustering of expertise at a single institution to solve problems.

Greg Petsko, who left the discipline-centered world of the Massachusetts Institute of Technology Chemistry Department for the interdisciplinary opportunities at the Rosentiel Center at Brandeis University, notes: "The most exciting areas of chemistry are where disciplines rub together. Things are moving so fast that the lines of disciplines are blurring"[11].

Although discipline-based research is still the basis for the vast majority of tenure decisions, as we will see in Chapter 14, "Insights on Tenure," Brandeis is rewarding a number of pretenure faculty for their interdisciplinary, multiinvestigator research.

Advances in interdisciplinary scholarship are also finding their way into the curriculum. Faculty cooperation across disciplines in research can often lead to similar cooperation in the development of interdisciplinary courses, and vice versa. One example is the increasing use of engineering project teams comprised of students from different backgrounds working together to design, build, and evaluate projects of interest to industry. In some cases, members of these teams are located at different universities, and are able to interact through the use of the communications and computer technologies referred to earlier.

Balance Between Basic and Applied Research

We may be seeing a shift in the balance between basic and applied research within the academic environment, although here again the signs are mixed. There is some indication that federal funding will move further toward the support of long-term basic research, with the idea that industry will pick up the cost of shorter term applied research. However, this approach may not be true with all agencies, nor is it clear what the consequences would be if it were. Susan Zolla-Pazner, professor of pathology at New York University Medical School, worries about a transformation from "university to research institute, to industrial subsidiary," noting that, "The promise of continuing industrial support is seductive but inevitably tied to commercial products and the bottom line" [17].

Certainly, industry's research priorities have changed, but what impact will this change have on its support of academic research? Will industry carry out short-term research on its own and support somewhat longer term research at universities, or will there be an unfunded gap in between? John H. Gibbons of the Office of Science and Technology Policy, for one, is concerned [19]:

> [Industry's] shorter product life-cycles and the international competitive squeeze causes them to have shorter and shorter vision in terms of where they make their future investments. And if they retreat toward the immediate market place, driven by the realities of international competition, and if the government retreats back to fundamental science, then what happens in the middle ground? It's been called the "Valley of Death" for good reason. If you separate fundamental research from its transformation and interaction with technology and subsequent markets, then you're in a real deep hole.

Balance Between High-Risk and Low-Risk Behaviors

There is some evidence that as a consequence of the funding problems outlined above, many faculty are less willing to do high-risk research. Some professors are playing it safe by concentrating on less risky projects that they feel have a higher likelihood of success. By "high risk," one usually is referring to ideas for which there is little preliminary data to support a proposal, a venture into a new field, an attempt to develop a new model or a new organism, or inventing a needed instrument [20, p.1].

Although it has always been true that faculty in general cannot afford to have too many student dissertations that result from inconclusive or negative findings, a recent report, *Science: The End of the Frontier?*, notes that [16, p.11]:

> As a consequence of the increasingly difficult search for funding, academic scientists are less willing to take chances on high risk areas with potentially big payoffs. Instead, they prefer to play it safe, sticking to research in which an end product is assured, or worse, working in fields that they believe are favored by funding agencies.

A professor of biology at Carnegie-Mellon University put it this way [16, p.11]:

> We are tending to do "safer" projects, avoiding the high-risk, but high-payoff projects. With the present climate we cannot afford to have experiments not work.

However, as with competition and cooperation, there is another possibility. Ursula Goodenough, president of the American Society of Cell Biology and a professor at Washington University in St. Louis, does not think keeping risky ideas out of proposals is such a good strategy. "I have seen reviewers say, 'This is fine, everything ought to work, but the answers are boring,' and they give it a middle priority," comments Goodenough. She thinks better advice for young scientists is to give a mix of some tried and true experiments and some neat and interesting ideas, with experiments that will address them [20, p.1].

Both the National Institutes of Health (NIH) and the National Science Foundation (NSF) have special sections to fund what NIH determines are "high-risk/high-impact" proposals [20, pp.8–9]. These "sugar grants," in NSF terminology, provide a small amount of money for a short period to allow an investigator to check out an interesting idea and develop it to the point where it could be submitted for a normal grant [20, p.9].

One thing is clear; we are seeing a shift away from stand-alone projects that do not contribute to a "stream of ideas," and movement towards programs in which the results of each project build on the efforts of previous work and are seen as part of a greater whole. Such programs do not always have to be large; they could involve just a few faculty or perhaps even just one faculty member. But they do need to build over time. They will need to be compelling overall, and not just interesting to the specific investigators. They do not all have to take place at Research I universities, particularly with the coinvestigator possibilities discussed above, but they *will* have to be cutting-edge.

Will the pressures outlined above compel faculty to take fewer risks in their teaching and course development or will the new technologies lead in the opposite direction? William Wulf of the University of Virginia believes that it will be the latter, and that the new information technologies will affect the education process in a profound way [1, pp.49–50]:

More fundamental is the opportunity to involve students in the process of scholarship rather than merely in its results. We like to say that we teach students to think, not merely to learn rote facts, but in truth we mostly limit them to thinking about what has been thought before. We can't ask them to explore new hypotheses because of the practicalities of access to sources and the sheer grunt work of collecting and analyzing data. Information technology eliminates those impediments.

I believe a number of faculty will respond to the above factors in ways that lead to an enhancement of the many forms of scholarship, that is, to an improvement in the discovery, integration, application, and dissemination of knowledge. This enhancement will not take place with all faculty at all schools, of course, and one of your tasks in choosing a graduate school, postdoc institution, and starting professorship position is to identify places where such outcomes are more likely. How to do so will be looked at more closely in Chapters 4–9.

Vignette #3	The Laboratory Without Walls

The following vignette describes an innovative, multiuniversity collaboration program made possible through current developments in information and computer technologies.

Paul Losleben
Stanford University

> I've been in this business for over 30 years, and I've never seen anything with the impact of the present revolution in computing and communications. It's like we've passed a threshold, and suddenly things are easy. The Internet is seeing a 15–20% growth rate per month. This is more than just an occurrence; this is a phenomenon.

So says Paul Losleben, senior research scientist in the Electrical Engineering Department at Stanford University. He is also the remote services manager of the National Nanofabrication Users Network (NNUN), a "colaboratory," linking Stanford University, Cornell University, Howard University, and the University of California at Santa Barbara in an interesting new form of research collaboration.

Losleben, who has had a hand in a number of innovations over the years dealing with the way research is conducted, is particularly excited about the Internet development. As he puts it, " It's the first opportunity for researchers all over the world to collaborate in ways never before possible, to not only do old things in new ways, but to do new things in new ways.

"When we think of the Internet or the Web, we think about getting information," notes Losleben. "But, you can also connect to people and to unique pieces of equipment at sites all around the world." Losleben likes to point out the impact such collaboration can have, not only for investigators at major research universities, but also for those at the smaller, less endowed ones. "Many of these smaller universities have first-rate people," he says. "What they often lack is a critical mass of colleagues with whom to collaborate. The Web changes all that. Just as the country doctor can now communicate with physicians at advanced facilities in metropolitan areas, professors at smaller institutions can collaborate, and even share facilities remotely, with their counterparts at larger universities."

Most researchers, of course, already communicate with each other. But they do so via conferences, transactions (publications), and personal correspondence. "Conferences usually take place annually so information from them is often a year old," comments Losleben. Transactions, in addition to having a low bandwidth, often reflect built-in delays between completion of research and publication of up to two years. Personal correspondence is good, but people are often hard to reach. "We particularly want to improve the personal communication part," says Losleben. "Increasing the efficiency here can lead to big productivity gains. By creating new opportunities for communication via the Internet, you expand the pool of ideas available to everyone."

Professors are often at a disadvantage if they only have access to equipment they can support out of their own research program. This disadvantage is particularly true as the cost of equipment and facilities continues to skyrocket. The goal of NNUN is to advance semiconductor technology by sharing ideas and facilities to manipulate and design materials at the atomic level. Equipment availability is critical in this area since the cost of doing research has become so high. As Losleben puts it, "If you can only buy what you can support out of your own research program, you lose; if your ideas are limited to the group at your university, you lose."

Such "laboratories without walls" are now springing up all over North America. Questions such as, "How do I find, evaluate, and connect to such operations?," "When can I use them?," "How much will it cost?," and "How do I pay for it?" can all be answered using the Web. According to Losleben: "Researchers now, or soon will, have unprecedented numbers of resources at their fingertips. You can transact business, engage in buying and selling, test computer models, and run experiments, all from different locations."

Losleben also notes that every published piece of research has a lot of information behind it, such as experimental data, benchmarks, and ideas not pursued. By putting this information on the Web, researchers can benefit from, and build on, the work of others. However, it is now possible to carry this information exchange a step further via what Losleben calls "active papers." Such an approach replaces much of the empirical research currently taking place, with computer models so that one cannot only read the paper, but "run the paper."

> We are also, he feels, on the verge of breaking the eight-meter rule, which states that most meaningful interactions take place among people who are physically within eight or so meters of each other. "With the Internet, you can probably access more resources more easily than at Bell Labs at the height of its glory," says Losleben.
>
> As noted above, Losleben is no stranger to innovation. Prior to coming to Stanford in 1986, he held various positions at the Defense Advanced Research Projects Agency (DARPA). There, among other things, he was responsible for the development of the Very Large Scale Integration (VLSI) fast-turnaround implementation service (MOSIS) that provided fabrication, parametric testing, and packaging of VLSI circuit designs produced by over 300 research and educational institutions geographically distributed across the United States. In some ways, MOSIS was a forerunner of the NNUN program. "But this 'laboratories without walls' collaboration is like nothing I've ever seen," says Losleben. "It's really going to change everything."

SUMMARY

We began this chapter with a look at some of the factors likely to bring about significant changes in teaching and research. These included the increasing use of communications and computational tools, the increasing focus on interdisciplinary programs, the prospects of decreased government funding, the increasing costs of doing research, and the changing role of industry in academic research.

We then discussed the implications these changes have for the broader area of faculty scholarship, as well as the balances faculty strike between cooperation and competition, basic and applied research, and high-risk and low-risk behaviors.

We concluded with a vignette describing a multiuniversity collaboration program using state-of-the-art computer and communication tools.

As tomorrow's professor, you need to know that the challenges and opportunities outlined in this chapter are here to stay. The impact of each will vary somewhat across disciplines and types of schools, but in one way or the other, they will affect all faculty. They are a part of the future academic reality, and are therefore a part of what you are going to be dealing with as a professor.

The question, as you begin to prepare for an academic career, is: "How can I gain an understanding of these issues as well as some experience with them in a way that will better prepare me for my role as a professor, and at the same time make me more competitive as I enter the job market for academic (and possibly nonacademic) positions?" We will look in detail at this important question in Part II, "Preparing for an Academic Career."

REFERENCES

[1] W. A. Wulf, "Warning: Information technology will transform the university," *Issues in Science and Technology*, vol. XI, no. 4, Summer 1995.

[2] P. DeHart Hurd, "Technology and the advancement of knowledge in the sciences," *Bull. Sci. Tech. Soc.*, vol. 14, no. 3, 1994.

[3] Committee on Science, Engineering, and Public Policy, *Reshaping the Graduate Education of Scientists and Engineers*. Washington, DC: National Academy Press, 1995, pp. ES-1–ES-9.

[4] Synthesis Coalition World Wide Web Home Page, http//:synthesis.org/synthesis.html, p.1, Mar.1996.

[5] G. Casper, *Stanford State of the University Address*, Stanford, CA, p.11, May 4, 1995.

[6] Symposium, "The third branch of science debuts," *Science*, vol. 256, pp. 44–64, 1993.

[7] R. Dutton, P. Losleben, and J. Plummer, "Computational prototyping for 21st century semiconductor structures," Research Proposal, p. 1, 1994.

[8] T. Abate, "Computer as big as a house to simulate nuclear blast," *San Francisco Examiner*, p. A-2, Sept. 8, 1995.

[9] D. Baraff, M. R. Cutkosky, S. Finger, F. B. Prinz, D. P. Siewiorek, L. E. Weiss, A. Witkin, and P. K. Wright, "Rapid design through virtual and physical prototyping," Research Proposal, p. A-1, 1995.

[10] Center for Advanced Electronic Materials Processing, World Wide Web Home Page, http//ww2.ncsu.edu, Mar. 1996.

[11] S. Benowitz, "Wave of the future: Interdisciplinary collaborations," *The Scientist*, vol. 9, no. 13, p. 4, June 26, 1995.

[12] W. Hanson, "Manufacturing leadership," unpublished manuscript, Massachusetts Institute of Technology, Leaders for Manufacturing Program, Mar. 1995.

[13] A. G. Kraut, "Science and the republican congress," *The Chronicle of Higher Education*, vol. XLI, no. 28, p. B-1, Mar. 24, 1995.

[14] W. J. Broad, "G.O.P. budget cuts would fall hard on civilian science," *The New York Times*, vol. CXLIV, no. 50,069, May 22, 1995.

[15] Committee on Science, Engineering, and Public Policy, *Reshaping the Graduate Education of Scientists and Engineers*. Washington, DC: National Academy Press, 1996, ch. 1, pp. 1–6. Copyright © 1996 by the National Academy of Sciences. Courtesy of the National Academy Press, Washington, DC. Reprinted with permission.

[16] L. M. Lederman, "Science: The end of the frontier?" supplement to *Science*, Jan. 1991.

[17] S. Zolla-Pazner, "The professor, the university and industry," *Scientific American*, vol. 270, no. 3, p. 120, Mar. 1994.

[18] R. P. Morgan, D. E. Strickland, N. Kannankutty, and J. Grillon, "Research on research, How engineering faculty view academic research," *ASEE Prism*, vol. 4, no. 3, p. 33, Nov., 1994.

[19] F. Hoke, "Presidential advisor Gibbons battles Capitol Hill to save research investments from the budget ax," *The Scientist*, vol. 9, no. 12, p. 12, June 12, 1995.

[20] B. Goodman, "Observers fear funding practices may spell the death of innovative grant proposal," *The Scientist*, vol. 9, no. 12, June 12, 1995.

Preparing for an Academic Career

CHAPTER 4

Your Professional Preparation Stategy

> During my graduate study at UCLA, and most of my three-year academic postdoc that followed, I thought I would end up in industry. But I see now that I was also preparing for an academic career. As an undergraduate at U.C. San Diego, I held some discussion sessions, and liked it. At UCLA, I was a T.A. for most of my time, and did very well (won many awards), and again, really liked it. Near the end of my postdoc, a position at San Jose State University (SJSU) was brought to my attention, and I decided to apply for it. I also applied to a number of other schools as well. I was looking for a place where teaching, research, and service were valued, and this seemed to be it. San Jose State had a very extensive national search going on, and they knew just what they wanted. I learned that they viewed me as more interested in research than teaching, so when I got back to UCLA to finish my postdoc, I arranged to teach a five-week section of a biochemistry course. I informed SJSU of this, and it made all the difference. They made me an offer, and I've been here ever since.
>
> *Pam Stacks, Professor and Chair of Chemistry,*
> *San Jose State University, San Jose, CA*

If you choose to pursue an academic career, you will be making what psychologists like to call a "consequential decision." Indeed, except for choosing a spouse or deciding to have children, it is hard to imagine a decision with a greater life-altering impact. When you consider that it usually takes up to 15 years from the beginning of graduate school, through a postdoc, to the possible awarding of academic tenure, you can see why such a decision must be made very carefully–and not too quickly.

To pursue an academic career, you will almost certainly need to pursue a Ph.D. since today it is difficult to have the first without the second. If you also consider that there are now more Ph.D.s and postdocs in science and engineering looking for academic positions than there are positions available, your decision becomes even more important, and more difficult. In some fields, such as electrical engineering and chemistry, where there has always been a path for Ph.D. graduates in industry, the oversupply is not as great as in other areas, such as civil engineering and high-energy physics, where there are fewer long-term options for Ph.D.s outside academia [1, p.2-3]. To make matters worse, in some fields of science and engineering, there is an oversupply of Ph.D.s in all sectors: government, industry, and academia [1, p.ES-8].

What lessons can be learned from the current situation that will help you decide what to do, and depending on that decision, enhance your chances of starting on the professional career of your choice? While avoiding unwarranted optimism, we must also guard against undue pessimism. In *Mercury, The Journal of the Astronomical Society of the Pacific*, Peter B. Boyce applies this to astronomy [2]:

> Realism, not cynicism, is the best response to employment trends in astronomy. Professors and teachers must tell students about the shaky job market, train them for a wide range of careers, and not lead them to believe that nonacademic positions are somehow inferior.

The key to your decision is the following:

- Follow your passion.

- Understand what you can and cannot do with a Ph.D., and whatever you pursue, be sure it is for the right reasons.

- Do *not* assume the current problems, with respect to available positions, will result in a drop in the supply of Ph.D.s or postdocs by the time you graduate.

- Do *not* assume there will be significant improvement in the demand for Ph.D.s or postdocs by the time you graduate.

- Given the above, if you still want to proceed with a Ph.D. and possible academic career, do so by adopting a preparation strategy that significantly increases your chances of getting the position you want.

The strategy proposed here is a three-pronged one of developing breadth as well as depth, pursuing multiple options, and at the same time thinking ahead, looking ahead, and acting ahead of your current stage in ways that establish your readiness for an academic career.

Before we examine this strategy in detail, let us look more closely at the decision to pursue an academic career.

THE DECISION TO PURSUE AN ACADEMIC CAREER

There are certainly many other things you can do with a Ph.D. besides becoming a professor, and we will look at some of them in the sections to follow. However, with very few exceptions, you can no longer be a professor in a four-year college or university without a Ph.D., or its equivalent. In all science fields, and in some engineering fields as well, your Ph.D. is almost always followed by a period of two–four years as a postdoc prior to seeking an academic position.

Earning such a degree is no small matter. It requires an exceptional capability and a significant commitment of time and resources. To be sure, it opens the doors to certain occupations, but it also closes the doors to others by making you appear unsuited or overqualified. What then, is a Ph.D. [3]?

> The doctor of philosophy degree is the highest academic degree granted by North American universities. Ph.D. programs are designed to prepare students to become scholars, that is, to discover, integrate, and apply knowledge, as well as to communicate and disseminate it.
>
> A doctoral program is an apprenticeship that consists of lecture or laboratory courses, seminars, examinations, discussions, independent study, research, and, in many instances, teaching.
>
> The first year or two of study is normally a probationary period, during which a preliminary or qualifying examination might be required. The probationary period is followed by an examination for admission to full candidacy, when students devote essentially full-time to completing dissertation research. This research, planned with the major advisor and the dissertation committee, usually takes 1–3 years, depending on the field. An oral defense of the research and dissertation before a graduate committee constitutes the final examination.

You should pursue a Ph.D. only if the things you want to do actually require such a degree. In addition to college or university professorships, the main other possibility is some form of research and scholarship, possibly followed by management of same in industry, government, or academia.

Do not pursue such a degree for the prestige and status it might bring, and certainly not for the job security you think it will provide. As David Goodstein, vice provost at the California Institute of Technology, puts it [4, p.1]:

Do it if you love it. Don't do it because the Ph.D. is your ticket to an easy life, because that's not true anymore. But if you love science and want to do research, you should still do it.

James C. Fleet, assistant professor at the Tufts University School of Nutrition, says it this way [5, p.1]:

> Others have suggested that Ph.D.s should consider alternative careers in areas of business, education or law, where scientific expertise may increase job prospects. While this may be realistic for currently underemployed Ph.D.s, this is not a plausible long-term strategy to help future Ph.D. candidates or graduates. The fundamental flaw in this proposal is that it ignores the motivations that bring people to study for a Ph.D. A love of science and an interest in discovery are the seeds that graduate schools nurture into Ph.D.s.

I recognize that passion is not the only consideration, and that we make career choices for a variety of reasons. One of these is often to impress others, particularly our parents. According to Peter J. Feibelman, author of *A Ph.D. Is Not Enough* [6], "A common theme in the minds of young scientists is impressing Mom and Dad. This strong motivation is to be cherished, of course, but only if it does not overwhelm one's ability to make rational decisions."

If what you want to do involves teaching at a secondary school or community, technical, or engineering technology college, working in most business settings, or performing much of the science and engineering in industry not requiring a Ph.D., then do not bother studying for it; stop at a master's degree. You will save yourself a lot of time, and will most likely be much happier for it.

Why People Choose, or Do Not Choose, Academic Careers

There are many reasons for wanting, or not wanting, to be a professor, and many possible paths to getting what you want. The following are just a few of those possible paths.

From graduate student to professor:

> I knew for a long time that I wanted to teach, but I also liked doing research. I got my Ph.D. in aeronautics and astronautics, but my real interest was in ships, in particular an area called unsteady free-surface flows. I also wanted my research to have some practical application. I applied to a number of schools right after getting my Ph.D. The University of New Orleans was building a tow tank, and I knew if I joined the Mechanical Engineering Department, I would also be around naval engineering faculty, which is what interested me the most. This access has provided me with some interesting opportunities for cross-disciplinary collaboration.
>
> *Norm Whitley, Associate Professor, Mechanical Engineering Department*
> *University of New Orleans, New Orleans, LA*

From graduate student to professor–twice:

When I was a young girl, I used to "play teacher," give little lectures in my attic, things like that. I also loved to read, and math came very easily to me. When I graduated from high school, I went to the University of Iowa with the idea of returning to high school as a math teacher. But, as soon as I took calculus, I realized that I didn't want to go back to high school and teach the factoring of polynomials. I went and got my master's degree at Creighton University, and in those days, you could get a college teaching job with just an M.S. degree. So I stayed on at Creighton and taught math full time. After awhile, I came to see that I really would be better off with a Ph.D., and Creighton even helped pay for me to do so at the University of Nebraska. Eventually, I got my Ph.D. at the University of Minnesota after following my husband around. I had no doubt as a Ph.D. student that I wanted to teach college mathematics because I had already done so.

Eloise Hamann, Professor and Chairman
Department of Mathematics and Computer Science
San Jose State University, San Jose, CA

From graduate student, to postdoc, to professor:

As an undergraduate, I got started in research and really liked it. During my doctoral studies at Colorado State University, I trained undergraduate researchers, and also helped get other graduate students started on their respective projects. These activities came naturally to me. As a postdoc in chemistry at Stanford University, I learned a great deal about what it takes to maintain a productive leading edge research group. I was prepared to go either way, industry or academia, after my postdoc, but the idea that I could also play a role in developing a teaching, as well as research, program had a lot of appeal to me.

Shon Pulley, Assistant Professor
Chemistry Department
University of Missouri—Columbia, Columbia, MO

From graduate student, to industrial scientist, to professor:

I had been a teaching assistant while working on my Ph.D. in electrical engineering at Rensselaer Polytechnic Institute. I really liked it, and always felt that I would return to academia after a period in industry. But I felt that to know the real world, you had to get into the real world, and there were certain things you could only learn in industry. I worked for IBM for ten years, and then decided to apply for professorships. It was a lot of work, and it took me over a year to land a position. I know that my industry experience helped me get the job, and that it has helped me in my teaching and research.

Kody Varahramyan, Associate Professor
Electrical Engineering Department
Louisiana Tech University, Ruston, LA

From graduate student, to government scientist, to professor:

I got hooked on research as a graduate student at M.I.T. After I got my Ph.D. in atmospheric chemistry, I took a job as a research chemist in the Aeronomy Laboratory of the National Oceanic and Atmospheric Administration (NOAA) in Boulder, Colorado. I had thought about teaching, but the Aeronomy Laboratory, with 30 Ph.D.s, gave me a lot of freedom to pick and choose the topics I wanted to work on. I worked there for about eight years, and then, on a leave of absence from NOAA, I served for two years as an associate program director at the National Science Foundation in Washington, DC. This brought me to the attention of people in academia, and I thought it was time for me to consider both teaching and research. My experience in the government, and particularly at NSF, certainly didn't hurt my application.

Mary Anne Carroll, Associate Professor
Department of Atmospheric, Oceanic and Space Sciences
and Department of Chemistry
University of Michigan, Ann Arbor, MI

From graduate student to industrial scientist:

I basically felt industry needed more Ph.D.s. When I graduated in the early 1980s, all my friends were accepting academic positions, but I felt there were real problems in industry that would benefit from people with Ph.D.s and the analytical skills accompanying such degrees. I really liked the industry pace and the immediate reward system. I haven't regretted it for a minute.

Cheryl Shavers, General Manager
Advanced Technology Operations
Intel Corporation, Santa Clara, CA

From graduate student, to professor, to industrial engineer:

I graduated from Wayne State in solid state physics, and I wanted to teach in the Northeast, really upper New England. I landed a job as an acting professor of physics at the University of Maine. The university was one of the lowest paying schools in the country, and I knew I needed to supplement my income. I obtained a grant from industry for summer research, but the university wouldn't let me earn the salary I put down on the grant even though the industrial sponsor was willing to pay it. They said it wasn't fair to the other faculty. That really upset me. I enjoyed my contact with students, but financially I just couldn't make it. So I decided to leave and take a good job in industry in New England, and it's worked out very well for me.

Roger Verhelst, Senior Engineering Manager
IBM Corporation, Essex Junction, VT

From graduate student, to professor, to entrepreneur:

I think the real problem for me was that tenure came too easily, and I began to see it as a trap, as a way to retire on the job, and I just couldn't do that. I was a professor of actuarial mathematics at the University of Manitoba. We were a very small department, doing the same thing over and over again. I was becoming obsolete. At age 46, an opportunity came up in California to consult on a big computer science project, and so I took a two-year leave of absence. After two years, I started my own company in Silicon Valley, and, of course, I didn't go back. It wasn't the weather or anything like that. I liked the university and I liked teaching, but I was getting stale and I had to do something on my own, and I couldn't do it where I was.

S. Amir Bukhari, Executive Vice President and Chief Technical Officer
Cardinal Technologies, Inc., Sunnyvale, CA

What can we learn from these stories? For one thing, they show us that there are many paths to an academic career, not just ones that go directly from a Ph.D., or even from a Ph.D. followed by a postdoctoral position. They also tell us that there are quite legitimate and rewarding careers with a Ph.D. outside academia, some of which can be achieved after a period as a professor.

Figure 4-1 shows possible paths one might take toward careers in science and engineering. Most students obtain a master's degree before deciding if they want to continue on for a Ph.D. The majority will decide not to continue, but rather to pursue careers in high school or community or technical college teaching, in government, or in industry. Still others may decide to work toward a professional degree in law, business, or perhaps medicine. Those who continue for a doctorate, either immediately or after a period in education, government, or industry, then have to decide what to do after obtaining their Ph.D. For science Ph.D.s, the almost universal path is that of a postdoctoral position for a few years prior to seeking a position as a professor or as a scientist in government or

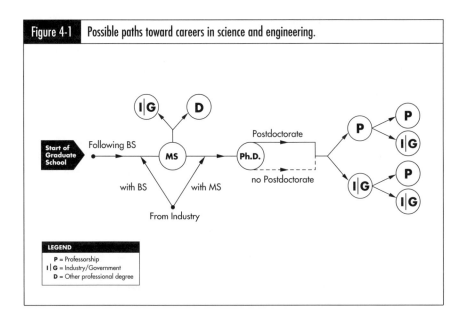

Figure 4-1 Possible paths toward careers in science and engineering.

industry. Some Ph.D.s in engineering will do postdocs as well, but their more common approach is to seek professorships or positions in government and industry immediately after obtaining their degree. As the figure shows, one could then remain as a professor or leave and go into government or industry, or vice versa. The figure is not meant to represent all possible options, but it does illustrate the most common paths.

In making a decision to study for a Ph.D. and, in the process, possibly preparing for an academic career, you need the right information. You can then factor this information (knowledge) into your assessment of your own interests, needs, capabilities, and strengths. We will look at this process more closely in a later section, but first let us see what is going on with supply and demand.

SUPPLY AND DEMAND—WHAT IS GOING ON HERE?

No one has a crystal ball. Predicting future job opportunities in any field is a little like predicting the stock market or the results of a horse race. However, from where we stand today, we project an abundance of engineering faculty positions well into the next century and a shortage of qualified candidates to fill them. For the past ten years there has been a widely publicized shortage of engineering faculty, and all indications suggest that this shortage will continue[1].

Job openings for college and university faculty will expand by 23,000, with the best opportunities for professors in business, engineering and science…There will be a shortage of 7,500 natural scientists and engineers with doctorates by the year 2,000[2].

Sounds terrific, but wait!

Ask graduate students about the job market. In scores of disciplines the answers will be much the same: They are finding that advertised positions at little-known colleges attract hundreds of applicants, the first 100 being from Ivy League post-doctoral students.

How can this be? As recently as 1990, we were reading reports written by distinguished educators predicting a shortage of professors throughout the 90's. Heeding these predictions we encouraged our best students to go to graduate school, and they followed our advice, swelling graduate enrollment to record numbers. Now they—and we—are reaping a harvest of bitterness and embarrassment[3].

---

[1] From *An Academic Career; It Could be for You*, by Raymond Landis, published by the American Society of Engineering Education, 1989.

[2] From *The 100 Best Jobs for the 1990s and Beyond*, by Carol Kleiman, Dearborn Financial Publishing, Inc., 1992, pp. 176–177.

[3] From *Chronicle of Higher Education*, by Shirley Hershey Showalter, Aug. 10, 1994.

Chapter 4 Your Professional Preparation Strategy

Down in the trenches they call it "The Myth." It's the idea, which started to make the rounds around 1987, that the nation faced a shortage of scientists. A wave of retirements in academia, plus burgeoning demand for scientists and engineers in high-tech industry, would create a short-fall of 675,000 scientists and engineers, crippling industrial competitiveness and threatening national security. Heeding the nation's call (and lured by a vision of recruiters beating down their dormitory doors) students labored through organic chemistry and differential equations to earn a bachelor's degree in science and in many cases, pushed on to graduate school. Now "The Myth" has met reality, and reality bites[4].

Figure 4-2 shows some recent headlines that capture the concern and frustration many feel about the present situation.

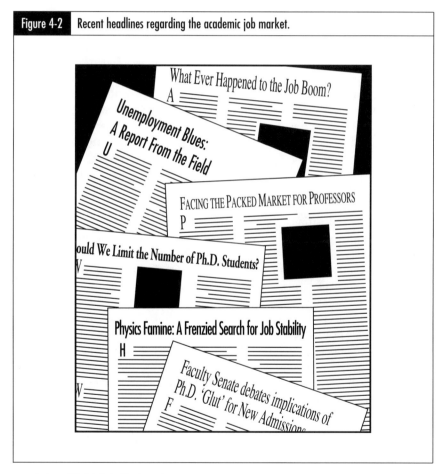

Figure 4-2 Recent headlines regarding the academic job market.

---

[4] From "No Ph.D.'s Need Apply," *Newsweek*, by Sharon Begley with Lucy Shackelford in Washington and Adam Rogers, December 5, 1994.

So what is going on here? Clearly, there has been a significant change in just a few years in both the supply and demand for full-time, tenure-track positions in science and engineering. How did this happen, and why did it come as such a surprise?

In a pure market economy, the demand for science and engineering professors would be proportional to the number of students taking science and engineering courses plus the number doing science and engineering research in graduate school. This demand, in turn, would depend on the number of students enrolled in higher education, the number of those required to take science and engineering courses, and the number who wish to do so because they are majoring in science, engineering, or related fields. It is among the latter group that there is the most variation. At the undergraduate level, this number ebbs and flows over a period of years, often in response to the perceived need for such graduates in industry. However, since it takes four or five years to obtain a bachelor's degree, the supply of science and engineering graduates is often out of phase with the demand. The number of graduate students choosing to pursue academic careers is also related to the perceived supply and demand for professors, and here the demand is also often out of phase with the actual supply.

Of course, a true market economy does not operate in an academic environment, and so the picture is more complicated. The increase or decrease in the number of science and engineering majors does not necessarily result in an increase or decrease in the funds available to support faculty. Even in private schools, the fraction of tuition going to supporting the institution is often less than half the total operating budget, and it is much less than this for public institutions. Furthermore, such tuition income is only roughly distributed to departments in proportion to the number of students taking courses in such departments. Add to this factor the matter of how faculty hiring contracts are set, tenure, faculty retirement rules, and distribution of funds among departments, and you have the case where faculty are not simply hired, or fired, in direct response to increasing or decreasing enrollments in a given field.

There are reasons to believe the supply, even back in the late 1980s, was not as insufficient as argued since the number of Ph.D.s has been increasing steadily over the last 20 years [5, p.10]. There is no question that it has increased over the last half dozen years, for both predictable and unpredictable reasons. The very predictions of a shortage in the late 1980s led to the expected result: an increase in the number of students entering Ph.D. programs. As David Goodstein, Vice Provost at the California Institute of Technology, puts it [5, p.10]:

> Even just the rumor that there might be academic jobs at the end of the decade prompted a large increase in the enrollment of American students in graduate school. The problem solved itself instantly if there was going to be a problem, but it was never going to be a problem.

Furthermore, as companies downsized in the early 1990s, demand for all types of degree holders dropped, and this drop encouraged some undergraduates to stay in school and continue for advanced degrees.

Supply can also be affected by events that, for the most part, could not have been anticipated. For example, when the Tiananmen Square uprising took place in April 1989, most of the relatively large number of Chinese mathematics Ph.D. students in the United States applied for, and received, political asylum, allowing them to stay in the United States indefinitely. The granting of political asylum resulted in a considerable bulge in the supply of mathematics graduates looking for professorships. Similarly, when the Soviet Union broke up, a number of senior Ph.D. mathematicians, some with 50 or more publications, became available. Many of these people were delighted to take jobs in the United States at assistant professor ranks and salaries. Both of these events had a smaller, although still significant, impact in other areas of science and engineering.

Change in demand can also manifest itself in unpredictable ways. The launching of Sputnik in the late 1950s and the Star Wars program of the mid-1980s are two examples. In the late 1980s, a wave of retirements was predicted based on the hiring of professors in the 1950s and 1960s. However, unanticipated changes in retirement laws resulted in a delay in the expected retirement of many of these professors. Yet, all of these people will eventually retire, die, or otherwise leave the profession. By the year 2008, nearly half of the 595,000 full-time college faculty members in the nation are likely to retire. These retirements should also coincide with an increase in college enrollments predicted by some demographers. For example, California alone predicts an increase in college enrollments of some 455,000 students by the year 2005 [7]. Yet, for the reasons outlined in Chapter 1, not all retiring teachers will be replaced by full-time, tenure-track professors [8]. Also, it is not clear what impact increases in productivity via advances in communications and other technologies will have on the demand for professors.

Another problem is that the ground is shifting in all areas of employment for science and engineering Ph.D.s. This fact is summed up in a recent report of the National Research Council's Committee on Science, Engineering, and Public Policy (COSEPUP) [1, pp. E-2–E-3].

> Hence, the three areas of primary employment for Ph.D. scientists and engineers—universities and colleges, industry, and government—are experiencing simultaneous change. The total effect is likely to be vastly more consequential for the employment of scientists and engineers than any previous period of transition has been.

In light of the above oversupply, some have advocated reducing the number of Ph.D.s by specifically limiting the enrollment of graduate students in science and engineering. A few schools have indeed instituted what Roman Czujko of the

American Institute of Physics calls "graduate student birth control," with Cornell University going the furthest by taking only 19 instead of its typical 40 physics Ph.D. students [9].

However, most schools do not have much of an incentive to reduce their graduate student populations. Kevin Aylesworth, theoretical physicist and founder of the Young Scientists Network, asserts that, "Because advisors depend on graduate students to put in many hours in their labs, they don't want to discourage graduate students from their narrow task of research" [4, p.10].

Others have argued against artificial limitations on supply, noting that it will not help those now seeking jobs, and that a better approach is to seek good advice followed by an application of the free market. The COSEPUP study referred to earlier concludes [1, p.ES-8]:

> Nevertheless, we see no basis for recommending across-the-board limits on enrollment for three reasons: First, conditions differ greatly by field and subfield. Second, we believe that an extensive, disciplined research experience provides valuable preparation for a wide variety of nontraditional careers for which scientific and technical expertise is relevant. Third, limiting actions would have little immediate aggregate impact even if they could be orchestrated effectively. Instead, we believe that our recommendations of greatly improved career information and guidance will enhance the ability of the system to balance supply and demand. When the employment situation is poor, better-informed students will be able to pursue options other than a Ph.D.; when the market is expanding, students will be able to move more flexibly and rapidly in the direction of employment demand.

It is also important to not make the mistake of assuming that just because supply exceeds demand, there is no demand. We are always going to need new professors. In fact, the latest predictions call for a constant academic hire rate of 5% for at least the next 15 years [10]. Yet, given the unreliability of any prediction, the approach for you to take is a conservative one that assumes there will be no decrease in the supply of graduates seeking academic positions and no increase in the demand for such positions.

Probably the best advice comes from Joseph S. Merola, director of graduate education for the Chemistry Department at Virginia Polytechnic Institute and State University. Speaking specifically of science, but in terms also applying to engineering, he notes [4, p.10]:

> I think science as a career is still a good choice. But if you view a Ph.D. in the same way that you view a vocational school—that it's going to give you some skills and those skills are going to be marketable—that's a big mistake. You have to go into science because almost from the day you were born you found yourself investigating, you found yourself being curious, you found yourself playing in the lab or building things, and this is exactly what you want to do with your life. So long as you have that internal motivation, science is a good career.

As noted above, there are plenty of interesting and worthwhile things you can do in science and engineering without a Ph.D., but there are some things, such as becoming a professor, for which a Ph.D. is almost always essential. If these are the things you think you want to do, then by all means, go for it! However, do so with versatility and flexibility so as to maximize your chances of success. Developing a strategy that will help you do just that is the subject of the next section.

THE THREE-PRONGED PREPARATION STRATEGY

A strategy, or overall plan, for achieving your goals is necessary because you have limited time, energy, and material resources. The plan should be flexible enough to allow you to explore different possibilities, and at the same time prevent you from running into too many dead ends. It should also allow you to assess progress toward your goals, and to make necessary adjustments along the way. A good strategy gives you a feeling of accomplishment as well as a reference point during your journey, often at times when you need it the most. Also, as noted in Part I, fundamental changes are taking place in academia with respect to teaching, research, and other forms of scholarship. Having a strategy that helps prepare you for these changes can be particularly valuable.

The strategy proposed here has three components: (1) Breadth-on-Top-of-Depth, (2) Next-Stage, and (3) Multiple-Option, as shown in Figure 4-3. Each approach complements the other, and all can be carried out simultaneously during your graduate student and postdoc periods. Let us take a look at each of these approaches in detail.

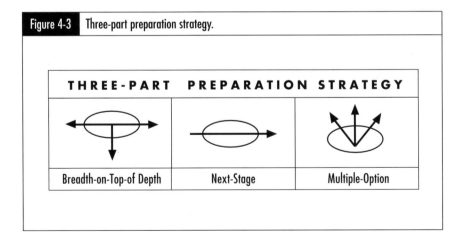

Figure 4-3 Three-part preparation strategy.

Breadth-on-Top-of-Depth

In the Breadth-on-Top-of-Depth approach, you seek to place your developing expertise in a broad context. By doing so, you are better able to see connections between your work and that of others, to make a more compelling case for your own contribution, and to be able to develop related areas of depth should the situation call for it.

One way to look at the concept is to imagine a capital "T." Here, depth is represented by the stem of the "T" and breadth by the cross bar. The first thing to understand about this concept is what it is *not*. Breadth-on-Top-of-Depth does *not* mean breadth *in place of depth*, nor does it mean breadth *over depth* in the sense that breadth is more important than depth. Breadth-on-Top-of-Depth means breadth *in addition* to depth. Developing depth, be it in a research area, another form of scholarship, or the teaching of a particular course, is essential to academic success. You need to be known for something, and that something needs to be both important and unique. The last thing you want to be is "a mile wide and an inch deep." However, there are at least three good reasons for developing breadth in addition to depth. First, by increasing your knowledge and exposure to related areas, you create the possibility of developing additional areas of expertise, "drilling multiple holes," as one faculty member put it. Second, by knowing what is going on in related areas, you increase the opportunities for collaboration in ways that can enhance your own scholarship. Finally, by placing your work in a larger context, you give it greater meaning and make it more compelling to a larger audience, which in turn makes it easier to justify and support.

As we will see in the next two chapters, the concept of Breadth-on-Top-of-Depth applies to all areas of research and teaching, not just to the choice of a specific research topic. By way of illustration, consider your choice of a research advisor. As we will see in Chapter 5, no matter who you end up "choosing" as your advisor, this one person will have strengths and limitations with respect to managerial style, knowledge of the field, and contacts with industry and government. In seeking Breadth-on-Top-of-Depth, you will want to identify "complementary" advisors, one or more of whom may be in industry or at another institution. These additional advisors can make up for deficiencies always found in any single advisor. Also, by choosing to work with complementary advisors, you broaden your experience and your exposure to opportunities that would otherwise not be possible.

Next-Stage

In the Next-Stage approach, you think ahead, look ahead, and to some degree act ahead of the stage you (and your future competition) are currently occupying. By doing so, you not only demonstrate your *willingness* to assume the role of the position you are seeking, but also your *readiness* to do so. Just as most of the best graduate students began taking graduate courses and/or conducting research as college seniors, you need to begin doing some of the things professors do while you are still a graduate student and postdoc. Today, it is not enough to be outstanding in your current job; you must also demonstrate that you can be successful in the next job for which you want to apply by actually performing in advance some of the activities and responsibilities that are part of that job.

Below are some areas in which demonstrating this "next-stage" competence would be important. As we will see in the next two chapters, no one expects you to demonstrate all of them. However, doing at least some of them will distinguish you from most of your competition, and within limits, the more you can do, the better.

- *Research*—In addition to having identified a dissertation or a postdoctoral research project that is compelling as opposed to just interesting, look for ways to engage in cross-disciplinary and multidisciplinary activities with faculty and students from other areas or departments. Obviously, you will want to pursue such activity with the support of your advisor.

- *Technical reviewing*—Find opportunities, both formal and informal, for you to review papers, grants, and proposals written by others.

- *Proposal writing*—In addition to reviewing the proposals of others and contributing sections to your advisor's proposals, write your own proposals and grant applications for research that you want to do as a professor.

- *Supervision of other students*—As you advance in your development as a graduate student or postdoc, find ways to play a more formal role in the supervision of other students, both undergraduate and graduate.

- *Publishing*—Coauthorship is fine, but make sure you publish at least one article in which you are the first author.

- *Presentations at conferences*—Establish a record of giving technical presentations at conferences in which faculty and industrial researchers are present.

- *Relations with industry*—Visit various research sites and give technical presentations, use equipment, samples, and other industry resources in your research, conduct joint investigations, publish with industrial collaborators, and consider internships and other forms of employment with industry or government laboratories.

- *Teaching*—Plan to acquire at least some experiences beyond those of a typical TA, such as giving lectures, covering sections of a class, or even taking full responsibility for a course.

The key steps in the Next-Stage approach are to ask questions (think ahead), make observations (look ahead), and acquire experiences (act ahead) by putting yourself in the right places at the right times and tuning your antenna to the gathering of the right information. You can do the above in a variety of settings, such as classrooms, laboratories, faculty offices, staff meetings, seminars (particularly with guest speakers from other schools), professional conferences, private discussions with students and faculty, and during visits to industrial and government R&D facilities. In all cases, the key question is: *Am I likely to encounter this situation as a professor, or future industrial scientist or engineer, and if so, what can I learn from it that will help to better prepare me for such a role?*

The Next-Stage approach involves actively seeking experiences that you are likely to encounter in the future, and we will look at a number of them in greater detail in Chapters 5 and 6.

Multiple-Option

In the Multiple-Option approach, you prepare concurrently for possible careers in academia, government and industry. There are four reasons why you should consider doing so:

1. At this point, you probably do not know enough about all the things you can do with a Ph.D. to zero in exclusively on any one of them.

2. By preparing for more than one possibility, you significantly increase your chances of professional employment after your graduation or post-doctoral experience.

3. By doing things that will make you more attractive to industry and government, you will, paradoxically, make yourself more attractive to academia. This increased attraction occurs because most colleges and universities want science and engineering faculty who can interact effectively with the other two sectors.

4. A corollary to (2) and (3) is that with the increase in part-time faculty positions, an industry/government career option can allow you to accept such part-time teaching while keeping open the possibilities of long-term academic positions at a later date.

While most beginning graduate students have little accurate knowledge of what it is like to work in the various employment sectors, many have preset ideas that prevent them from considering options that might be quite beneficial. By exploring multiple options and not making up your mind too soon, you avoid the mistake of not pursuing an academic career when, if you had additional information, you would have chosen to do so. You also avoid the reverse: choosing to pursue an academic career when, if you had additional information, you would have decided otherwise.

As someone considering an academic career, you have a particular advantage. You have seen your future profession in action throughout your undergraduate and graduate study. However, what you have seen is only a portion of the professional life of a faculty member, and one purpose of the three-pronged strategy is to help you learn as much as possible about the rest before making a final decision.

In describing the rewards of an academic career, Ray Landis, dean of engineering and technology at California State University, Los Angeles, sent a survey to the nation's engineering deans, asking this question: "If you were to talk with one of your best undergraduate students, what would you tell him or her are the rewards of a faculty career?" The responses, ranked in order of their frequency, were [11]

1. Joys of teaching/Rewards of working with students

2. Freedom/Flexibility

3. Work environment

4. Rewards of research

5. Variety of work

6. Financial rewards

7. Lifelong learning

8. Job security

It would have been interesting had Landis also asked the deans what they thought were the least rewarding aspects of a faculty career.

Richard Bube, former chairman of the Materials Science and Engineering Department at Stanford University, thinks that much of the above is pure myth.

As he puts it [12]:

> An idealized view of a career as an engineering or science professor at a major research university involves quickly earning tenure, spending time helping young minds develop, and measuring personal success by the maturation of one's students. One participates in a community dedicated to truth and does research in its pursuit, studying problems of personal interest. Safe in an "Ivory Tower," one has time to think and be absorbed by scholarly pursuits, enjoying the chance to work one-on-one with students.

Even though Bube's comments apply to research universities, and Landis's results cover a broader spectrum of schools, the two contrasting views raise important questions about what is real and what is rhetoric in statements about the life of science and engineering professors.

Similar misunderstandings can apply to positions in government and industry. In some fields, such as computer science, electrical engineering, chemistry, geology, and certain areas of biology, there is a history of Ph.D.s accepting positions outside academia, and consequently a greater understanding of what these positions are like. In other science and engineering fields, industry positions for Ph.D.s are much less common, and attitudes about such options reflect this lack of experience. As William Jaco of the American Mathematical Society notes: "It is important to change the traditional view that the only job worth having is in academia. The culture of the science and math community considers anything short of academic employment a failure. We have to change that" [9].

As one industrial research manager recently observed [1, p. 2-20]:

> Most recent graduates, particularly those who have not summer-interned, do not have the foggiest idea of what industrial research is all about. Some even think that using or developing technology to do something useful is not research and if it is a product that makes a profit, is even slightly dishonorable.

However, Ph.D.s are increasingly finding employment outside universities, and more and more are in types of positions that they had not expected to occupy [1, p.6-3]. Figure 4-4 contains some recent headlines that make this point.

With the Multiple-Option approach, you are encouraged to gain a variety of skills applicable to many sectors of Ph.D. employment. According to the Committee on Science, Engineering and Public Policy report, this greater versatility can be promoted on two levels [1, p.ES-4].

> On the academic level, students should be discouraged from over-specializing. Those planning research careers should be grounded in the broad fundamentals of their fields and be familiar with several subfields. Such breadth might be much harder to gain after graduation.

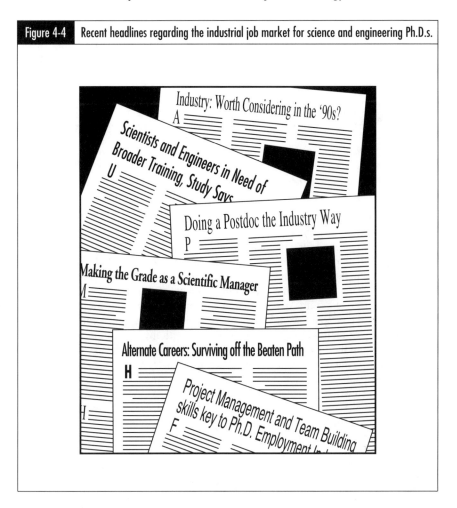

Figure 4-4 Recent headlines regarding the industrial job market for science and engineering Ph.D.s

On the level of career skills, there is value in experiences that supply skills desired by both academic and nonacademic employers, especially the ability to communicate complex ideas to nonspecialists and the ability to work well in teams. Off-campus internships in industry or government can lead to additional skills and exposure to authentic job situations.

As noted earlier, one advantage of the Multiple-Option approach is that by making yourself attractive to industry, you simultaneously make yourself more attractive to many academic institutions. At first, this dual attraction may seem counterintuitive. How can industry with its focus on shorter term applied research be compared with academia and its focus on longer term theoretical understandings? In spite of the tensions created by such differences, industry and academia need each other more than ever. Having faculty with a knowledge of industry who can work at the intersections of these domains is becoming more, not less, attractive to academic institutions, including many at the Research I and II levels.

Vignette #4   A Ph.D. Career in Industry

Academic positions are not the only possibilities for those with Ph.D.s in science or engineering. In the following vignette, we look at a path that has led to a very successful career in industry.

Cheryl L. Shavers
Intel Corporation

"I haven't regretted for a moment my decision to go into industry," says Dr. Cheryl Shavers, the general manager of the Advanced Technology Operation in the Technology and Manufacturing Group at Intel Corporation in Santa Clara, CA. But that is certainly not what all her friends were doing with their Ph.D.s from Arizona State University (ASU) in the early 1980s. "Most of the people I went to school with wanted to become professors. I saw many of them putting their lives on hold with low paying postdocs, in effect taking the low risk, easy way out," comments Shavers. "That wasn't for me. I wanted to get going, get back to industry where I could make things happen."

Growing up in the black community on Phoenix's South Side, Shavers came to realize that of the few women she knew who went to college, most became either nurses or teachers; that was the expectation. However, even at an early age, only doing what was "expected" was not one of Shavers's characteristics.

After observing how the police investigated a tragic homicide in her neighborhood, Shavers became intrigued with the possibility of becoming a forensic scientist. She did extremely well in math and science in high school, and after graduation enrolled in a criminal justice program at a local community college. She soon discovered, however, that to actually work in a crime laboratory you needed a background in science, particularly chemistry, and so she switched her major. "This was a life-saver for me," she says. "Chemistry was a lot harder than criminal justice, but it made all the difference in the world in terms of my options." Shavers also discovered, after a summer internship with the Phoenix Police Department, that she did not want to work solely in an environment that imposed so many restrictions and provided such limited promotional opportunities.

After earning her bachelor of science degree in chemistry in 1976, she took a job at Motorola's Semiconductor Sector in Phoenix, AZ, where she had a set of experiences that strongly impacted her future career. In the mid-1970s, Motorola required new college graduates like Shavers to take graduate courses at a local university while they were working for the company. Shavers began by taking MBA courses, but found them less than challenging. For intellectual stimulation more than anything else, she took a graduate course in thermodynamics in the Chemistry Department. Her professor soon recognized her potential, and offered her a fellowship to study toward a doctorate in solid-state chemistry.

By this time, Shavers was also noticing a situation at Motorola that would impact her decision to return to industry after completing her doctorate. Most of the people she worked with were either young, enthusiastic, but naive start-ups like herself, or very much older employees who seemed to lack the energy and drive of her younger colleagues. There were few experienced, intellectually strong mentors with advanced degrees who younger science and engineering graduates could look up to. Shavers wanted to become such a person while still making a contribution to technology.

So, she left her $15,000 a year job at Motorola in 1978 for a $3300 a year fellowship at ASU. "Most of my colleagues thought I was crazy," she says, "but I saw this as a temporary move, as a way to get the credentials I needed to return to industry and have the influence I wanted." Shavers loved the graduate student experience, but wanted to get through quickly. Three and a half years later, she left ASU with her Ph.D.

Shavers then took a job as a semiconductor process development engineer at Hewlett-Packard Company in Cupertino, CA. A few years later, this led to a job at Hewlett-Packard headquarters as a patent agent. Subsequently, Shavers held positions as a factory manager at Wiltron Company in Mountain View, CA and as a thin films application manager at Varian Associates in Palo Alto, CA. "Varian taught me a lot about being a manager and about a high-pressure business environment," comments Shavers. "I did well, but it left me emotionally drained."

In 1987, Shavers was recruited by a Varian customer, Intel Corporation, as a member of the technical staff of the Components Research Group in Santa Clara, CA. In her current position as general manager, she investigates future generation devices for PC platforms, as well as peripheral chipsets that fit into Intel's strategic wafer investment objectives. She also participates in numerous university outreach, as well as community, programs. "Now," she says, "I am in a position, and at a time in my life, where I can fulfill my original goals of mentoring younger employees in the technical and managerial challenges of high-technology companies." "One of my personal obligations," says Shavers, "is to provide industrial soft landing pads for students and interns who come to Intel." She works with these new employees to help them learn how to navigate the ropes, to see that exciting contributions can be made in industry by people with Ph.D.s who are not that much older than themselves.

Shavers does not like the term "role model," although as the only senior-level, black female Ph.D. in the company, being seen as one is inevitable. She does consider herself an example of what is possible for bright, ambitious college graduates. And it is clear that, while Shavers may have decided not to become a professor, if her fellow ASU graduates could look at her now, they would certainly see a teacher.

SUMMARY

We began this chapter by pointing out that the decision to pursue an academic career is a consequential one with long-term implications. It must be seen in the context of the more basic decision to study for a Ph.D., since such a degree is a prerequisite for virtually all academic positions in four-year colleges and universities. We examined the basis for an academic career decision, including a detailed look at the supply and demand situation in science and engineering. For those of you who wish to pursue a Ph.D., we proposed a three-pronged strategy that will prepare you for an academic career while maintaining options for careers in government and industry. The three elements of this strategy, Breadth-on-Top-of-Depth, Next-Stage, and Multiple-Option, can be applied to all aspects of your preparation activities. We concluded the chapter with a vignette describing a successful career path in industry for a woman with a Ph.D. in chemistry.

REFERENCES

[1] Committee on Science, Engineering, and Public Policy, *Reshaping the Graduate Education of Scientists and Engineers*. Washington, DC: National Academy Press, 1996. Copyright ©1996 by the National Academy of Sciences, Courtesy of the National Academy Press, Washington, DC. Reprinted with permission.

[2] P. B. Boyce, "Should we limit the number of astronomy students?", *Mercury, The Journal of the Astronomical Society of the Pacific*, vol. 23, no. 5, p. 8, Sept.–Oct. 1994.

[3] "The Doctor of Philosophy degree: A policy statement," in *Reshaping the Graduate Education of Scientists and Engineers*. Washington, DC: National Academy Press, 1996, p. 1–3. Copyright © 1996 by the National Academy of Sciences, courtesy of the National Academy Press, Washington, DC. Reprinted with permission.

[4] R. Finn, "Discouraged job-seekers cite crisis in science career advice," *The Scientist*, vol. 9, no. 11, May 29, 1995.

[5] J. C. Fleet, "Young researchers' disillusionment bodes ill for future of science," *The Scientist*, vol. 9, no. 11, May 29, 1995.

[6] P. J. Feibelman, *A Ph.D. Is Not Enough! A Guide to Survival in Science*. Reading, MA: Addison-Wesley, 1993, p. 13.

[7] "Colleges face revenue gap," *The San Jose Mercury*, p. 3B, June 5, 1995.

[8] S. Mydans, "Part-time college teaching rises as do worries," *New York Times*, p. A17, Jan. 4, 1995.

[9] S. Negley with L. Shackelford and A. Rogers, "No Ph.D.'s need apply," *Newsweek*, p. 25, Dec. 5, 1994.

[10] E. Goldman, "Fac sen: Grad students be wary of poor market," *Stanford Daily*, vol. 207, no. 60, p. 6, May 19, 1995.
[11] R. B. Landus, *An Academic Career, It Could Be For You*. Washington, DC: American Society of Engineering Education, 1989, pp. 4–7.
[12] R. Bube, "Expectations vs reality in engineering faculty careers," *Engineering Education*, vol. 79, no. 1, pp. 33–36, Jan./Feb. 1990.

CHAPTER 5

Research as a Graduate Student and Postdoc

Have the breadth to see the problems, and the depth to solve them.

Anonymous

You will want to acquire a variety of research experiences before becoming a professor. Some of these, such as choosing an advisor and completing a dissertation, are common to all Ph.D. students. Others, like writing your own research proposals and supervising other researchers, although technically not required for your Ph.D. or postdoc assignment, can make all the difference when it comes to your first years as a professor.

The key is to see your graduate student and postdoc experiences as the first steps to a life of scholarship, not as the last hurdles to the completion of your degree or postdoc assignment. This approach should be the case no matter what type of academic or industrial position you ultimately occupy. Seen in this context, you will go beyond the minimum requirements and take additional steps that can have significant returns for you down the road. The experiences we will discuss in this connection include:

1. Choosing a graduate school or postdoc institution.

2. Choosing a research topic.

3. Choosing a dissertation advisor/postdoc supervisor.

4. Writing your own research proposals.

5. Carrying out your research.

6. Publishing.

7. Attending conferences and other professional meetings.

8. Giving talks on your research.

9. Supervising other researchers.

10. Managing research projects and programs.

We will take a look at each of these activities in the context of the three elements of the preparation strategy outlined in Chapter 4, Breadth-on-Top-of-Depth, Next-Stage, and Multiple-Option. Figure 5-1 shows the relationship among these three elements and the ten research activities listed above. Of course, most of these activities do not stand alone. For example, choosing a graduate school or postdoc institution, a research topic, and a research supervisor are clearly related. However, for discussion purposes, we will examine them one at a time.

CHOOSING A GRADUATE SCHOOL OR POSTDOC INSTITUTION

> Where you go to graduate school determines where you do your postdoc, and where you do your postdoc determines where you become a professor.
>
> *Shon Pulley, Assistant Professor*
> *Chemistry Department, University of Missouri—Columbia*

While it may not be quite that simple, it is certainly important to look ahead when choosing a graduate school and postdoc institution. In so doing, keep in mind that the institutions where you gain these experiences are not the places where you are likely to start as a professor, so be sure you choose them for the right reasons.

In the last chapter, we mentioned the importance of following your passion in all matters related to your research. The research experience is too demanding and so full of setbacks and frustrations that if it is not accompanied by a desire, even a zeal for the subject, it is unlikely to be sustained over time. That said, it is also important to realistically match your passion and aspirations with a level-headed assessment of your skills and present level of achievement.

Consistent with the above factors, you should choose a graduate school with the reputation of the university, and particularly the department of interest uppermost in your mind. Being able to attend a school with a strong reputation is more than just a statement about your capabilities and undergraduate preparation. The

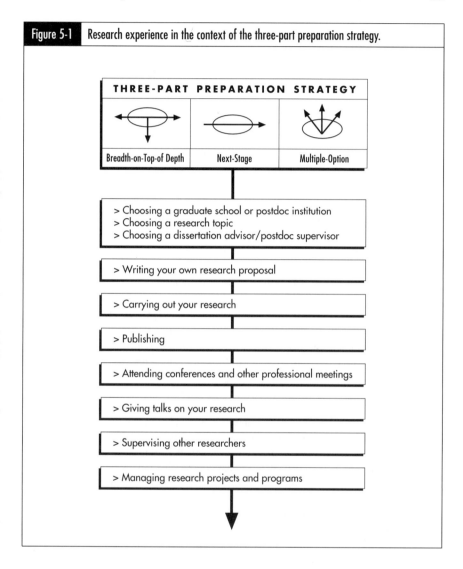

Figure 5-1 Research experience in the context of the three-part preparation strategy.

reputations of graduate schools are built on their faculties and on their graduate students and postdocs. Each attracts and enhances the other. The better the combination, the more likely you are to acquire the experiences that will make you attractive to other institutions as a future professor.

Remember, however, that the reputation of a specific department may not always coincide with the reputation of the institution as a whole, and in some cases, the former can significantly exceed the latter. This difference is particularly true with departments in less common specialties, such as astronomy, agricultural sciences, and atmospheric sciences. For example, on average, the Georgia Institute of Technology's engineering research doctorate program ranks 12th out of approximately 250 institutions. Not bad, to be sure. Its Industrial Engineering Department, however, ranks number one [1]! You should look first at the reputation of the department, and then at the reputation of the school. The best source for this information in the United States is the current edition of the *Research–Doctorate Programs in the United States—Continuity and Change*, available in your library or from the National Research Council Office of Scientific and Engineering Personnel, 2101 Constitution Ave., Washington, DC, 20418.

For postdocs, all of the above comments apply, but in addition, it is particularly important to pay attention to the reputation of your potential supervisor. For new Ph.D.s, the first question often asked is, "Where did you get your degree?", but for recent postdocs it is, "Under whom did you postdoc?" This important aspect will be discussed in greater detail in the section "Choosing a Dissertation Advisor or Postdoc Supervisor."

In addition to reputation, you want to consider schools that will expose you to a wide range of opportunities. Look for institutions with research programs in a number of fields, not just in an area of interest to you at the moment. Such schools will give you more choices as you seek to develop a research topic and choose a research advisor. They will also give you the opportunity to work with students and faculty in areas adjacent to the one in which you choose to develop or enhance your depth.

> Siemens's Ralf Brinkmann did his dissertation on the rings of Saturn, or, more specifically, the kinetic theory of planetary ring structures. Such structures are an example of what are called plasmas, sets of particles in a confined space under the influence of a gravitational or electromagnetic field. During his research, Brinkmann was exposed to work underway in another department on high temperature plasmas used as etchers in the preparation of semiconductors. This exposure led to a request from a visiting researcher for Brinkmann to apply his knowledge of plasmas gained in his dissertation to a plasma problem in the semiconductor industry, and eventually to a research position at the Siemens Corporation in Munich, Germany.

You should also pay attention to schools in which there is some evidence of interdisciplinary collaborations across units or departments. Remember, it is at the intersection of disciplines where scientists are finding some of the most interesting problems. You will probably find more such interdisciplinary collaborations taking place at larger research universities, but do not automatically rule out smaller institutions. Sometimes smaller size and fewer resources may actually encourage interdisciplinary interactions. Brandeis University, a school of about 3000 students, has a number of interdisciplinary programs, including those of the Rosentiel Center in which 16 faculty from a number of departments share office, meeting, and laboratory space. One example of such collaboration is that taking place between physicists and biologists on the development of faster and more sensitive X-ray detectors [2].

In addition, keep in mind that as interuniversity interactions increase, there should be greater opportunities for smaller schools to collaborate with larger institutions.

In looking ahead to acquiring Next-Stage experiences, pay attention to institutions providing graduate students and postdocs with opportunities to write research proposals, publish papers, give presentations, interact across disciplines, supervise other students, manage research efforts (particularly for postdocs), and acquire teaching experiences beyond those of a standard teaching assistant. A given school may not provide all of these opportunities, but the one you choose should offer most of them.

With respect to Multiple-Option experiences, look for schools where there is a history of interaction with industry. This interaction is most likely to occur at schools located in metropolitan areas surrounded by industrial enterprises, but here again, there are plenty of exceptions. Purdue University in West Lafayette, IN has developed extensive ties with a number of industries across North America. Schools with well-developed undergraduate, or master's, co-op programs are particularly likely to support such interactions, and to be able to provide contacts for graduate students and postdocs. But, again, there are exceptions. Until recently, Stanford University had no undergraduate co-op programs, and yet, since the 1950s has had extensive interactions with local, national, and international companies.

Do not automatically assume that all schools or departments will provide opportunities or encouragement in your preparation for an academic career. If at all possible, visit the institutions you are considering. Ask to see the facilities and to meet with faculty, graduate students, and postdocs. In particular, seek out students and postdocs who are considering academic careers, and find out what types of experiences they have had that support their interests. Pay attention to how people interact, their level of enthusiasm, and whether or not they would repeat their decision to attend the institution. Read between the lines. Ask about faculty tenure and morale, as well as graduate student morale. Finally, try to get a sense of whether or not the department really wants you, as well as the other way around. If they do, and most of your other criteria are met, then it is probably the place for you.

CHOOSING A RESEARCH TOPIC

It is really important to do the right research as well as to do the research right. You need to do "wow" research, research that is compelling, not just interesting.

George Springer, Chairman
Aeronautics and Astronautics Department, Stanford University

By choosing a research topic, I do not mean selecting a subject from possibilities presented by a professor, or worse, having such a topic assigned to you, attractive as this may first appear. Choosing means going through the process of finding and then developing a topic with all the initial anxiety and uncertainty such a choice entails. It is a way of developing your capacity for independent thought. If, like many undergraduates, you spent most of your time solving assigned problems rather than asking open-ended questions, the idea of searching for areas or problems to investigate can seem a bit daunting. On the other hand, if you have had some research or special study experience as a senior, master's student, or industrial researcher (Next-Stage approach), the task will seem less formidable. It is important that you take the time to find a problem that is interesting to *you*, significantly complex, and compelling.

James Plummer, professor of electrical engineering at Stanford University, believes that when students struggle through the process of defining and developing a topic, they will be more excited and will feel a greater ownership of the effort than if the problem, no matter how interesting and important, were handed to them by others. Plummer tells his students to take the time—up to a few months if necessary—to read as much as possible in their field of interest as well as in related fields. He also encourages them to talk to people at seminars, conferences, during industrial visits, and on the Internet. As he puts it, "I want my students to sift through the grains of sand to find those that can possibly make a significant difference. In the process, they usually end up throwing 99% of the sandbox away, but this is what they have to do to get to the real gems."

Plummer believes that it is perfectly okay for his students to feel inadequate for awhile. As a postdoc and as a professor, you will have to define your own programs and projects, as well as supervise other students, so you need to get some experience with this early on. Plummer wants each of his students to define and develop a project, publish at least one article on it in a refereed journal, and give at least one presentation about it at a professional conference before graduation.

The research you do as a graduate student will set the stage for much of your research to follow as a postdoc and as a professor. While it is unlikely that your later research will be straightforward extensions of your dissertation, it is also unlikely that it will be completely outside your field. Stories to the contrary are the exception, not the rule. Thus, the knowledge, expertise, and skills that you gain early on will form the foundation for later investigations. By choosing the right topic now (Breadth-on-Top-of-Depth), you help ensure your research viability in the future. As Cliff Davidson and Susan Ambrose of Carnegie-Mellon University put it, "The most successful research topics are narrowly focused and carefully defined, but are important parts of a broad-ranging, complex problem" [3, p.101].

There are a number of reasons for considering a given area. Some, of course, have to do with your particular interests, capabilities, and motivations. Other considerations center around areas that will be of greatest interest to both the academic and private sector. Fields that fall into this category include information technology, industrial technology and automation, biotechnology, and health and medical technology [4, pp.36–37]. But, of course, commercial application is only one consideration. There remains a huge need for compelling research of a fundamental nature in all areas of science and engineering.

Chemistry professor and author, Robert Smith, in his book, *Graduate Research: A Guide for Students in the Sciences* [4, p.36], lists 11 points to consider in finding and developing a research topic. They are as follows:

1. Can it be enthusiastically pursued?

2. Can interest be sustained by it?

3. Is the problem solvable?

4. Is it worth doing?

5. Will it lead to other research problems?

6. Is it manageable in size?

7. What is the potential for making an original contribution to the literature?

8. If the problem is solved, will the results be reviewed well by scholars in your field?

9. Are you, or will you become, competent to solve it?

10. By solving it, will you have demonstrated independent skills in your discipline?

11. Will the necessary research prepare you in an area of demand or promise for the future?

I would add to Smith's list problems that have both a basic and applied component, the potential for interdisciplinary collaboration, and, particularly in areas with large infrastructure needs such as observational astronomy, high-energy physics, and semiconductor processing, the potential for interuniversity interaction.

As noted earlier, it is important to have a problem that you can enthusiastically pursue (1), and that you can sustain interest in over the long haul (2). Much of research is just that, re-search. At times, it will be quite mundane and will surely be frustrating. Experiments will not go right, equipment will fail, data promised from other sources will not arrive on time (or at all), other researchers who pledged assistance will not come through as expected, and still others will do work that competes with your contribution. During these times, you need courage and fortitude. Having developed a topic that you are truly excited about can make a big difference.

> Sanford Dickert is a graduate student working on the problem of resonance in robotic arms, which is a subset of a number of problems in the field of control theory. Dickert's studies have implications for robotic use in medical surgery, manufacturing, and even planetary exploration. In addition, his resonance investigations have implications for studies on the movements of large structures such as buildings and bridges. By placing his robotic arm studies in a larger context and courting interest and financial support from a number of sources, such as NASA and the Bechtel Corporation, he has been able to sustain his efforts through many frustrating periods.

Choosing a problem that is solvable in a reasonable period of time (3), that will lead to further research (5), and that is manageable in size (6) is a particular challenge for most graduate students and postdocs. Initially, there is a tendency to take on more than is necessary to complete a dissertation or obtain another publication or two. In such cases, it is important to have the right supervisor, part of whose job it is to help you understand what is and is not an original, publishable contribution (see next section). There will be plenty of time for further work after you complete the current stage (see vignette at the end of this chapter).

Whether or not a problem is worth solving (4), will make an original contribution to the literature (7), and, if solved, will have results that will be viewed well by scholars in your field (8) is at the heart of what we mean by choosing compelling topics leading to a meaningful "stream of ideas." As Davidson and Ambrose point out, "The most desirable projects are challenging, intriguing, and important (e.g., the results can contribute to improving our quality of life in some way or are intellectually interesting)" [3, p.100].

One way to tell if a subject is compelling is to note the varying attendance at seminars or symposia on different research topics. It is rarely the same for each session. In some cases, attendance may be up for big-name speakers, but often it is because the work presented is also of broader interest. Pay attention to these seminars as clues to possible research directions and topics. Of course, going into areas where there are too many other researchers has its drawbacks, but watch the opposite extreme of being one of a kind with little complementary interest or support.

Another way to tell if your idea is compelling is to imagine giving a talk on your work to a group of faculty and graduate students not in your area of specialty. Do you understand their problems well enough to give a talk that would be understood and appreciated by such an audience? In other words, do you recognize the applicability of your ideas to other fields? If so, you are probably on the right track.

Generally speaking, problems rooted in theory or broad concepts are more compelling than are those specific to a particular time, place, or activity [5].

Meteorology professor Alison Bridger's research on the winds of Mars is interesting particularly when seen as a way of predicting dust storms during future unmanned landings. It becomes compelling, however, when it provides insights about the behavior of all planetary atmospheres, including Earth's.

Materials science professor Brian Love's research on contact materials for dental crowns is interesting. It becomes compelling when seen in the context of the broader theories of surface science, and applications to such areas as artificial implants and electronic packaging.

Mechanical engineering professor Fritz Prinz's research on the deposition of metal droplets is interesting. It becomes compelling when seen as a way to construct unique metal artifacts, some of which are machine tool prototypes built in hours instead of months.

Biologist Rick Hopkin's research on radio telemetry is interesting. It becomes compelling when seen as a unique way to use portable, low-power electronic systems (placed in collars) to study the home-range characteristics of wild animals.

A further test of whether or not a topic is compelling is if industry shows an interest. This test does not apply in all situations, of course, but it is a way of seeing if your work can be put in a broader, applied context.

Your competency to solve the problem (9) will, of course, depend to some degree on your innate abilities. However, your ability to develop basic knowledge and technical understanding, computer skills, and experimental expertise will also depend on such things as the availability of certain courses and seminars, library materials, independent study opportunities, and most importantly, other students, postdocs, faculty, and even industrial scientists and engineers with whom you can interact. While all of these people do not have to be in your department, or even at your university, many should be accessible to you in one way or another through the Internet, guest seminars, and visits to industrial sites.

Developing independent skill in your discipline (10) is part of what is meant by acquiring depth in an area. To acquire such skills, you must define and develop a problem that is sufficiently robust, the solution of which requires a fundamental understanding of certain phenomena or behaviors, and in some cases experimental techniques. However, it is important that your *focus* be on problems, and not on techniques or specialized tools. The latter come and go, and as a researcher, you want to be able to shift your approaches as needed to solve the more fundamental problems. Comments Peter Feibelman in, *A Ph.D. is Not Enough!: A Guide to Survival in Science* [6, p.46]:

> It is the researchers who focus on a significant problem, and who are willing to bring to it whatever resources are necessary, who give the most interesting talks, write the most significant papers, and win grant support most easily. I strongly recommend that you try to teach yourself to be "problem-oriented," to plan your research projects so that they address important scientific issues regardless of what techniques you and your coworkers will need to use.

Developing a research effort in an area that will be in demand and hold promise in the future (11) can be tricky. Some fields, such as semiconductor physics and fiber optics, may have been compelling for some time, but are now approaching maturity and shifting focus, and are likely to be in less demand in the future.

Other areas, such as telecommunications and biotechnology, are quite popular. However, their very popularity may have oversaturated the field. In such cases, large numbers of investigators often compete for limited financial and experimental resources.

Some areas serve as technology drivers for other fields, and may be in a better survival position as specific applications shift. You need to look at emerging fields and see if your work can impact these areas in some specific way. For example, work on amorphous silicon may apply to the emerging fields of flat panel displays, which in turn are part of an even broader field of low-power portable communications systems.

While breadth-on-top-of-depth does not mean breadth-in-place-of depth, being able to move around within a field can be quite helpful. You will always need to maintain an area of depth, but your depth can shift over time. In the words of Prasad Raji, a research scientist at the Hewlett Packard Company Research Laboratories in Palo Alto, CA: "We want people who have demonstrated a depth so we know they can do this kind of work. But we don't want them to be married to a particular depth. We want them to be able to develop a new depth in a related field somewhere within their overall breadth."

Ph.D. students who have spent time in industry before returning to school often have a special perspective on what research problems would be of interest to the commercial sector. These typically more applied research topics are becoming more acceptable as dissertations in some academic areas. The Multiple-Option approach points to either considering such a topic for your own research, or at least looking for connections between your work and that of students and postdocs from industry.

Postdocs in particular need to look for research experiences beyond the specialty they developed in their dissertation. Here, again, Feibelman makes the point that [6, p.103]:

Without at all wanting to argue that you should strive to be broad and shallow, or that you should spread yourself so thin that you are unable to make progress in any area, I suggest that by having your fingers in several pies you are more likely to prosper scientifically. As one area loses its scientific appeal, another, with which you are already familiar, may increase in importance. The clever ideas you learn, or develop, in one area may be applicable in another. This can be an extraordinarily efficient way to make progress.

While doing research as a graduate student and postdoc, look ahead to doing research as a professor, and consider such things as start-up and maintenance costs. Can you replicate the very expensive laboratory that your advisor has acquired over many years? Is it necessary to do so? Martin Ligare, associate professor of physics at Bucknell University, is a case in point.

> Ligare received his Ph.D. from Columbia University in atomic and molecular physics with a specialty in laser spectroscopy. His experimental research required a fair amount of laboratory space and equipment. After graduation, Ligare received a National Research Council postdoctoral fellowship, ostensibly with the National Bureau of Standards (NBS) in Gaithersburgh, MD, but, in fact, he spent the entire time (18 months) in residence at the Massachusetts Institute of Technology. He then spent four years at the City College of New York (CCNY) where he continued his experiments, first as a postdoc and then as an assistant professor. While at the NBS, M.I.T., and CCNY, Ligare was part of an experimental laser group requiring large amounts of grant money and infrastructure support.
>
> A voracious reader of physics material outside his immediate field, Ligare was more able than most of his colleagues to see the bigger picture. He began to take on the role of a person who could place the work of other individuals in a broader context, and who understood the theoretical underpinning of his field. His papers on laser spectroscopy became more theoretical.
>
> Ligare decided to leave New York for Bucknell University, a liberal arts school in Lewisburg, PA. Had he stayed at CCNY, he would have had to proceed with his experimental work and continue to try and bring in large NSF grants. At Bucknell, he had freedom to work on projects that were still in the spectroscopy area, but which were of a more theoretical nature. Such work did not require the infrastructure support of his experimental studies, something Bucknell could not have provided. Ligare still goes to the same conferences, sees the same colleagues, but plays a different role, one more suited to his own, and his institution's, interests.

Finally, you need to pay attention to the interdisciplinary implications of your work, and to the possible appeal such work has to both academia and industry. Just as breadth-on-top-of-depth does not mean breadth-instead-of-depth, interdisciplinary does not mean "instead of disciplinary."

As chemistry professor Robert Smith notes [4, p.35]:

Interdisciplinary research is no substitute for good disciplinary training during the greater part of a graduate career. It is advisable, however, to seek exposure to interdisciplinary activities in graduate as well as postdoctoral training since most researchers engage in interdisciplinary research during their professional careers.

An illustration of this interdisciplinary approach can be found in the work of Renate Fruchter.

A lecturer and research associate in the Civil Engineering Department at Stanford University, Fruchter has been working on the development of computer support for collaborative teamwork and software communication tools that enable architects, structural engineers, and construction managers to collaborate in the design of buildings, bridges, and other large structures. This development in itself represents a significant interdisciplinary effort. However, Fruchter was interested in exploring the generality of her concept and her collaboration technology. She developed tools that could be used by computer scientists and mechanical, electrical, and spacecraft engineers in the design and manufacture of what are called mechatronic, or electro-mechanical, systems. Such systems, which include everything from laser printers that can operate nonstop for up to a year at a time, to automobile power windows that automatically reverse if an object is caught in their path, are of particular interest to the manufacturing industry.

In looking for ways to expand her work beyond the civil engineering and architecture communities, Fruchter came to the attention of the Stanford Integrated Manufacturing Association (SIMA). She developed a proposal that engaged graduate students and other postdocs from a number of departments. The proposal was funded on a pilot basis by SIMA. Subsequently, a group of SIMA's industrial members who had a particular interest in her work provided additional support, and Fruchter is now developing a larger proposal for submission to the National Science Foundation.

CHOOSING A DISSERTATION ADVISOR/POSTDOC SUPERVISOR

A rabbit is sunning himself outside his house when a fox comes along and tells him that he is going to eat him for lunch. The rabbit explains, rather smugly, that the fox cannot eat him because he is working on his dissertation, the subject of which is the superiority of rabbits over foxes and wolves. The fox laughs, but the rabbit persuades him to come into his house and examine his dissertation with the understanding that if the fox did not agree that the title was correct, he could eat the rabbit for lunch. The fox follows the rabbit into the house and never emerges.

A few hours later, the rabbit is out sunning himself again when a wolf comes by. The above scene repeats itself with the same result.

Later in the afternoon the rabbit is outside again when a squirrel comes by and comments on the satisfied look on the rabbit's face. The rabbit explains that he is indeed satisfied because he has just completed his dissertation on the superiority of rabbits over foxes and wolves. The squirrel is skeptical, but agrees to follow the rabbit into his house to examine his dissertation.

In the house is a computer on which appears the completed dissertation. On the floor on one side of the room are the bones of a fox. On the other side, the bones of a wolf. In the corner sits a lion.

The rabbit smiles and says to the squirrel, "You know, it doesn't really matter what your research topic is, as long as you have the right advisor."

Anonymous message on the Internet

Well, not quite, although choosing the right advisor is certainly of comparable importance to choosing the right research topic. As S. E. Widnall, a former president of the American Association for the Advancement of Science, notes, "The advisor is the primary gatekeeper for the professional self-esteem of the student, the rate of progress toward the degree, and access to future opportunities" [7].

Choosing an advisor and determining a research topic are clearly linked, and the first does not always precede the second. Beginning Ph.D. students, particularly those without industry experience, are usually less sure than are postdocs of what research topic, or even what field, to pursue, so it is a good idea to look for opportunities to explore various possibilities before making a commitment. Some departments even have formal rotating internship programs that allow you to spend a period of time, usually up to six months, with different faculty to explore mutual interests and possible fits.

Your freedom to choose an advisor decreases as you focus in on a research field, which makes it all the more important to consider what you want in such a person. In addition to a primary advisor, you will most likely select one or two secondary advisors, and what one is lacking in experience and temperament can often be found in the others. One advisor may be from industry, another from a different department, and another with strengths in a particular specialty.

Our Breadth-on-Top-of-Depth approach suggests that you look for an advisor with connections to other faculty, and one who will give you "wiggle room" in choosing a topic, although with tighter funding constraints, the latter becomes more difficult. The Next-Stage approach calls for an advisor who will give you opportunities to do some teaching, write research proposals, supervise other students, and in other ways let you begin doing some of the things you will do as a postdoc or professor. The Multiple-Option approach calls for an advisor who is open to your pursuing possibilities in academia, government, and industry.

According to Smith [4, p.30], the choice of an advisor should be based on the person's:

- Accomplishments in teaching and research

- Enthusiasm for advising students

- Experience in directing graduate students

- Management and organization of his or her research group

- Reputation for setting high standards in a congenial atmosphere

- Compatible personality.

Smith refers to three types of advisors: (1) the collaborator, (2) the hands-off, and (3) the senior scientist [4, pp.26–27]. The collaborator type is more likely to be young and hungry for results. His/her success depends to a larger extent on yours, so he/she has a vested interest in how well you do. Often, this can mean rapid progress toward your degree, but be careful. In such cases, topics are often chosen more by the advisor than by the student. The topics may be less risky, and the advisor may want more than the appropriate share of credit. The key with young faculty is to see that your interests overlap with theirs.

Young faculty usually have quite a bit of energy. While they lack experience in supervising graduate students and postdocs, they remember more clearly what it was like to be in such a position. Also, how well such professors supervise graduate students, or at least how many they graduate, may be a factor in tenure decisions that can be to your benefit. Of course, the existence of this factor can also be a problem in terms of the pressure on you to perform. As one graduate student with such an advisor noted, "No laid-back six months to browse the literature in my situation."

The hands-off advisor is generally a midlevel academic with other responsibilities, but in Smith's words, "less greedy for results." Such a person can be a source of wise counsel, and might let you choose areas of greater risk and significance.

The senior scientist type is a well-established faculty member with varying amounts of time. Smith thinks the quality of attention may be the best of all because of their extensive experience. However, while older faculty may not compete with you, as might younger faculty, they may also think they know it all, are less likely to help you learn the ropes, and may not be as available. Also, their energy level may be lower, they may be out of date, or living on past glories.

Joanne Martin, a professor of organizational behavior in the Graduate School of Business at Stanford University, has also studied the advisor–advisee relationship in some detail. She, too, likes to classify advisors into three types, in this case, (1) authoritarian, (2) coach, and (3) laissez faire [8].

Martin thinks in terms of management styles. The authoritarian advisor is likely to set the goals and lay out tasks for the research, usually in some detail. He/she will want weekly meetings to assess progress, and will give more objective criticism than praise. "Expect lots of red marks on the drafts," notes Martin. Authoritarian advisors welcome conflict, expect you to speak up, and are active throughout the research process.

The coach, on the other hand, will seek to set goals jointly with the advisee. There may be a lot of guidance in the beginning or planning phase, but not much during the research itself. "Active in the planning stage, passive during the process, and active in the evaluation stage," is the way Martin puts it. You will be expected to take over the research and writing, and there will be no micromanagement. Coaches give good feedback and try to draw out criticisms from their student. The style is somewhat impersonal and task-oriented, and conflict is okay.

The laissez faire advisor is friendly and constantly supportive. If you do not like criticism, this is the one for you. You set the goals. The advisor encourages you, "keep up the good work," and avoids conflict. Such advisors are friendly, but it is not certain that you will learn much from them. They will be relatively inactive on the research task unless you take the initiative, but supportive throughout, and generally available. Attractive as they may first appear, going with a laissez faire advisor is a high-risk strategy, and is only likely to work if you have strong research skills, are independent, and know what you want.

Unlike Smith, Martin does not see a correlation between her categories and whether or not an advisor has tenure. Although it is tempting to match Smith's collaborator, hands-off, and senior scientist with Martin's coach, laissez faire, and authoritarian, the real point here is not to force-fit people into categories; after all, everyone has some of all six characteristics. In fact, some advisors are able to combine elements of all three styles as the student evolves. However, each advisor probably fits one description more completely than any other, and you need to be aware of these differences and make decisions and adjustments accordingly.

First, keep in mind that you want an advisor who is in a research area in which you have, or seek to develop, a strong interest. This element is essential, given all the inevitable difficulties you are going to experience. At the same time, you want a professor who understands that "finding the right problem" is half the battle, and that he or she needs to give you the encouragement and time to do so. Martin says this quite eloquently when she notes:

> Topics you approach with passion lead to the best research. You want to get out with something significant, something you care deeply about (rather than the purely pragmatic) choice, so select a topic that is a window into your soul.

Look at the styles of various professors, how they treat their graduate students, how many students they have, and what has been their track record in getting students out. Consider how much direction and supervision you want and need. Be willing to push your boundaries, but not to an extreme. You need to be somewhere in your comfort zone, but not too comfortable. Above all, you want at least one advisor who will be demanding. In the words of a graduate student in chemistry, "You want your anxiety level to go up a little when you get an e-mail message from your advisor."

If possible, talk to the students and postdocs of potential advisors. Ask lots of questions. Do the students and postdocs see the big picture? Are any of them doing compelling work? Do they have high expectations for themselves and others in the group? How often does the research group meet? If possible, attend one or more of these meetings, and pay attention to the interactions. Consider what your relationship will be with the other students and postdocs in the group. Who will give you survival tips? Who will mentor you? Are there postdocs involved, and what is their relationship with respect to the professor and to graduate students? Where are recent graduates and former postdocs now working?

It is also important to look at things from the other direction, and consider what is in all of this for your advisor. It is rare for an advisor to take you on to simply fulfill a departmental obligation to supervise a certain number of students or postdocs. Most advisors already have enough students, and one more will not make much of a difference in this regard. Faculty supervise graduate students and postdocs for two basic reasons: it helps them with their own research agenda, and it gives them a chance to do what is called "deep teaching." Often, the two go together.

Helping faculty with their own research may involve more than just working on existing projects. Professors often want to explore new areas, and having a graduate student or postdoc who is so interested may prove particularly attractive.

> Karen Huyser, a graduate student in electrical engineering, was interested in exploring the then new field of neural networks. Professor Mark Horowitz, whose specialty was in microprocessor design, had a burgeoning interest in this area, and thus took Huyser on as a graduate student. Because it was such a new field, Huyser had to reach out to faculty in such areas as biology, computer science, psychology, and even philosophy. This reaching out by Huyser, in turn, brought Horowitz into contact with professors who he would never have met had it not been for Huyser's interests.

By now, it should be clear that not all teaching takes place in the classroom. Working with graduate students and postdocs over a period of years, not months, can be particularly rewarding. As organization professors John Darley and Mark Zanna point out, "Training graduate students is the one aspect that makes a professor's career sharply different from almost all other jobs" [9]. To engage future colleagues on a deep level, and to see them evolve in their understandings, capabilities, and degrees of responsibility is something a faculty member does not experience in the classroom. As not only a recipient of this kind of teaching, but as a future provider of it, you can benefit more than most from the process.

WRITING YOUR OWN RESEARCH PROPOSALS

> If I had my life to do over again, I would have written half a dozen proposals *before* I got my Ph.D., instead of just contributing to one or two. Once you have a faculty position, chasing dollars is the number-one activity; the more experience you bring to it, the better off you will be. Use your research advisor to help you learn to write proposals, *before* you start sending things in blind to funding agencies.
>
> *Michael Reed, Associate Professor*
> *Department of Electrical Engineering, Carnegie-Mellon University*

Writing proposals is not just for those heading for Research I universities. No matter what type of faculty position you take, be it at a baccalaureate, master's, doctoral, or research-granting institution, or even a research position in government or industry, you will soon have to start writing proposals. In academia, if you want any scarce resources such as funding from an internal or external agency, equipment for research or course development, a fellowship for your graduate student, or even just permission to perform a procedure, you must first write a proposal. All proposals, from two to three page statements seeking small amounts of seed funding from inside an institution, to lengthy documents with extensive budgets that can take weeks or months to prepare, must be written with care.

The objective in this chapter is not to teach you how to write proposals, although I will say something about doing so in Chapter 12, "Insights on Research." In addition, you can take advantage of the many workshops, seminars, books, and other writings readily available in academia to help you with such preparation. My purpose here is to persuade you that writing proposals is something you need to get started on *before* you become a professor. It is a perfect example of the Next-Stage and Multiple-Option strategies.

Writing represents a very high level of thinking, much higher than conversation. That is why it is so difficult, and why, when it is done well, so valuable. It does get easier with practice, so you need to get started. Look for every opportunity to put your case in writing and to receive the critical feedback your efforts will bring. Completing such an important and difficult task brings many rewards. Again, Smith captures the point best by noting [4, pp.233–234]:

> Writing proposals and applying for grants require the discipline and insights that will help throughout a professional career. Commitment, thoroughness, and patience are essential components of grant getting. Tolerance of failure is also required, because

many proposals are rejected. Overcoming the trauma of rejection is an important lesson for the professional. I know researchers who have received dozens of grants during their lives. This has often meant confronting five times that many rejections. Researchers have to develop emotional strength to survive. Survival skills develop with the self-confidence that comes from research accomplishments, and these skills are useful in one's personal life as well as work.

Preparing your own proposals helps you confront the Breadth-on-Top-of-Depth challenge by requiring you to place your work in a larger context. Science writer Anne Simon Moffat discusses the keys to successful grantsmanship with an interesting example from biology [10]:

> A proposal to isolate and characterize a new gene or to describe a new molecular structure may not be enough to snare a grant, unless an explanation is offered on how these discoveries relate to larger problems, such as gene regulation or structure/function relations. Take protein crystal structures, which are now being solved by crystallographers at a rate of about one per day. The techniques and materials are expensive so proposals have to go beyond "another new one." There's got to be "a good story behind the protein," says crystallographer Joe Kraut of the University of California, San Diego.

Most graduate students and postdocs begin by reading and critiquing the proposals of others, and then moving on to contributing drafts of sections to a proposal written by a principal investigator. This type of contribution is fine, yet it is not a substitute for putting together your own complete proposal.

Perhaps the best place to start is with your dissertation proposal. Some departments require all students to produce such a document, but even if yours does not, you should do so anyway. Not only will it give you much needed experience, it will help ensure that you are undertaking an effort that has the support of your advisor and other key faculty. It also helps you understand clearly what you are, and are not, committing to, and what is at least the initially projected time frame for the completion of your work. The same thing can be done as you begin your postdoc work. Here, it is particularly important to have expectations about what you can accomplish and when, and one way to help ensure this understanding is through a written proposal that you and your supervisor agree on before you start. The postdoc period is also a time to develop proposals that can help launch your academic career (see the vignette at the end of this chapter for a specific example).

Show drafts of your proposal to as many people as possible, including professionals who are in your general field, but not in your particular specialty. Be sure to consider faculty at other institutions and interested researchers in government and industry. These are the people most likely to recognize excessive jargon, obscurities you have missed, and connections you may have failed to identify. The fact that these are also the people who will be your future colleagues, and that some of them may even be in a position to help you with your first professional position, is something I am sure you will consider as well.

Writing your own proposal does not mean you have to be the sole investigator. Indeed, as a graduate student, or even postdoc, it may be necessary for you to do your work in conjunction with a principal investigator employed by the institution. It does mean that you are taking the overall responsibility for the development of the document, and it is "your" piece of research that is being proposed.

What the above discussion boils down to is this: the time to get started on writing your proposal is *now*. Free-lance writer Robert Burroughs puts it best when he says [11]:

> As with anything else, proposal writing improves with practice, and persistence can be key. If you can focus your attention on a grantor's interests, show how your research matches that interest, and then demonstrate that your research has a wider theoretical application, you will have made a good start down the road to funding.

CARRYING OUT YOUR RESEARCH—AN EXAMPLE

As noted earlier, the research process has its moments of excitement, and even drama. But much of it is also mundane work punctuated by periods of frustration, delay, and anxiety. At these times, it is particularly helpful if you can incorporate the three elements of the preparation strategy outlined earlier. Below is one example of how this strategy can work. It is taken from an organization I am involved with at Stanford University called the Center for Integrated Systems (CIS). The Center seeks to foster multidisciplinary research in microelectronics and computer science through the financial support of its industrial members.

> Michael Farn, a graduate student in electrical engineering, was interested in exploring a new field called "integrated optics," which involved attempts to produce Fresnel (essentially flat) lenses on semiconductor wafer-like materials. The ability to mass-produce such lenses using semiconductor technology could be of significant interest to industry.
>
> Joseph Goodman, a professor at Stanford with a specialty in optics, was interested in serving as Farn's advisor. However, the topic represented a new area for Goodman, who had just become chairman of the Electrical Engineering Department, and thus had a limited amount of time to supervise a student in an emerging field.
>
> Farn realized that if he were going to explore this area, he would need additional resources, both material and intellectual, to complement what was available from his department. Texas Instruments, headquartered in Dallas, TX, a member of CIS, was interested in the possibility of such lenses for two reasons. If they could be made using semiconductor techniques, then the company, a leading semiconductor

manufacturer, would be in an excellent position to produce them. Furthermore, Texas Instruments had a division that made infrared detectors for the military and law enforcement, and it was possible that such lenses might replace the bulky glass ones currently in use in such devices.

A relationship was established among Farn, Goodman, and scientists at Texas Instruments in which Farn, who was given a Texas Instruments fellowship, used some of the company's facilities in Dallas while consulting with their industrial scientists. All of this was done with Goodman's enthusiastic approval and involvement.

Farn quickly took on the role of facilitator during his bimonthly visits to Dallas. It was Farn who brought together, for the first time, semiconductor scientists and the optics engineers who, prior to this project, had no particular reason to talk to each other.

According to Graddy Roberts, Farn's chief contact at Texas Instruments, "If all he did was to bring these two groups together to talk about mutual problems and opportunities, it would have been worth it from our point of view." But Farn did much more. By placing his research in the context of its possible application to semiconductors and infrared optics (Breadth-on-Top-of-Depth), bringing together people who needed to deal with the challenges of mass-producing such devices (Next-Stage), and making contacts with scientists in a nonacademic institution (Multiple-Option), Farn covered all the bases while pursing research that lead to his Ph.D.

The results of Farn's initial efforts were promising enough that Farn and Texas Instruments wanted to continue their relationship. After graduation, Farn accepted a postdoc at M.I.T. Lincoln Laboratories where he continued his work in integrated optics under a contract partially funded by Texas Instruments.

PUBLISHING

To succeed you must make your talents well known and widely appreciated. Publishing provides you with an important way to accomplish that. Your papers, available in libraries around the world, represent not only your product but also your resume.

Peter J. Feibelman
Sandia National Laboratories [6, p.39]

When it comes to publishing, having the right perspective is essential. If you view publishing as a burden and as a requirement imposed by the system, then you will do it reluctantly, if at all, and are unlikely to feel proud of the outcome. On the other hand, if you view publishing as a natural part of your scholarship, as a way to take some of the credit for what you have done and to communicate the results of your hard work to your colleagues and beyond, then you are likely to set standards that will help make you a professional success. The important thing is to get in the publishing habit *now*, and to treat publishing as a forethought, not an afterthought. You can do this by creating the expectation in your own mind that you will do publishable work.

Publishing produces a permanent record, so you want to be sure what you produce is of high quality. Even though the peer review process rejects far more than it accepts, a lot of not-so-memorable work gets through. This reality means that ultimately it is your own standards that should determine what to do. If you plan to do publishable work right from the beginning, it will help set the baseline for the research that follows. Many people find it helpful to get in the habit of writing as they go along, not just after they complete their scholarship. Each (writing and scholarship) reinforces the other.

Most graduate students and postdocs will end up as coauthors on publications with more senior researchers and/or faculty. This junior–senior coauthorship raises some interesting issues having to do with the proper appropriation of credit, which we will discuss in Chapter 13, "Insights on Professional Responsibility." In most circumstances, you should be open to sharing credit with others, even if you feel you have made a disproportionate contribution to the research. Contributions can be made in different ways and at different levels. However, one important test of coauthorship is whether or not everyone listed would be able to give a talk, and answer subsequent questions, on the paper at a professional conference or symposium. Be certain that this ability at least applies to you.

As noted earlier, our purpose in these sections is not to give specific details on how to carry out the activities that are discussed, but rather to motivate you to start on the activities (Next-Stage approach) while a graduate student and postdoc. There are plenty of resources available in most institutions to help you with the details. (See also "Writing Research Papers" in Chapter 12.) One particularly useful reference is by the editors of *ASEE Prism*, who have developed an excellent number of tips on how to avoid publishing pitfalls [12]. We reproduce one especially good set of pointers here that were adapted from sociologist B. B. Reitt's "An Academic Author's Checklist [13]."

Is the article, as written, complete?

- What is the problem (question, issue) the article sets out to solve?

- Does it achieve a solution (resolution)?

- If a full solution is not reached, why not?

- Is the solution reached the best one among the possible answers?

- Is the significance of the solution apparent to the reader?

- Is any information in the article unnecessary?

Is the article, as written, authoritative?

- Is the solution (information, hypothesis) presented based on a proper mixture of the writer's and others' work?

- Is the solution based on an appropriate mix of primary and secondary sources?

- Does the writer give evidence of wearing discipline blinders?

- Are all the sources cited genuinely relevant and necessary?

- Is the balance between published and unpublished sources good?

- Are all the acknowledgments of indebtedness made? Has permission to acknowledge personal assistance been obtained?

Is the article, as written, singular?

- Does the article provide new information or hypotheses, or does it contribute corroborative evidence to an existing body of knowledge? Which elements, exactly, in the article are unique, original, completely new?

- Which elements, though not new to this field or specialty, are new to other groups of readers who have been overlooked before?

- What kind of information does the article provide its readers? What mixture is there of new and established information?

- Is the information timely or timeless?

- How specialized is the information?

- Who needs this information? For what purposes(s)?

- Is the information already available to readers?

ATTENDING CONFERENCES AND OTHER PROFESSIONAL MEETINGS

As a professor, you will certainly attend your share of professional conferences. Attending such meetings is also something you will want to do as a graduate student and postdoc, particularly as you near the end of your tenure. Indeed, one of the functions of a good advisor is to encourage such attendance, and where possible, to introduce you to interesting colleagues. While conferences can be expensive, student discounts are available. Also, conferences tend to change locations from year to year, so within a two- or three-year period, a conference in your field is likely to take place relatively close to your current institution. In many fields, it is at such conferences where job contacts are made, and in some cases, where preliminary interviews take place. We will look at this important aspect in more detail in Chapter 8, "Applying for Positions."

The purpose of this section is to make you aware of the advantages of attending conferences and other meetings while you are still in the early stages of your research. Conferences are ideal places to find out what is hot, and not so hot, in your field, observe the various debates and controversies underway, meet interesting people, make contacts for future interactions, and in general participate in the milieu of the professionals in your field. All aspects of your preparation strategy can be carried out to some degree at professional conferences.

Conferences can range from as few as 100 or so participants for very specialized or local events, to many thousands for annual meetings such as those of the American Chemical Society, the American Institute of Biological Sciences, the American Society of Mechanical Engineers, and the American Physical Society.

Conferences can often seem overwhelming, and it is easy to be intimidated by the list of speakers and attendees, all of whom seem to know more than you. As astronomer Timothy Ferris noted during a recent cosmology convention [14]:

> ...there was hope of learning something useful, especially given the roster of speakers which included Edward Witten, the string-theory wizard widely said to possess the most acute scientific intellect since Einstein; Andrei Linde, the Russian cosmologist who theorizes that our universe is just one among many; and Allan Sandage, the astronomer who has done more than anyone else alive to chart the size and age of the universe.

The key that I have found is to treat the whole thing as a bit of a game in which you ask specific questions, make certain observations, and do particular things, all designed to make the experience worthwhile and even fun. Here are some suggestions:

Questions to ask

- What can I learn from my three days at this conference that I could not pick up in any other way?

- What would I do differently if I were giving the talk I am now attending?

- Am I ready to step into this situation (talk, discussion, or meetings), and if not, why?

- Who are the people asking the "big" questions? Who are the leaders in the field?

- Are there new directions developing in my field, and if so, who are the emerging leaders?

- Who are the professors whose behavior and actions I wish to emulate and why? Are they associated with a particular school or field?

- Who are the recent alumni from my institution with whom I can make contact, seek information, and share experiences?

- Who are the professors who might review my work, collaborate with me, and possibly help me with future connections?

Observations to make

- Note which professors are being offered positions at other academic institutions or in industry.

- Note which professors, students, and postdoc, seem comfortable interacting with professionals from industry.

- Note which professors take the time to introduce their graduate students and postdocs to colleagues in academia, government, and industry.

- Note which professors are sought out by the media and why.

- Note hallway conversations, political and job opportunity discussions, who is talking to whom, and what kinds of alliances are being formed.

Things to do

- When appropriate, give a presentation on your research, and seek as much formal and informal feedback as possible (see next section).

- Introduce yourself to speakers who you found particularly interesting. Exchange business cards.

- Briefly (five minutes maximum) describe your experiment, dissertation, or postdoc research topic to a variety of people, and note their reactions.

- Share with others a list of the courses you have taught or lectures you have given, as well as a statement on your teaching philosophy (see Chapter 6).

- Attend presentations in your own and related fields. Look for contacts that could lead to expanded research interactions.

- Attend specialized meetings and social gatherings where you can meet contemporaries from other institutions.

- Talk with other graduate students and postdocs about their plans, and exchange information.

- Obtain names and business cards of professors from schools with whom you want to have additional interactions. Do the same for scientists and engineers from government and industry.

- Follow up with a note and promised abstracts or publications.

- Get on the mailing lists of publishers for the latest textbooks in your field.

Remember that you cannot possibly do or see everything, or meet everyone. But you can become engaged in the process, and set the stage for a host of future interactions that will benefit you in the months and years to come.

PRESENTATIONS

When it comes to talks on your research and related activities, you need to do two things: (1) give them, and (2) obtain honest feedback on how well they were received. The second does not always automatically follow from the first, so you may need to take special steps to ensure that it does. As with proposal writing and publishing, giving talks on your work to various audiences is something you will be called on to do throughout your professional career. According to Cliff Davidson and Susan Ambrose of Carnegie-Mellon University [6, p.59]:

> You will deliver many talks on your research in a variety of settings throughout your career as a professor. The goals of these presentations might vary widely, e.g., to convey highly focused information to an audience of specialists, or conversely to cover a broad range of issues appropriate for the general public. To cover the spectrum between these two extremes, imagine discussing your research in front of the following audiences: experts in your discipline at a professional conference, students and faculty in your department, a government committee composed of technical and nontechnical people, and a local community volunteer organization. The purpose, amount of detail, and rhetoric of your presentation will vary in these situations. Nevertheless, there are general guidelines that you can use to prepare and deliver effective talks for all of these audiences.

An important addition to the above list for future professors is the "academic job talk," given during most job interviews, that can make or break your chances for a faculty position (see Chapter 8, "Applying for a Position," for further details on this particular talk).

Becoming comfortable with giving talks about your work is essential. Here, again, practice is the key. There is certainly a relationship between giving good talks on your research and giving good lectures, and one would assume that if you are thinking about becoming a professor, you will look for opportunities to do both (see Chapter 6, "Teaching Experiences Prior to Becoming a Professor"). Yet, many students and postdoctorates are reluctant to give talks on their research, particularly to their peers and to faculty. Usually, it is because they do not feel they are ready. We often do not think that we have enough to say when what we really mean is that we have not taken the time to develop the right presentation. As long as you start with the correct expectations, it will rarely be the case that you do not have something worthwhile to offer.

An informal presentation to your research group of a half dozen researchers requires a different degree of completion than does a selected presentation at a professional conference. There are times, particularly in a seminar or in a presentation to senior undergraduates, where a talk on how you came to choose your research topic would itself be an interesting and appropriate subject.

In all cases, however, you want to obtain critical feedback on how well your talk was received. Colleagues who show no reluctance to give very frank feedback on papers and proposals are often much less willing to do so with oral presentations. You can certainly pick up clues in the number and kinds of questions asked and in informal comments heard after the talk. However, you owe it to yourself to probe beyond comments such as, "nice talk," or "enjoyed your presentation." Acknowledge the compliment, and then ask the person giving it if they have any suggestions on how you might improve the presentation in the future. With such an opening you will often receive some very specific pointers that can be quite useful.

We will not go into detail here on the specifics of presenting good talks since, as with proposal and grant preparation, there are resources available at most academic institutions to help you with presentations. In some cases, it is even possible to have your talks videotaped for additional feedback. In giving talks, it is worthwhile, however, to keep in mind Feibelman's dictum, "never overestimate your audience [6, p.28]." We have all suffered through presentations in which we thought we were the only ones who did not understand what was being said, only later to discover we were far from alone. Why is it that faculty who have no difficulty giving talks at the right level in their courses feel the need to impress their colleagues with presentations loaded with jargon, unnecessary equations, and obtuse references? You lose nothing, and gain much, by taking the time to place your work in the proper context, present interesting examples, use clear illustrations and graphics, and even include some humor and a good story or two to make your point. You can always provide more detail during the question and answer period.

An example of an activity incorporating our three-pronged preparation strategy with respect to graduate student and postdoc presentations is the Stanford Center for Integrated Systems' Student–Partner Information-Exchange program (S.P.I.E.). In this program, teams of advanced graduate students—those within one–two years of completing their Ph.D.s—give presentations on their research to the Center's industrial sponsors. As many as 75 students, representing over two dozen research areas in applied physics, chemical engineering, computer science, electrical engineering, materials science, and mechanical engineering, comprise a pool from which companies choose teams of four–five students to give talks at their research and development sites. Students may end up visiting companies across town, across the country, or in some cases across an ocean. To participate, each student must develop a half-hour presentation that can be understood and appreciated by all of the other students on the team. Thus, someone working on relational databases in the Computer Science Department must give a talk that can be understood by a student working on fiber optics in the Applied Physics Department, and vice versa. This requirement puts the students in good stead when they then give their talks to scientists and engineers at the visited companies. The program also gives the students an opportunity to put their work in a broader context, to look for connections to other research areas, and to expose themselves and their work to industrial companies.

SUPERVISING OTHER RESEARCHERS

As a professor, you will certainly be supervising the work of others, be they undergraduates doing independent study projects, master's students doing theses, Ph.D. students doing dissertations, or in some cases, postdocs managing projects of their own. In addition, you will most likely have some supervisory responsibility for support staff. Any experiences you can acquire along these lines before you become a professor will redound to your benefit, not only in the way it helps you with your current responsibilities, but in how it demonstrates your readiness for the responsibilities of a professorship.

The level of supervisory responsibility you undertake will depend on your willingness and readiness for such responsibilities, and the number of players on your team needing such supervision. In the beginning, you are, of course, the one being supervised. Use this time to observe your own relationship with your supervisor. Also note the relationships that other students have with graduate students, postdocs, and faculty supervisors. What do you like and dislike about the relationship, and in particular, what would you do differently? (As we will see in the next chapter, you will do the same with respect to teaching.)

One of the many challenges of supervision is to figure out how to do it in a way that enhances your own activities, while at the same time benefiting the person being supervised. In some cases, the people you supervise will work directly for you and will help you on your particular research project. In most cases, however, their work will be indirectly related to yours, part of your general area, but not directly supportive of your research. This indirect relationship is fine if it gives you a chance to broaden your understanding and to connect to areas adjacent to your own.

An illustration of how such supervision works can be seen in the case of Elizabeth Drotleff, a Ph.D. student in chemistry at the University of California, Santa Cruz.

> Drotleff began supervising undergraduates as well as another graduate student in the second year of her Ph.D. studies. "It's been an important experience for me," says Drotleff, "there's a lot more to pay attention to than I thought, but with the help of my postdoc and faculty supervisor, I'm learning to handle many things."
>
> Drotleff is part of a small team that includes herself, her faculty advisor, a postdoc, another Ph.D. student, and two senior undergraduates. They are organic chemists working in the area of natural product synthesis. Drotleff and her colleagues seek to synthesize naturally occurring compounds from marine organisms that may have potential benefit in the treatment of disease. Once they are able to reproduce these compounds, which often have complex chemical structures, in the laboratory, others can experiment with creating derivatives of the compounds that might be of particular interest to medical scientists.
>
> The task of synthesizing the compounds is broken up and assigned to different researchers depending on their knowledge, experience, and skills. The faculty advisor oversees the entire project, and the postdoc works on the most difficult synthesizing tasks. Drotleff is learning to synthesize particular parts of the molecule, and then help bring these parts together. She supervises another graduate student,

who at the moment has little formal laboratory experience, and who is attempting to learn a number of synthesis methodologies. The undergraduates she supervises produce the starting materials that she and others need for the compound synthesis. Says Drotleff:

"I'm excited about the research project I'm on; it's quite challenging and it's great to apply the chemistry I've learned in classes. Overseeing less experienced members of the group has initially been very difficult, especially because of the problem of learning how to manage my time. Not only am I responsible for myself: my research, preparation for seminar presentations, and reading the literature, but I'm also partly responsible for keeping these people busy and interested in their own chemistry as well as teaching them laboratory techniques and safety. Overall, I feel very fortunate to have this experience before I acquire a postdoc position or a job."

At the postdoc level, supervisory responsibilities can be quite significant, and are often part of the broader responsibilities of managing a research project or program (see next section).

Taking on supervisory duties is clearly part of the Next-Stage approach. It is also the kind of activity that will help you in maintaining multiple employment options. As we saw earlier, industry and government are looking for people who can communicate well with their colleagues and who can eventually manage projects and programs. Therefore, everything you do in the area of supervision will prepare you for positions in all three employment sectors.

MANAGING RESEARCH PROJECTS AND PROGRAMS

You have to be able to generate quality independent ideas, plan experiments to carry them out, and manage the people who will help you do the work. My three-year postdoc helped me do this. I had thought about going out [on the job market] after two years, but I am really glad that I waited an extra year. It was critical to my being able to make it when I got to UCSD.

Elizabeth Komives, Assistant Professor of Biochemistry
University of California, San Diego

The step up to actually managing a research project or even a small research program is something more likely to be done as a postdoc than as a graduate student. In fact, most postdocs do not even go this far, but the ones who do are sure to make themselves much more competitive. Doing so requires you to think more broadly (Breadth-on-Top-of-Depth), gives you obvious Next-Stage experiences, and in most cases brings you in contact with opportunities in two or more employment sectors (Multiple-Option).

Opportunities for such experiences can come about in unexpected ways:

A few weeks after arriving at the Stanford Linear Accelerator Center (SLAC) as part of his University of Colorado postdoc assignment, Nety Krishna ran into, as he put it, "a few problems." His principal investigator back in Colorado abruptly resigned and moved to France, essentially leaving Krishna and the investigator's one graduate student on their own. Krishna thought he was going to spend a year or two working on the front-end electronics of what are called drift chamber particle detectors before moving up to full responsibility for the chamber and a host of experiments it was designed to support. Now, he had to do both, and much more quickly than he imagined. If he failed to obtain data from the first run of the experiment, it might be up to a year and a half before the effort could be repeated. Such a failure would have been hard enough on Krishna, but it would have been devastating for "his" graduate student. And, of course, there were problems with the chamber. As Krishna describes it:

"I was faced with a 'nightmare' situation when a 25-micron wire (as thin as a human hair) split and rendered the whole drift chamber inoperable. I was forced to consider unconventional techniques of dealing with the problem because the detector was embedded between tons of lead and iron. At the time, I was undergoing knee surgery for an injury sustained during a frisbee game. I noticed that the surgeon was using an instrument that functioned as a remote retriever (arthroscope). I used this idea and performed a successful 'surgery' on the detector and restored the chamber to full working order."

This initial experience brought Krishna to the attention of the senior staff at SLAC. From then on, he was given increasing levels of management responsibilities that included overseeing the operation and schedule of all the detector experiments, supervising graduate students, functioning as an official liaison to up to 250 physicists, and serving as the youngest member of the "commissioning team" that set policy and made decisions for accelerator experiments. These management experiences have made Krishna much more attractive as he explores a variety of opportunities in government, industry, and academia.

The reasons for stepping up to these Next-Stage experiences are summarized nicely by Guy Blaylock, assistant professor of physics at the University of Massachusetts:

As you go from one stage to the next, the role you play changes. As a student, I was expected to excel in a specific subject, and I was judged by my individual contribution. As a postdoc, I was expected to be more of a team player, to contribute to the researchers around me and review their papers, as well as run a complete project and write my own proposals. As a professor, things changed again, and now I have a dozen balls in the air. Given all of this, I strongly recommend that as a graduate student, you do some of the work of a postdoc, and as a postdoc, you do some of the work of a professor. Not only does it make things easier when you get to the next stage, but it also separates you from the rest of your competition.

NETWORKING

Many of the activities discussed above are illustrative of a more general activity called "networking," the process of connecting with others with whom you have a mutual interest. The term has gotten a bit of a bad name of late because it is overused and/or used in the wrong way. However, networking is absolutely essential to your strategy of obtaining contacts that can make a difference in your path to becoming a professor.

Most of us want to believe that our good work will be enough to get us what we want, but it will not. Recognition and opportunity come to those who also have good contacts. I began to feel good about networking when a colleague whom I respect a great deal confided to me that he spent 60% of this time doing the best damn job he could, 20% of his time making sure that everyone else knew what a good job he was doing, and another 20% of his time looking for something better to do. You may quarrel with the percentages, but the reality is that you are going to have to make contacts and market yourself as well as do substantive work in order to get the recognition and opportunities you want and deserve.

The beauty of networking is that, if done right, almost everything you do will rebound to your benefit if you end up choosing an academic career. In the open environment of academia, all your contacts in industry, government, and at other colleges and universities will be important to you. In one way or another, all of these contacts will be your future colleagues. Remember, every visitor to your department or laboratory, those you meet at conferences, at other schools, and during visits to industrial and government sites are both current and future resources for you.

We will return to this important subject in Chapter 7, "Identifying the Possibilities." Now is the time, however, to start developing a personal database of people you want to network with now and after you become a professor. As you find yourself in various situations, remember to introduce yourself to interesting individuals, obtain their business cards, follow up with a note, and enter the information into your database for future reference.

Vignette #5	The Research Continuum

Applying the three-pronged preparation strategy throughout your undergraduate, graduate, postdoc and job search periods is the best possible approach. In the following vignette, we show how this strategy took hold for Professor Shon Pulley of the University of Missouri—Columbia.

Shon Pulley
University of Missouri—Columbia

Shon Pulley began thinking seriously about research while an honors undergraduate chemistry major at Utah State University. Since then, he has moved through graduate school and a postdoctoral appointment, to an assistant professorship at the University of Missouri—Columbia, all the while expanding on his initial research interests. Although perhaps not fully aware of it at the time, Pulley applied much of the three-pronged preparation strategy outlined in this book in his path toward an academic career.

From the start, Pulley was interested in the organic synthesis of natural products. As an undergraduate, he worked on a number of projects including the synthesis of polymer supported reagents. During this period, he was able to coauthor four publications with his undergraduate advisor. At the same time, he was also paying a lot of attention to the way research was being conducted. According to Pulley:

> The experience with my undergraduate advisor, while he was starting as an assistant professor, provided insights into getting a group started and developing undergraduate and new graduate students into productive independent researchers. I was able to build on these insights later on as a graduate student and postdoc.

At Colorado State University, Pulley expanded his interests in organic synthesis to include the synthesis of natural and unnatural peptide fragments using optically active chromium carbene complexes. Here, he published two papers, one coauthored with his advisor, and the other with his advisor and two other researchers. And, again, he was involved in more than just his own research. As he notes:

> During my doctoral studies, I trained undergraduate researchers and helped new graduate students start on their respective projects. These experiences were very helpful as I began my own academic career.

In addition, Pulley served as a laboratory instructor for general chemistry classes, which gave him further insights into the interests of undergraduates, particularly those who did not want to become scientists.

As an American Cancer Society Postdoctoral Fellow at Stanford University, Pulley directed his research toward the asymmetric total synthesis of natural products using enantioselective palladium catalysis. During this time, he continued his supervision of undergraduate and graduate students, and in his words, "These experiences demonstrated the commitment required to maintain a productive leading-edge research group, which I fully intend to draw on to develop a vigorous research and teaching program as a professor."

One of the most important things Pulley did as a postdoc was to develop a series of research proposals reflecting possible areas of interest as a future professor or research scientist in industry. He began with a one-page statement, reproduced below, that places his interests in a broad context, making a compelling case for further study. Note how the first paragraph establishes the applicability of his work, the second paragraph his approach and reasons for carrying it out, and the final paragraph his suggestions for future research directions. The statement is written for an organic chemistry audience, but is general enough to be comprehensible to all chemists, thereby placing his work in a broader context while effectively introducing his plans for further study.

The Development of Synthetic Organic/Organometallic Methods and Future Interests

Recently, a synthetic organic renaissance has changed the way we plan synthetic strategy. Governmental regulations demand cost minimization and reduction of hazardous waste streams. The use of enantiomerically pure drugs in chemotherapy is necessary not only to realize enhanced specificity, but also to avoid possible side-effects caused by the other enantiomer. Furthermore, the elucidation of biological processes through structure activity relationship (SAR) studies depends heavily on organic synthesis to identify clinical compounds and improve pharmacological profiles. The development of synthetic methods that meet the regulatory and commercial needs of the chemical industry, especially pharmaceutical interests, requires the training of students in organic synthesis.

In light of these requirements, my research program concentrates on transition metals as a means of achieving efficient and cost-effective organic synthesis. The use of transition metals to effect a desired transformation has several advantages over classical organic methods. First of all, metals can effect reactions catalytically ultimately leading to reduced waste and more cost-effective syntheses. Second, enantioselective processes occurring on a metal center containing chiral ligands will afford enantiopure compounds. Finally, the mild and chemoselective reactivity of transition metals allows a more convergent approach to complex organic molecules without the need for cumbersome protection/deprotection strategies. The following projects develop novel synthetic methodologies using transition metals and examine their scope and limitations, the ultimate goal being the efficient and economical asymmetric synthesis of clinically interesting compounds.

Using the methodological studies described below as a foundation, I envision my program expanding into bioorganometallic chemistry as a method of achieving selective chemical transformations. For example, transition metal-catalyzed processes using ligands capable of molecular recognition should be useful as models for naturally occurring metalloenzymes. The design of peptide and carbohydrate-based ligands that will impart selectivity as a result of distinctive molecular associations is an area with enormous potential and I present some of my initial interests toward this end in the last proposal of this section. This represents long-term research interests that will allow my group to use its knowledge of organic and organometallic synthesis to make valuable contributions to the field of bioorganometallic chemistry.

This statement was followed by five proposals, four–five pages each, outlining possible areas for further investigation. Each proposal contained the following categories: (1) Specific Aims, (2) Background, (3) Significance, (4) Experimental Design and Methods, and (5) References. In Pulley's mind, there is no question that it was the careful thought put into these proposals and a well-prepared interview presentation that separated him from the pack when it came to the three or four candidates who were called to the University of Missouri for interviews.

There was one more thing that Pulley did as a postdoc. He kept his options open by interacting with industry while doing his research. As he notes:

> I was prepared to go either way, industry or academia after my postdoc, but the idea that I could also play a role in developing a teaching, as well as research, program had a lot of appeal to me.

SUMMARY

There are a number of research-related activities you need to complete prior to becoming a professor. Some, such as choosing a research topic and identifying an advisor, are required of all Ph.D. students and postdocs, while others, such as writing proposals and supervising other researchers, although not specifically required, will nevertheless put you ahead of most of your competition looking for academic, government, or industry positions.

By applying professional preparation strategies to these activities, you enhance your attractiveness still further. For example, in seeking Breadth-on-Top-of-Depth, you demonstrate a capability and flexibility for future tasks as interests and needs change. You are also better able to make a compelling case for the work you are currently doing.

However, it is no longer enough to be very good at tasks expected of someone at your present stage of development. You must also look ahead to tasks expected of those who are in positions you wish to occupy in the future, and then perform some of those tasks now so you can demonstrate your readiness for such responsibilities. Such demonstrations are the essence of the Next-Stage approach.

If you then take steps that expose you to people and activities in industry and government as well as academia (Multiple-Option approach), you expand your options while at the same time enhancing your attractiveness as a college or university professor.

In all of the above, you need to identify and maintain contact with individuals through a personal database that can be called upon as you set about applying for academic positions.

There is one additional thing that you need to do in your preparation for an academic career. You need to acquire teaching experiences beyond those of an ordinary teaching assistant. It is to this subject that we turn in Chapter 6.

REFERENCES

[1] M. L. Goldberger, B.A. Maher, and P.E. Flattau, Eds., *Research–Doctorate Programs in the United States—Continuity and Change.* Washington, DC: National Research Council, 1995, p. 472.

[2] S. Benowitz, "Wave of the future: Interdisciplinary collaborations," *The Scientist*, vol. 9, no. 13, p.4, June 26, 1995.

[3] C. I. Davidson and S. A. Ambrose, *The New Professor's Handbook: A Guide to Teaching and Research in Engineering and Science.* Bolton, MA: Anker Publishing, 1994.

[4] R. V. Smith, *Graduate Research: A Guide for Students in the Sciences*, 2nd ed. New York: Plenum, 1990. Copyright © 1990 by Robert V. Smith. Reprinted with permission.

[5] R. F. Adler and T. J. Baerwald, "How to plunge into the research funding pool," *The Professional Geographer*, vol. 41, no. 1, p. 2, Feb. 1989.

[6] P. J. Feibelman, *A Ph.D. is Not Enough!: A Guide to Survival in Science.* Reading, MA: Addison-Wesley, 1993.

[7] S. E. Widnall, AAAS presidential lecture, "Voices from the pipeline," *Science*, vol. 241, pp. 1740–1745, 1988.

[8] To appear in P. Frost and S. Taylor, Eds., *Rhythms of Academic Life.* Newbury Park, CA: Sage, in press.

[9] J. M. Darley and M. P. Zanna, "The hiring process in academia," in *The Compleat Academic: A Practical Guide for the Beginning Social Scientist.* New York: Random House, 1987, ch. 1, p. 139.

[10] A. S. Moffat, "Grantsmanship: What makes proposals work?", *Science*, vol. 265, p. 1922, Sept. 23, 1994.

[11] R. Burroughs, "Getting your research funded," *ASSE Prism*, vol. 2, no. 1, p. 31, Jan. 1995.

[12] Editorial staff, "How not to write a scholarly paper," *ASEE Prism*, vol. 3, no. 6, pp. 22–23, February 1994.

[13] B. B. Reitt, "An academic author's checklist," *Scholarly Publishing*, vol. 16, no. 1, pp. 66–72, Oct. 1984. Copyright © 1984 by Scholarly Publishing. Reprinted with permission.

[14] T. Ferris, "Minds and matter," *The New Yorker*, vol. LXXI, no. 12, pp. 46–47, May 15, 1995.

CHAPTER 6

Teaching Experiences Prior to Becoming a Professor

> Almost every body does TAships. The important question is: Who taught in summer school, at a local community college, or as a sabbatical replacement? Research and teaching experiences are what will get you the job.
>
> Martin Ramirez, Assistant Professor of Biology
> Bucknell University

In the previous chapter, I discussed how the three-pronged preparation strategy (Breadth-on-Top-of-Depth, Next-Stage, and Multiple-Option) could be applied to your research experiences while a student and postdoc. Teaching in some form prior to becoming a professor is also a part of this strategy. Consider the Breadth-on-Top-of-Depth component. It is often said that if you really want to learn something, take the time to try and teach it to someone else. This statement is true whether you are talking about your current research to a group of Ph.D. students or about basic physics concepts to beginning undergraduates. Your depth of understanding only increases as you seek to explain what you know to students who are able to ask you questions in the process. At the same time, your breadth of understanding also increases. As you seek to place your topics in a meaningful context, look for examples illustrating the concepts you wish to teach, and seek connections between what you know and the backgrounds and subject matter interests of the students, you invariably broaden your own understanding of the material.

Clearly, teaching prior to becoming a professor is an example of a Next-Stage activity. However, it is also illustrative of the Multiple-Option approach. We see this in the comments of chemistry Ph.D. student Kriste Boering [1]:

> One of the things I had thought would be important would be a separate category on my resume with something I called "Teaching and Leadership Experiences,"

because I consider those to be related to one another. I find invariably I am asked about my teaching skills by industrial interviewers such as large chemical or oil companies [as well as by potential academic employers]... I think my teaching experience shows that I am able to communicate. This is very important in any field you'll be in, anywhere, in any kind of job setting, academic or industrial. If you're a very technical person, you still need to communicate to the marketing people in your company about what you're doing, and what it's good for. Well that's exactly what I tried to do in my general chemistry classes and it's exactly what companies are looking for beyond your professional training: organizational and communication skills. You can develop both of these qualities in teaching.

WHY TEACH AS A GRADUATE STUDENT OR POSTDOC?

In addition to enhancing your communications skills, the benefits of acquiring teaching experiences prior to becoming a professor include:

- Confirming in your own mind that teaching is an important part of what you really want to do

- Helping you prepare for your first teaching assignment as a professor

- Giving you a significant leg up on your competition in your search for an academic position.

On occasion, such experiences can also provide needed supplemental income.

Chapter 6 begins by discussing the above benefits in greater detail. It then looks at a variety of teaching possibilities open to graduate students and postdocs, some of which may not have occurred to you. Information will then be provided on how to locate the appropriate opportunities and how to prepare for them. Tips on how to prepare a teaching portfolio to capture and present your successful teaching experiences is discussed next. The chapter concludes with a vignette describing the teaching experiences of a chemistry postdoc at a large Canadian university.

Confirming That Teaching Is What You Want To Do

You have observed teaching for as long as you can remember, and now it is something you think you would like to do. Yet, as you may already suspect, it is one thing to observe teaching and another thing to actually do it, and do it well. Teaching assistantships (TAships) expose you to certain aspects of teaching, and they are a good place to start. However, nothing compares with giving lectures that are engaging and informative, dealing with student questions on the spot, and preparing examinations and homework assignments that connect with your lectures. As a geology graduate student who had just taught an introductory earth sciences course remarked, "The idea that I was the only one responsible for what my students were getting, that my course might determine who went on to major in the earth sciences, that I was the one who determined their final grade, and that the University counted what I did just like they did a 'real' professor, was quite sobering." For all of you, including those heading for Research I and II institutions, teaching will be an important, if not central, activity during your years as a professor. Now is the time to find out if this is what you really want to do.

Helping to Prepare You for Your First Teaching Assignment as a Professor

Wanting to teach is important, but, of course, wanting alone is not enough. As someone once said, "Those who think all you have to do to be a good teacher is to love to teach, have to believe that all you have to do to be a good surgeon is to love to cut." Indeed, the medical model is quite instructive.

> Your health maintenance organization has assigned you to a new physician who is just starting his practice. You make an appointment to see him, complaining of pain in your lower right side. He confirms it is appendicitis, and says he wants to operate right away. In answer to your anxious questions, he informs you this will be his first operation, and that he will probably be a bit nervous. But he also tells you not to worry. He has studied the appendix extensively, and he even did some research on it while a medical student. Furthermore, he has dreamed of being a surgeon for as long as he can remember. He has observed hundreds of operations, although it has been some time since he witnessed an appendectomy. However, before completing his degree, he assisted a well-known surgeon by handing him the scalpels during a heart bypass operation. But no, he has not yet performed any surgery on his own.

I suspect that under such circumstances, you would say thank you very much, head for the door (as best you could), and start shopping around for another physician.

Of course, nothing like the above is actually going to happen because it is not the way surgeons are trained. All future surgeons go through an extensive period in which they work side by side with experienced surgeons. At first, they just observe. Then they assist the experienced surgeon, and only then, with an experienced surgeon at their side, and usually dozens of other physicians looking on from above, they perform their first operations. It is quite some time before they are actually performing operations in which they are the only physician on the scene.

I recognize that teaching does not involve life or death situations as is often the case with surgery. Yet, we insist that all future professors obtain training as researchers prior to entering the profession, even though half will do little or no subsequent research, but we do not do the same with respect to teaching. Is it because we think that, compared to research, teaching is "easy" and does not require much preparation? Or is it because we think that the cost of poor teaching, at least in the beginning, is not that great? Or is it both?

Good teaching is not easy, and the cost of doing it poorly is high. By actually teaching part, or in selected cases all, of a course as a graduate student or postdoc you will be better prepared and have greater confidence when you start teaching as a professor.

But does not such an approach create an obvious problem? If one of the arguments for obtaining teaching experience while a graduate student or postdoc is that the students you first teach as a professor will be better off because of it, then what about the students you teach before you become a professor? Are not they being shortchanged by being your first students? *Not necessarily.* We just touched on the medical training model in which the resident surgeon does his or her work under the direct supervision of an experienced physician with many other physicians observing and with plenty of feedback from everyone after the operation. A similar process takes place in the training of high school and elementary school teachers. Here, "student teachers" teach one or two classes under the direct supervision of a master teacher. The student teacher often goes over the lesson in advance with the master teacher, and there is considerable review of the teaching, homework assignments, and examinations by the master teacher.

A similar system could be used in the preparation of professors. It might involve a multistep approach beginning with teaching assistantships, followed by guest lectures, and/or the teaching of class modules in courses in which the regular professor is present. These experiences might then be followed by team teaching with an experienced professor, and/or the teaching of a special course such as one taught through a university extension, in a field in which the graduate student or postdoc has clear subject-matter expertise. Finally, under certain circumstances, and still *under the mentorship of an experienced professor*, an advanced graduate student or postdoc could take full responsibility for teaching an entire course.

Some schools have begun to take steps in this direction. One example is a pilot program at North Carolina State University, called "Preparing the Professoriate." In focus-group discussions, it was found that doctoral students wanted "opportunities to prepare more fully for the academic life of a professor... to learn to teach in the same way that they learn to do research in a significant and extensive advising atmosphere" [2]. According to North Carolina State [2]:

> The program uses "mentoring pairs," each of which teams a doctoral candidate with a current or emeritus professor. Throughout an academic year, the mentors work with their graduate students ("teaching associates") to develop individualized plans

for substantive teaching experiences; these range from course preparation and planning to final course evaluation.

Another example is a project sponsored jointly by the Association of American Colleges and Universities and the Council of Graduate Schools. Known as Preparing Future Faculty, it has awarded 17 doctoral institutions a total of $1.8 million from the Pew Charitable Trusts to create special teaching programs for future academics. Says Bianca L. Bernstein, dean of the graduate college of Arizona State University, one of the participating schools, "For too long, the only skills that doctoral programs have sought to impart involve research. Too many Ph.D.'s wind up learning teaching skills on their first faculty job." The Preparing Future Faculty program seeks to change this situation through mentoring programs, and seminars, and by providing opportunities for students to explore all aspects of teaching, including developing a course from scratch [3].

Giving You a Significant Leg Up On Your Competition in Your Search For An Academic Position

You can bet your Ph.D. that evidence of teaching experience will be looked for in your application for an academic position. You can also be sure that you will be asked about such experiences, or lack thereof, during your job talk and campus interview. In fact, having such experiences can go a long way toward determining if you are one of the three or four, out of perhaps hundreds of applicants, who even gets a job interview. Remember, also, that getting a job offer from a school you are interested in is only the start. You want to be able to negotiate the best possible set of initial conditions in terms of teaching, research, type of position, and available resources. To do so, you need to be the most competitive, and teaching experience is one thing that can help make you this way.

We will discuss this matter in greater detail in Part II, "Finding and Getting the Best Possible Academic Position." The point to note here is that, in spite of statements about research expectations you see in academic advertisements, the primary reason for hiring most new faculty is to have them teach classes. Positions usually become available because a professor has left the department and a replacement is needed to teach his or her classes. On more rare occasions, a position opens up because the department is expanding into new areas. In both cases, the primary need is to fill teaching slots, and search committees will want to know how you can help to do so.

Over the last few years, there has been a shift in the statements about teaching experience in academic job announcements for assistant professors in science and engineering. A few years ago, most announcements sought "outstanding *potential* for excellence in teaching and research." Today, many more announcements seek "*demonstrated skills* in teaching and research," and a number clearly want *"evidence* of teaching excellence," accompanied by a statement of teaching philosophy and the presentation of a teaching portfolio. Here is an example of an advertisement for a beginning assistant professor position that appeared recently in the *Chronicle of Higher Education* [4].

> **Industrial/Systems Engineering**: The Industrial and Systems Engineering Department (ISE) at Virginia Polytechnic Institute and State University (Virginia Tech) seeks an entry-level tenure-track person to continue advancing the program in Systems Engineering with an emphasis on Systems Life-Cycle Engineering, the Systems Engineering Process, Economic Analysis, Design, Evaluation, and Logistics Engineering. This position will be available in the Fall, and requires teaching in ISE at both the graduate and undergraduate levels. Candidates must demonstrate a potential for creative research and the capacity for significant contributions with the Systems Engineering Design Laboratory. *Demonstrated teaching excellence is expected.* Preference will be given to applicants with the ability to lecture via distance learning media, to collaborate across disciplines, to interface with systems oriented agencies and firms, and to publish. At least one degree must be in Industrial and Systems Engineering (Ph.D. preferred), with a preference for all degrees in engineering. Salary is commensurate with qualifications and experience. Applicants should send a complete resume including personal data, education, publications, research, and professional experience, together with names of at least three references to....(italics added).

TYPES OF TEACHING EXPERIENCES

A variety of teaching opportunities are available to interested and qualified graduate students and postdocs. (See Figure 6-1.) Some have existed for years and are well established, while others are not so common and may require special permission from a professor or the department. Possibilities in addition to teaching assistantships include guest lecturing, teaching class modules, team teaching with a professor or another student, teaching a regular course at your own or another institution, teaching extension courses, and teaching via distance learning. The key is to customize your experiences to fit your particular interests and capabilities, and to do so in stages that are accompanied by the oversight of an experienced professor.

Teaching Assistantships—a Good Place to Start

In some schools and departments, serving as a teaching assistant in return for partial tuition credit and/or stipend may be required whether or not you are considering an academic career. One advantage of a teaching assistantship is that the infrastructure for doing it is already in place. Mechanisms for obtaining these positions are well known, although this does not mean that getting a specific

Figure 6-1 Types of teaching experiences.

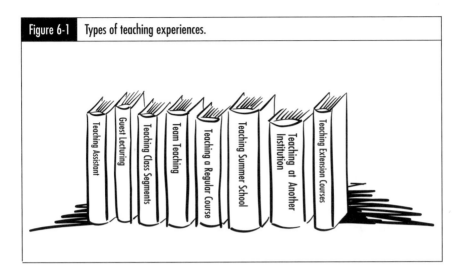

position is automatic. In most departments, there is at least some written material on teaching assistantships, and in many cases, there is at least some training provided. You are surrounded by "colleagues" who have done or are doing teaching assistantships, and they are available to give you advice and support.

Teaching assistants handle many routine, but essential, teaching responsibilities such as preparing problem sessions, writing problem sets and examinations, holding office hours, preparing laboratory experiments, writing up handouts, giving late/early examinations, grading, and answering electronic mail. As a teaching assistant, you will begin to experience some of the essential nonlecturing responsibilities that are part of being a professor.

The key to successful teaching assistantships is leveraging. You need to see this activity as preparation for the teaching experiences that follow. Imagine that the course is one you will later teach as a graduate student, postdoc, or beginning professor. You will want to keep copies of the handouts, homework assignments, laboratory exercises, and examinations. You may even want to ask the professor for a copy of his or her lecture notes. Most teaching assistantships require you to attend the class lectures given by the professor. Even if they do not, be sure to do so. In class, imagine that you are the professor giving the lectures. What would you seek to emulate? What would you do differently? How would you answer questions, change the handouts, overheads, and examinations? Finally, try to identify a few lectures that you would particularly like to give, and make a note to talk to the professor about doing so in a later semester.

Also, take the initiative to discuss the design of the course with the professor. As a teaching assistant, you may have wider latitude than you think about what you can contribute to the course, and you might want to talk to the professor about your role, especially if you want to have independent responsibility for reviews or developing handouts or visual aids. He or she may welcome more assistance/contributions from you.

Guest Lecturing

The next logical step after serving as a teaching assistant is to start giving guest lectures. To do so, you need only the permission of the professor in charge of the course. One type of guest lecture takes place when you, in effect, substitute for the professor and seek to give, with perhaps some modifications, the lecture the professor would have given. This approach may not seem all that interesting or innovative. However, for your first time out, it will do just fine. The novelty of the experience will be exciting enough.

Another type of guest lecture occurs when you give a presentation on your particular interest or specialty. In this case, your talk complements the regular course material. You have greater flexibility in what, and how, you present your ideas since you are the expert on the subject matter. On the other hand, you will probably have to develop your presentation from scratch with little or no help from the instructor.

Keep in mind that just because you are talking about your research does not mean your lecture has to be given only to an advanced class. Your talk can go over very well in an introductory course, provided it is presented at the right level and in the appropriate context. Here are just a few examples of guest lectures given recently by graduate students or postdocs to introductory classes:

- An astronomy graduate student at the University of Arizona, working with charged coupled devices (CCDs) to enhance images from telescopes at the Kitt Peak National Observatory, gave a lecture on this subject, accompanied by photographs, to an introductory astronomy class.

- A computer science graduate student at the University of California, Los Angeles, studying parallel computing architectures, gave a talk to an introductory meteorology class on the use of such computers in weather forecasting.

- A materials engineering postdoc at Virginia Polytechnic Institute and State University, studying how different surface materials behave when in contact with each other, gave a lecture to the school's first-year dentistry students on how his work applies to problems with dental crown adhesion.

- A biology graduate student at the University of California, Berkeley, studying implantable telemetry systems in deer, gave a talk on his subject to an introductory ecology course at San Jose State University.

Teaching Class Segments or Modules

After you have given a few guest lectures, you might want to step up to teaching longer segments of a few weeks or more. One reason for doing so would be the scheduled or unscheduled absence of the regular professor. In this case, you will probably try to replace the professor in much the same way as you did in the first type of guest lecture.

Another reason for teaching a segment or module would be your expertise or specialty in an area to be covered in the course.

> The first time a graduate manufacturing course was taught at Stanford University, it focused on two manufacturing processes: metal forming and semiconductor fabrication. Feedback from the students indicated that exposure to additional processes would be desirable, so the next time the course was taught, the professor invited a postdoc to teach a three-week segment on composite materials manufacturing.

Obviously, teaching a segment or module of a course takes more time than one or two guest lectures. In addition to planning a sequence of presentations, you will probably need to develop related examination questions and homework assignments. However, doing so can be a nice extension of your earlier lecture and teaching assistantship work, while still having you operate under the guidance of a regular professor and within an existing class structure.

Team Teaching

Team teaching, in which you and someone else such as a professor or perhaps another graduate student or postdoc, share responsibilities, is yet another possibility. Under these circumstances, you would probably become a paid university employee. This situation is more likely to occur when a new course is being proposed, and you and your partner are working together to both develop and teach the course. Both of you are likely to be present at all the lectures, although specific duties may be divided along the lines of individual interests and expertise. This approach is a great way to obtain both teaching experience and experience in developing a new course, something you will most likely do in your first year or two as a professor.

Teaching a Regular Course

Teaching a full course entirely on your own is the ultimate preparatory experience. Here, you are the "professor," and will be legally responsible for all aspects of the course. In most cases, you will probably be replacing someone who is on leave or otherwise not available. This way, the university will not have to create a new slot for you.

Teaching a full course is a very time-consuming activity, so you want to think carefully before making such a commitment. You may want to do it near the end of your graduate student or postdoc experience. True, this period is also the time when you are most engaged in your research. However, teaching a course while doing research can provide a welcome balance in your life. It is also something you will almost certainly have to do as a beginning professor; therefore, showing this Next-Stage capability in advance could be a smart move.

Many introductory science and engineering courses are offered in multiple sections at various times in a given semester. You might want to teach one of these sections. I am referring here to full teaching responsibility of a section that has about 30 or so students, and not a teaching assistantship in which you work with a study section of a larger lecture course. Under these circumstances, you will likely stay in step with the professors in the other sections, use the same textbook, and give the same homework assignments and examinations. Again, for your first time out, such an assignment will be just fine.

> Tava Lennon, a professor of industrial operations and engineering at the University of Michigan, attended the University of Auckland in New Zealand. After doing her honors work in mathematics, and before coming to the United States, she taught a section of a statistics course for nonscience majors. Other sections were taught by other professors. They shared their examinations and homework assignments. "They were my mentors and guides," says Lennon, who credits this experience with setting her on the road to becoming a professor.

Developing a new course from scratch that you teach yourself will certainly take the greatest amount of time, but it can also bring the greatest rewards.

> Martin Ramirez worked for the Academic Excellence (ACE) program while a graduate student at the University of California, Santa Cruz. ACE is an honors program for minority students in the natural sciences that seeks to challenge them academically and motivate them to take an active, cooperative role in learning through group projects and group discussions. As an ACE staff member, Ramirez developed and taught an introductory biology course especially for these students that met four hours per week. "It was hard work, but an invaluable experience," says Ramirez. "I used this approach later at Pomona College in Southern California, and now at Bucknell University in Pennsylvania."

Teaching Summer School

Finding courses to teach during the summer may be easier than during the academic year because fewer regular faculty are available at that time. This way you do not have to wait until a slot opens up from a professor who might be going on leave. However, teaching summer school presents its own challenges. The much faster pace, the possibility of fewer resources being available, of not having as many regular professors around to interact with, and in some cases having students who are not part of the regular student body or who are not expecting to work as hard are all factors to consider.

Yet, summer school is often a time when you can experiment more freely with different courses of nonstandard length and format.

> Nancy Kelly, of the Massachusetts Institute of Technology, taught a one-week summer course she developed in "Speech Spectrogram Reading," which began with a half dozen graduate students from electrical engineering, but quickly expanded to dozens more from other departments and from industry. "I eventually packaged it as a one hour per week, ten week course for Voice Processing Corporation," remarked Nancy.

Teaching at Another Institution

Your teaching possibilities need not be confined to your graduate institution. If you have a specialty that is in demand and are willing to teach at less popular hours, such as early morning or in the evening, then a nearby institution may present an ideal opportunity.

> Princeton University has a program that allows interested science and engineering graduate students to teach basic math and science classes in the evening at local community colleges. This arrangement works nicely for the community colleges, which need part-time, temporary teachers in the evening, and it works well for the Princeton graduate students who are looking for teaching experience and extra income.

By looking outside your graduate institution, you increase your teaching possibilities, not only because you are looking at more schools, but also because the type and number of courses available will be much greater. Teaching at another institution, particularly if it is different from your own, i.e., Master's or Baccalaureate, can help you decide if this is a type of school you would be interested in as a professor. This Multiple-Option approach broadens your portfolio of experiences and references, and increases your overall competitiveness.

Teaching Extension Courses

Many colleges and universities offer extension or continuing education programs. Some of the courses offered in these programs are regular university classes taught at a different time and location, but many others are specially designed to meet the needs and interests of the local community. In the San Francisco area alone, over 3000 such courses are taught each semester. Some extension courses are taught by regular university professors. Most are taught by "specialists in the field" who are usually working professionals or graduate students and postdocs who have a particular area of expertise. A recent notice in the University of California, Santa Cruz extension catalog put it this way [5]:

Teaching a UCSC Extension course not only represents an important service to one's community, it can also be a personally challenging and rewarding experience. If you have a related bachelor's degree or higher, five or more years industrial experience and would like to be a UCSC Extension instructor, please send your resume and names of courses you would like to teach to...

Some courses that are part of certificate programs attest to the students, knowledge and success in a specialized area of professional education. Examples include:

- Advanced environmental management
- Computer engineering
- Managing the development of technical information
- VLSI design engineering.

Mark Gittler, manager of financial planning at Syntex Corporation, teaches an intermediate accounting class through the UCSC Extension. Tom Adams, who works for Apple Computer, teaches a Supply Chain Integration course through the same system.

Other courses, many of which are in the sciences, are taught to those seeking personal growth and enrichment.

Andrew Fraknoi, former executive officer of the Astronomical Society of the Pacific, teaches a Saturday course on "Black Holes and Other Deep Space Mysteries" to over 100 students at San Francisco State University.

Bruce Elliot, a director at the California Department of Fish and Game, teaches a short course on Wildlife of Northwestern California through the University of California extension.

In certain situations, particularly with courses that are part of a certificate program, instructors are provided with syllabi, homework assignments, and examinations. In other cases, the courses are developed by the instructors from scratch. In almost all cases, whether or not the course is actually taught, and you actually get paid, depends on a minimum student enrollment. On the positive side, your course is likely to be filled with adults who want to be in your class.

A Note of Caution:

Teaching, with all its preparation, takes a great deal of time. And the more complex the tasks, the more time they take. More time is required to prepare and teach a three-week segment than to teach one or two guest lectures. Much more time is required to teach a full course than to teach a segment. In all cases, it will take much more effort than you first imagine. A good rule of thumb is to estimate as best you can how long it will take, and then multiply by three.

Also, unlike much of your research, teaching is something you have to do on schedule. You cannot postpone it to a later time if something more pressing comes up. In some cases, you will have to sign a contract that will be even more specific about expectations and responsibilities.

It is very important not to overcommit yourself. If time concerns mean stopping at a teaching assistantship, then do so. If it means giving a few guest lectures rather than a sequence of lectures, or a sequence of lectures rather than a full course, then do so no matter how enticing the more time-consuming opportunity may appear. You want to do something that you will be proud of, and that you can include in your teaching portfolio.

HOW TO FIND THE RIGHT TEACHING OPPORTUNITIES

There are a variety of ways to identify or create teaching opportunities. The most obvious is to simply announce that you want such opportunities. Tell as many professors, students, and postdocs as possible. Often, faculty declare that, had they known students wanted to give guest lectures, teach class segments, or even teach a full course, they would have encouraged and supported this desire.

Next, find out what is currently going on. Who is doing what you might want to do? By definition, the possibilities discussed in this chapter are temporary, which means the people now doing them will not be doing them one, two, or at most three years from now. Now is the time to get in line for some of these opportunities. Start with the area closest to home: your advisor, your research group, and then move on to your department. Do not rule out other departments, particularly when it comes to guest lecturing. Remember, some courses are cross-listed in more than one department.

Arrangements at other institutions will probably require personal referrals. Many faculty at your institution know colleagues at nearby institutions, so start with them. Get to know faculty from other institutions who are visiting your campus to give seminars. Here, again, the main thing is to put out the word that you are interested in such possibilities. If you do, and if you remain flexible, then you are likely to end up with many more good opportunities than you can handle.

PREPARING FOR A SUCCESSFUL EXPERIENCE

Obviously, you want to do all you can to make your teaching experiences a success for both you and your students. It is important to remember, however, that nobody gets it completely right the first time, and even the best teachers are always improving. It is very likely that your first lecture will be in a course you have already taken, or have observed as a teaching assistant. In any case, since you are almost certain to have some advance notice before giving your lectures, be sure to sit in on the lectures leading up to the one you will give. If you cannot observe such lectures live, see if you can watch a videotape of the professor teaching the course.

Most departments have orientations for teaching assistants. If not, ask to have set one up. At the very least, talk to other teaching assistants, in particular the person who was doing the same class last year.

In preparing your lectures, you need to consider the objectives of your presentation, how they relate to the objectives of the course, and the backgrounds and interests of the students. Your best source of information in this regard is the professor who has taught the course before. Do not be afraid to ask for advice and suggestions from those who have already done what you want to do. It does not reflect poorly on you; just the opposite. It will help you avoid costly mistakes, while saving you significant amounts of time.

In addition, there are numerous written materials that can assist you in preparing lectures, examinations, laboratory exercises, and homework assignments. Three sources I have found particularly useful are *Teaching Engineering* by Philip Wankat and Frank Oreovicz [6], *The New Professor's Handbook: A Guide to Teaching and Research in Engineering and Science*, by Cliff I. Davidson and Susan A. Ambrose [7], and *The New Faculty Member*, by Robert Boice [8].

Obtaining feedback on your teaching is essential. At most universities, a course evaluation is passed out at the end of the semester just before the final examination. However, feedback need not take place just at the end of the course. You can hand out a midcourse evaluation form to your class anytime you wish, and thus make changes before the course is over. Another form of feedback is the students' reactions to what you do. Look at their faces when you speak. Are they bored? Are they keeping up? Often, a well-posed question with a healthy pause afterwards can give students a chance to speak up, and give you an opportunity to gauge how much they are learning.

A word of caution is in order here. Missing in any on-the-spot evaluation is a measure of the lasting value of what is taught, and how it is taught. Some courses that were highly rated by students at the time they were taken were thought to have been of much lesser value by those same students three or four years later. The reverse is also true. I know of a chemistry teacher whose course was much more highly regarded by his former students after they moved on to medical school than it was by those same students at the time they actually took the course. Only longitudinal evaluations can provide such an assessment. Unfortunately, only a few colleges ask for such evaluations even at tenure time, much less for former student teachers.

If your school has distance learners, you may want to meet with the instructional television network set up on your campus to support them. Typically, students watching the class remotely have special needs. For example, you may need to give problem sets and handouts to be distributed to remote students several days before the actual day of the class. Alternatively, you may need to schedule a special room with camera equipment for the final examination review session.

Finally, be aware of other resources at your disposal. Many colleges and universities have centers for instructional development that assist faculty and graduate students in improving their teaching. There are now several hundred such centers, with the number growing each year. Examples include: the Program for Instructional Excellence at Florida State University, the Faculty and T.A. Development Center for Teaching Excellence at Ohio State University, the Teaching Excellence Center at Rutgers University, and the Center for Teaching and Learning at Stanford University. Most of these centers have workshops on the keys to successful lecturing, and many have videotape libraries containing presentations by award-winning professors. These centers can also help you obtain feedback on your teaching. They have everything from lecture and course evaluations that can be given to students, to observers who can sit in on your classes, to ways of videotaping your presentations for playback and evaluation at a later time. Not only are such records helpful to you in improving your teaching, but they provide a way of documenting your successes in your teaching portfolio.

YOUR TEACHING PORTFOLIO

You have put a lot of time and effort into acquiring various teaching experiences. Now you need to capture your successes in a record for later use. Since your teaching ability is not easily shown on your curriculum vitae or during a job interview, a "teaching portfolio" illustrating such experiences can serve as a very useful addition to your application for an academic position. The intent of the teaching portfolio is to capture the intellectual substance and actual samples of teaching methods that an academic interview, vitae, or application letter cannot. Such capture is particularly important when applying for positions where teaching is stressed in the job description. In addition, teaching portfolios are being used with increasing frequency in faculty retention, tenure, and promotion (RTP) decisions. For this reason, the portfolio you start working on now can be the one you add to on a regular basis after becoming a professor. (See also Chapter 11, "Insights on Teaching and Learning".)

A useful guide to teaching portfolios is *The Teaching Portfolio—Capturing the Scholarship of Teaching* by Russell Edgerton, Patricia Hutchings, and Kathleen Quinlan. According to the authors, the notion of the teaching portfolio "lies in the analogy to portfolios kept by architects, designers, painters, and photographers to display their best work" [9, p.3].

Think of a teaching portfolio as a special insert in your curriculum vitae under the heading of "Teaching." As in the case for artists, it is your best work, not everything you have done. According to Shore *et al.*, it is "… a summary of a professor's major teaching accomplishments and strengths. It is to a professor's teaching what lists of publications, grants, and academic honors are to research" [10].

In fact, Edgerton *et al.* make an interesting observation about the different way we look at the evaluation of research and teaching, and how it is connected with the teaching portfolio [9, p.5]:

> When it comes to research, faculty take for granted that it is their responsibility to present evidence of accomplishment. In the case of teaching, however, evaluation often appears to be something that happens *to* faculty—be it through student course ratings or obligatory classroom visits by chairs or deans. Portfolios place the initiative for documenting and displaying teaching back in the hands of the person who is *doing* it; they put the teacher back in charge…selecting, assembling, and explaining portfolio entries that accurately represent actual performance.

Your portfolio should begin with background information, such as your curriculum vitae, a description of the school and setting in which you taught, and the nature and prerequisites of the course, followed by selected entries. A typical portfolio might include [9, p.9]:

Work samples from current or recent teaching responsibilities, such as

- Course materials prepared for students such as exams, handouts, discussion questions
- Essays, field or lab reports, and other student works with TA critiques and feedback
- An edited videotape or written case study of a classroom teaching experience
- A reflective memo on the course syllabus, if you developed the syllabus yourself or collaborated with others in developing it.

Documents of one's professional development as a teacher, such as

- Records of changes resulting from self-evaluation
- Evidence of participation in workshops, seminars, and professional meetings intended to improve teaching.

Information from others, such as

- Statements from colleagues who observed your teaching
- Invitations to teach from outside agencies.

These suggestions are taken from a more complete listing that appears in Appendix A, "Possible Items for Inclusion in a Teaching Portfolio."

Increasingly, colleges and universities are asking that your teaching portfolio contain a statement of your personal philosophy of teaching and learning. This statement is something you should think carefully about during the period leading up to the time you apply for your first professorship position. An example of such a statement appears in Appendix B, "Statement of Personal Philosophy Regarding Teaching and Learning."

Finally, here are some tips for developing your teaching portfolio [11]:

- Start compiling samples for your portfolio as soon as possible.
- Form the habit of filing away samples of work which demonstrates your teaching.
- Select those items which you deem to be the best examples of your work demonstrating teaching quality.
- The format of your teaching portfolio will vary, depending on intended use.
- Be sure the format is well organized and presents your work with care, neatness, and creativity.
- After you secure a job, plan to continue to retain copies of your work.

Vignette #6	Teaching as a Postdoc

Obtaining teaching experience while a postdoc can be critical to your future academic success. In the following vignette, we see how such experiences are helping a chemistry researcher at a large Canadian university.

Ulrike Salzner
Memorial University of Newfoundland

> I knew there were a number of things I needed to do to make myself more attractive to academia, and obtaining teaching experience was clearly one of them.

So says Ulrike (Uli) Salzner, a postdoc in the Chemistry and Physics Departments at Memorial University of Newfoundland (MUN), a Doctorate I university of about 18,000 students in St. John's, Newfoundland. The approach taken by Salzner is illustrative of both the Breadth-on-Top-of Depth and Next-Stage components of our professional preparation strategy.

Salzner's undergraduate and graduate studies took place in the Federal Republic of Germany, culminating in a Ph.D. in theoretical organic chemistry (1993) from the University of Erlangen–Nuremberg. After graduation, she accepted a two-year postdoc in the Chemistry Department at Northern Illinois University in DeKalb, IL. "This position gave me the opportunity to continue my work in theoretical chemistry, and to be in an environment where I could improve my English," comments Salzner. As her appointment came to a close, she applied for a number of faculty positions in the United States and Canada. She was on the short list at a few schools, but did not receive an offer. According to Salzner:

> My foreign status may have been part of it. The fact that I had concentrated in one area of chemistry also narrowed my chances. And then there was my lack of teaching experience. In German universities, Ph.D. students almost never have the opportunity to teach classes. I was able to conduct small seminars, and the experience showed me I could explain things well to others. I also came to feel that teaching and research were things I would like to do.

Since Salzner did not have a faculty position, she decided to apply to other universities for a second postdoc. Now, however, she had some new objectives in mind. "While I couldn't do anything about my visa status, I realized I could do something about my research and teaching. I wanted to broaden my research horizons, and at the same time obtain real classroom teaching experience," she explained.

The position at MUN offered her the opportunity to do research in an entirely new area. As Salzner puts it:

> When doing a postdoc, you have to be careful. It is easy enough to get a position advancing the work of your supervisor, but what does this do for your career? You want to be sure to do what is good for you as well. I needed to try another area, one that had a broader appeal within the field as a whole. The MUN offer would have me doing research in a more applied materials science area involving the development of intrinsically conducting organic polymers. I would be part of a team consisting of myself, an electrochemist, a theoretical chemist, and a physicist, and this appealed to me as well.

However, Salzner had another problem that had nothing to do with research and teaching. She needed more money. She was a single mother with a small son, and the standard postdoc salary simply was not enough. MUN really wanted her; yet, the only way they could provide her with a higher income was if she agreed to teach a chemistry class as part of her postdoc duties. This "requirement," of course, fit perfectly into her plans, and she eagerly accepted the position.

Her teaching assignment took place during the first semester she was in St. John's. Comments Salzner, "I had a freshman chemistry class with 84 students. It met four hours a week for lectures, and another three hours a week for laboratory. In addition, I supervised two other laboratory classes of another professor. This assignment was basically full-time, so I put my research somewhat on hold during the first four months."

Salzner's course was taken by students who had not had chemistry before or who had done poorly in such a course. They included nurses, some future science majors, and those who were taking it solely to fulfill a general education requirement. In Salzner's words:

> It certainly made for a challenging experience. For the most part, I had to assume my students didn't know anything, not even the basic structure of the atom. I had to figure out how to bring science to them in ways they could understand, while also making it interesting. I also found I had to read the textbook carefully myself in order to organize the material in a didactically useful way.

> For me, the most enjoyable part was when some of my students really caught on, when they started sitting in the front of the class, and when they came to see me in my office to discuss things further. I really feel I reached some of them in a significant way.

> At the same time, I know there were others I couldn't reach. Some stopped coming to class, and eventually dropped out. Around 20% of my students failed the class, which is about average for the course, yet I still felt badly about it.

Salzner had a lot of independence in terms of what, and how, she taught. As she puts it:

> To a certain extent, I was surprised at how much they trusted me. On the other hand, they were aware of my background, and knew I could give well-organized talks. Also, we went out for pizza every week, and I asked a lot of questions, so they knew I was putting effort into my teaching.

Salzner is now fully engaged in her research at MUN, but may teach again in the second, and final year of her postdoc. She is clearly delighted at having had the opportunity to do so in her first semester, summing up her feelings this way: "Not only has it closed an important gap in my background, but it has convinced me that more than ever I want to become a professor."

SUMMARY

We began this chapter with a discussion of the benefits of obtaining teaching experience prior to becoming a professor. These included: confirming in your own mind that teaching is what you want to do, better preparing yourself for your first full-time teaching assignment, and giving yourself a leg up on your competition for a faculty position. These experiences fit nicely into our three-pronged, Breadth-on-Top-of-Depth, Next-Stage, and Multiple-Option preparation strategy.

We then examined a number of teaching arrangements starting with teaching assistantships, followed by guest lectures, the teaching of class modules or segments, team teaching, teaching a full course during the academic year or during the summer, teaching extension or continuing education courses, and teaching at another institution. We discussed how to obtain one or more of the above, followed by a look at some of the ways to prepare for your teaching assignments and how to obtain useful feedback on your teaching. We then introduced the idea of the teaching portfolio, and how it can be used to capture the successes in your teaching experiences for presentation to potential employers. The chapter concluded with a vignette describing the teaching experiences of a chemistry researcher at a large Canadian university.

Having discussed the various research and teaching experiences that you need to acquire prior to becoming a professor, we are now ready to turn in Part III to the important task of finding and getting the best possible academic position.

REFERENCES

[1] "TA talk," *Stanford Teaching Assistant Newsletter*, vol. 2, no. 2, p. 4, Winter 1991.

[2] Committee on Science, Engineering, and Public Policy, *Reshaping the Graduate Education of Scientists and Engineers*. Washington, DC: National Academy Press, 1996, pp. 3–5.

[3] M.C. Cage, "Learning to teach," *The Chronicle of Higher Education*, vol. XLII, no. 22, p. A19, February 9, 1996.

[4] *The Chronicle of Higher Education*, vol. XLI, no. 28, p. B12, Feb. 10, 1995.

[5] *UCSC Extension Course Catalog, Spring 1995*. Santa Cruz, CA, vol. 28, no. 2, Feb. 1995, p. 110.

[6] P. Wankat and F. Oreovicz, *Teaching Engineering*. New York: McGraw-Hill, 1993.

[7] C. Davidson and S. Ambrose, *The New Professor's Handbook: A Guide to Teaching and Research in Engineering and Science*. Bolton, MA: Anker Publishing, 1994.

[8] R. Boice, *The New Faculty Member*. San Francisco, CA: Jossey-Bass, 1992.

[9] R. Edgerton, P. Hutchings, and K. Quinlan, *The Teaching Portfolio— Capturing the Scholarship of Teaching*. Washington, DC: American Association of Higher Education, 1991.

[10] B. M. Shore *et al.*, *The Teaching Dossier: A Guide to its Preparation and Use*, rev. ed. Montreal, Quebec: Canadian Association of University Teachers, 1986, p.1.

[11] *Teaching Portfolio Preparation.* Stanford, CA: Stanford University Career Planning and Placement Center, Oct. 1994, p. 1.

Finding and Getting the Best Possible Academic Position

CHAPTER 7

Identifying the Possibilities

Consider this:

You have been single for quite a while. You are secure with yourself and enjoy your independence, but you are starting to feel like you are ready for a relationship. You have done your share of dating and have had a few "trial runs," but now you want to really find "the one."

So where do you look for him or her? You could simply hope he/she finds *you*, or you could figure out some ways in which your chances of having an encounter with a compatible person are greatly enhanced. One way may be to determine, and then select, certain *places* you would be more likely to meet people with interests similar to your own.

Until now, you have met people mostly by chance or at social gatherings, in other words, in a very general setting. While you know you cannot "shop" for Mr. or Ms. Right, you do think, and your nonsingle friends have told you, that looking in the right places can make a big difference.

You are aware that first you must decide what type of person you are looking for, including what kind of values and goals you would want them to have. In order to determine this, however, you must first take a closer look at yourself and be secure with what you find. You need to draw on your earlier explorations and experiences, and connect them with your particular interests and desires.

Once you have gotten a mental picture of the person you want to find, then think of the places you would most likely meet him or her. Are they the same places you like to go? Is your image of this person very general or is there something unique that you are looking for? The image you have could make a big difference in where you look and how much "competition" you might encounter. If what you want is not all that special, it probably will not take long to meet someone, but chances are the person you do meet will only have some of what you are looking for. However, if you are seeking someone really special, it may be more of a challenge to find, and "get" him or her, but much more rewarding in the end.

Your chances of finding the right person at the right place improve tremendously if you take the time to decide what your best options are before you set out on the hunt. This way, when the time comes, you will know where *not* to look as well as where to look. You will know the kinds of people you were seeking, and perhaps could even modify your approach to increase your chances of finding the "one." You could also ask other nonsingle friends about their experiences in finding a mate. The information they provide would not substitute for your own exploring, but it could give you valuable insights on where to focus your efforts. Such insights would be helpful in spite of the fact that each person would have his/her own preference for a particular partner and areas in which to look.

EXPLORE NOW, SEARCH LATER

In seeking an academic position, as in seeking a mate, it is essential that you explore before you hunt. You need to compare what is available (types of institutions, positions, and locations) with what you need and want (capabilities, interests, and values). Only then will you be in a position to seek out—apply for—specific jobs. Your resources (time and knowledge) are limited, your competition is fierce (many other applicants), and you have to make every opportunity (application and job interview) count. By taking the time to explore first, you will be in a position to zero in on specific options with a targeted effort that increases your chances of success. Exploring exposes you to various institutions and people without the risk of rejection. It allows you to measure, or benchmark, yourself against certain settings and situations without expending limited resources that you will need later for the actual job search.

This chapter is about exploring and identifying the possibilities, Chapter 8 is about searching and applying for positions, and Chapter 9 is about closing the deal and getting the results you want.

The approach we will take in Part III, as shown in Figure 7-1 involves:

- Exploring (what is out there).

    1. Deciding what you want; your values, interests, needs, capabilities, and strengths in relationship to academic possibilities.

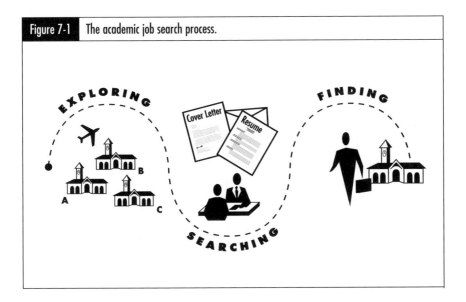

Figure 7-1 The academic job search process.

2. Researching what is out there; background reading, discussions, and visits to other schools.
3. Preparing for the search; putting (1) and (2) together.

- Searching (for specific opportunities).

    4. Setting the stage; how academic positions are established, what departments look for in new faculty, what jobs are available, and the time frame for applications.
    5. Preparing your application materials; cover letters, CVs, letters of recommendation, and teaching portfolios.
    6. Applying for positions; conferences, campus visits, and the academic job talk.

- Finding (the position that is right for you).

    7. Negotiations; principles for responding to job offers.
    8. If you do not get the job you want; the Multiple-Option approach revisited.

The biggest mistake job seekers make is to go immediately to steps (5) and (6), while often skipping steps (1) - (4) altogether. Let us begin, then, with a look at what you want.

DECIDING WHAT YOU WANT

In examining your values, interests, needs, capabilities, and strengths in relationship to various academic possibilities, keep in mind that you are trying to figure out how you want to live your life, not just how you want to manage your career. One way to address these important matters is to ask yourself specific questions about the types of institutions, types of academic positions, and types of locations or settings in which you would be willing to spend the next several years.

You need to be honest in your answers to these questions, and not deny/ignore what is truly important to you. At the same time, you are unlikely to get all the things you want, and therefore need to be clear about what is really nonnegotiable (such as academic positions in the same city for a two-professor couple) as opposed to what is merely desirable (such as teaching in your specialty area versus teaching general introductory courses). Exploring gives you a chance to identify your boundary conditions: what is negotiable, and what is not worth further time and effort.

Your Type of Institution

First, consider *types* of institutions and departments. The vast majority of candidates with Ph.D.s received them from Research I and II universities, with the remaining doing so from Doctoral I and II institutions. However, the majority of all Ph.D. candidates attended, as undergraduates, smaller liberal arts colleges or larger public universities granting bachelor's and master's, but not doctorate degrees. In other words, some of you have had exposure to more than one kind of institution, while others have spent all your student and postdoc time at a research university. The various types of colleges and universities were discussed in some detail in Chapters 1 and 2. Now, you need to ask how these types of schools fit with *your* particular values, interests, needs, capabilities, and strengths.

> Alison Bridger knew her options would be limited with a major in atmospheric sciences. She wanted to teach meteorology, and there were only a small number of schools in North America with such departments. Teaching was a priority, but she also liked doing research, although as she put it, "I knew I wasn't going to be turning out a major paper in my field every year." San Jose State University (Master's I) turned out to be the ideal place for her to achieve both of these objectives.

Of course, even institutions of the same classification can vary considerably in terms of such things as student selectivity, the emphasis placed on various forms of scholarship, and the likelihood of obtaining tenure. Mary M. Heiberger and Julia M. Vick, authors of *The Academic Job Search Handbook* [1], have developed a list of contrasting institutional possibilities. In considering an academic position they ask, Do you want to work for:

- A university with a large enrollment in doctoral programs where tenure depends on research (and increasingly, at least adequate teaching), or a public or private institution where both teaching and research are highly valued?
- A school with a distinctive personality, such as a strong religious affiliation, or one with a liberal education approach?
- An institution that caters to heavy involvement in the life of the school, or one in which you primarily identify with the department?
- A school which is highly selective, or one which offers educational opportunities to a broad section of the community?
- An institution where there are many graduate courses to teach, or one where most of the teaching is at the undergraduate level?
- An institution with a history of involvement with industry, via co-op programs for example, or one that is more "academic" in orientation?
- An institution where there are good prospects of getting tenure, or one that is more challenging and selective?
- A department where socializing with other faculty is expected, or one where more professional involvement is encouraged?
- A hierarchically structured department, or one that places an emphasis on participatory decision making?
- A department in which you would be the first person of your gender, social, or ethnic background, or one in which you feel most others are like you?
- A department where the faculty age distribution suggests a large turnover in the next few years, or one in which there is a more uniform distribution of age and rank?

No single institution will have all the characteristics you desire. However, identifying your initial preferences will give you a reference point from which to make various tradeoffs as your exploration deepens.

Your Type of Appointment

Now, let us consider various kinds of appointments. Most of our discussion will center on full-time, tenure-track positions at the assistant professorship level. Such appointments are of greatest interest to most recent Ph.D.s or postdocs. However, tenure-track positions bring with them tenure requirements and conditions, which may not suit everyone's interests and needs. Many other kinds of appointments exist, one of which may be more suited to your situation. Examples include temporary, or fixed-term appointments, part-time appointments, consulting appointments, and adjunct and research professorships, to name a few. Such appointments can provide experiences that eventually lead to permanent positions, meet special needs and circumstances, or both.

> For a variety of reasons, Bezhad Razavi wanted to teach and do research at one of only a half dozen research universities. At the time of his graduation from Stanford, there were no openings in his field at any of these institutions, and so he decided to go to work for a major industrial research laboratory (AT&T Bell Laboratories), and accept part-time teaching appointments at local universities as a way of maintaining his "readiness" should a full-time position open up at one of the schools of interest to him. Razavi eventually landed a tenured associate professorship at the University of California, Los Angeles, in Los Angeles, CA.

In exploring various types of positions, you should ask:

- Is a tenure-track position the only appointment I would consider?
- Are there circumstances in which a full-time temporary appointment (three years, for example) would make sense?
- If I mainly want to teach, would a part-time appointment satisfy my interests and needs?
- If I am waiting for my spouse to finish his/her education, or if the right tenure-track position is not yet available, is a part-time or temporary appointment appropriate in the interim?
- Must the appointment be in the same department as that from which I earned my degree or might it be in a related department?

Your Setting

Location and setting are certainly considerations for most job seekers, but particularly so for tomorrow's professors. As noted in Chapter 1, there is often at best one or two academic institutions of a given type in any town or city. Because of the way academia works, professors do not move from school to school in the same way scientists and engineers often do in industry. Many faculty positions end up being decades-long or even life-long appointments, and for this reason, location and setting take on added importance.

In exploring various locations, you will obviously consider the physical and cultural environment. Personal preferences and family considerations may also play a part. In addition, you should look at the relationship between the institutions you are considering and other local colleges and universities, industrial companies, and national laboratories, particularly if these entities are ones with which you will want to interact.

> Norm Whitley, a professor of mechanical engineering at the University of New Orleans in New Orleans, LA, liked the idea of living in the South, and not too far from his wife's family in Oklahoma. He was also looking for a school where teaching and research were given about equal weight. In addition, he wanted to live

in a city large enough to provide opportunities for his wife's social work specialty. In this case, New Orleans fit the bill.

For Mark Hopkins, a professor of electrical engineering at Rochester Institute of Technology (R.I.T.) in Rochester, NY, being close to certain kinds of industries was very important. He had had little industrial experience prior to obtaining his Ph.D., and he wanted to be at a school where he could spend some of his time doing research at a nearby company. R.I.T. provided such an opportunity with a program that allowed him to teach at R.I.T. while also doing research at the Xerox Wilson Center for Research and Technology in nearby Webster, NY.

Of course, all of the above may take a back seat when it comes down to two professors who just want to be together.

For Noel and Kirk Schulz, anything would have been an improvement over their arrangement of the last few years, in which Noel was a Ph.D. student in electrical engineering at the University of Minnesota in Twin Cities, MN, and Kirk was a professor in chemical engineering at the University of North Dakota in Grand Forks, ND, some 325 miles away. When, upon graduation, Noel was not offered a position at the University of North Dakota, she and Kirk set out to find the right school that would welcome both of them. Michigan Technological University in Houghton, MI, with its balanced emphasis on teaching and research, turned out to meet their needs.

RESEARCHING WHAT IS OUT THERE

Now that you have a better idea of what you want, you can move on to learning more about what actually exists. You have already begun to do so through your reading of this book, informal discussions with faculty from other institutions, and in some cases, visits to other colleges and universities. Your efforts must now become more focused and purposeful, but remember, you are still exploring. Searching for specific positions will come soon enough.

There are three approaches to take in your explorations of other institutions. The first is background reading, the second is discussions with students and faculty familiar with these institutions, and the third is occasional visits to other colleges and universities.

Background Reading

Just as would-be travelers can learn important things about a distant place through books and travel references, you can learn much about specific academic institutions from resource guides, directories, college catalogs, and the Internet pages. Your library and bookstore have dozens of guidebooks that give brief summaries about various colleges and universities. Many of these are written primarily for students, but they can also be a source of information for prospective faculty. The best general guides to graduate programs are the three Peterson's graduate programs guides in biological and agricultural sciences [2], physical sciences and mathematics [3], and engineering and applied sciences [4].

Many professional societies have directories and other publications describing programs in their respective fields. Examples include the annual *American Society of Engineering Education (ASSE) Engineering Graduate Studies and Research Directory* [5], and the *Graduate Student Packet*, produced jointly by the American Physical Society (APS) and the American Institute of Physics (AIP) [6].

College catalogs are a source of important information about specific courses, background on faculty, characteristics of the student body, and institutional structure. The *Chronicle of Higher Education (CHE)* is a weekly periodical covering all aspects of higher education. It also includes an extensive classified section. It is well worth the $75 annual subscription fee, but if this is too steep, you can always find it in your library. You should get in the habit of perusing recent copies; it is a great way to keep up with what is going on in academia. The *CHE* also produces an almanac in September of each year, summarizing annual data on students, faculty, and resources for all of higher education. The almanac is an excellent resource for all present and future professors.

"Surfing the Net" is also a good way to explore other institutions, their programs, and faculty. You will want to use these resources later when you are zeroing in on specific institutions (searching), but for now, use them as an interesting way to explore. A good guide for exploring on the Internet is *Using the Internet in your Research—An Easy Guide to Online Job Seeking and Information* [7]. The URL (Uniform Resource Locator) addresses for the seven sample schools introduced in Chapter 1 are as follows:

Bucknell University	http://www.bucknell.edu/
Memorial University of Newfoundland	http://www.mun.ca/
Rochester Institute of Technology	http://www.rit.edu/
San Jose State University	http://www.amalthea.sjsu.edu/
Stanford University	http://www.stanford.edu/
University of Michigan	http://www.umich.edu/
University of New Orleans	http://www.uno.edu/

Talking to Others

Reading, of course, can do just so much. You also need to talk with those who have direct experience at other institutions. As noted in Chapter 4, it is important to initiate contacts with such people early on, and not just in the last year of your graduate study or postdoc. By now, you should have a database of contacts that include:

- Former students and postdocs currently working in academia
- Faculty at your current institution, especially your advisor
- Faculty you met through seminars and other visits to your institution
- Faculty at other institutions with whom you have corresponded about your research
- Faculty you met at conferences and other professional meetings
- Industry and government contacts who may have connections to academia.

Now is the time to draw on your network for information about different institutions, while at the same time letting them know of your interests in an academic position. Do so through personal contact, not through mass mailings to everyone on your list. Telephone calls or personal appointments are best, but you may want to start with electronic or hard copy mail. Be sure to personalize your communication. Make references to your past interactions, and say that you will followup with a telephone call. Do not send other information such as your curriculum vitae via electronic mail; it is both irritating and presumptuous. Call first, then followup in the appropriate way with additional information if requested. Remember, you are seeking information, not applying for positions.

Visiting Other Institutions

Undoubtedly, your travel budget is quite limited, so you cannot fly off and visit lots of other schools. However, you can do more in this regard than you might think if you look for leveraging opportunities. Start by contacting the people you know at institutions in your area. Even if these institutions are not among those to which you are likely to apply, just visiting another campus will give you a different perspective. Of course, some of these schools might well represent the *types* of institutions to which you will want to apply.

You can also take advantage of trips made to cities or towns for other purposes. Vacations, visits to family and friends, attendance at conferences, and industrial sites all provide opportunities to spend an extra day or two at a local college or university. Contact people you know at these places (as far in advance as possible), tell them you are going to be in the area, and ask if you can stop by for a chat. Assure them that you are simply trying to learn more about their *type* of institution, rather than asking to be considered for a specific position.

In Chapter 4, we talked about questions to ask and observations to make with respect to your current institution. It is a relatively simple matter to now do the same with the institutions you are visiting. Russ Hall, senior acquisitions editor for Prentice-Hall, has, by his own estimate, visited over 300 institutions of higher education in North America. Hall looks at such visits in the same way as he looks at visits to someone's house. "Every house has a story to tell about the people who live there," he says. "You just have to look for it in the right way." The size of faculty offices and the physical proximity of departments and buildings to each other is one useful piece of information. "You can also tell a lot about the 'haves' and 'have nots' on campus just by looking at the buildings," he notes.

Other things to look for include: How rushed are the professors and students? Do people in other departments, or even the same department for that matter, know each other? Are there social events that bring people together within and across departments, or do most people function independently and in relative isolation? Answers to these questions can tell you a lot about a particular institution and department, including the direction in which it might be heading.

The skills you have acquired in obtaining your Ph.D., such as an ability to learn quickly, observe carefully, and generalize in many situations, are just the ones to apply in your explorations of other institutions.

PREPARING FOR THE SEARCH

Your purpose in exploring is to discover new possibilities, some of which you might not have considered before. By comparing what you have learned about your interests, values, and needs with the types of institutions that exist, you can now focus on a plan of researching and applying to specific schools that have what you want.

It is a bit like the 20-question game we all played as youngsters. When I played this game with my son, we only used ten questions, the limit of both of our attention spans. I would tell him I was thinking of an object in the house, and he could ask me up to ten "yes/no" questions that might lead him to it. In the beginning, he had no strategy at all. He just started guessing. Once in a while, he would guess right, but most of the time, he would use up all ten questions without getting to the object. As time went on, however, he began to develop a crude strategy. He would ask if the object was upstairs, and would be happy if I said "yes" and unhappy if I said "no," not realizing he had the same amount of information in both cases. Eventually, he got to the point where he would ask about attributes, i.e., "Is it touching the floor?," "Is it made out of wood?," and "Is it alive?." This approach would usually get him into the right room, knowing a lot about the characteristics of the object. He could then use his last two or three questions to try, often successfully, to identify it.

The purpose of your exploring is to use the information you have gathered to both eliminate possibilities like research universities and the northeastern part of the United States, for example, as well as to identify possibilities, for instance, master's and baccalaureate institutions in the western United States and Canada. Once you have an idea of what you want in an academic position, as well as what may actually be available, you can bring this information together in a plan that can lead to desired job offers.

How to formulate such a plan is the subject of Chapter 8. If you do the exploring outlined in this chapter, plus the suggestions for applying for positions described in the next, you will have a good chance of obtaining at least one good academic job offer. Yet, in spite of your best efforts, and perhaps for reasons beyond your control, it is also possible that you will not be successful in your first attempt at finding the desired position. Therefore, you need to have a plan in place that allows you to continue on while reviewing the situation and deciding if you want to apply again at a later date.

In Chapters 4–6, we employed a three-pronged preparation strategy, one component of which is the Multiple-Option approach where you prepare simultaneously for careers in academia, government, and industry. There were two reasons for this approach:

1. In the highly competitive and uncertain Ph.D. market, it makes sense to pursue more than one kind of possibility.
2. Connections made with industry will rebound to your benefit, even if you end up accepting an academic offer.

As you begin your academic job search, you need to decide if you want to also apply for nonacademic positions in government and industry. You should do so only if you feel some genuine enthusiasm for such opportunities. These positions do not have to be your first choice, and indeed they can be a temporary one. However, just as with academic positions, a lack of interest will be detected quickly, and thus your chances of success will not be high. The contacts you have made and the relationships you have established should now be part of a database that you can tap into for industrial and government possibilities. We will discuss this approach in greater detail in Chapters 8 and 9.

Vignette #7	From Industry to Academia

Not all paths to an academic career go through a Ph.D., or occur right after graduation or a postdoc. The following vignette describes a return to academia after a number of years as an engineer in industry.

Joseph Reichenberger
Loyola Marymount University

The most common entry into academia is through the doctorate degree, either right after graduation or after a few years as a postdoc. Yet, this is not the only option. Increasingly, a number of engineers and scientists who have worked in industry for many years are thinking about returning to academia as professors. The reasons for wanting to make such a return, the ease with which it can be done, and the benefits to the faculty member and the institution are often quite different from what they are for those who become professors right after graduate school or a period as a postdoc.

After a successful 30-year career in industry, and at age 51, Joseph Reichenberger became a full-time, tenure-track associate professor of civil engineering and environmental science at Loyola Marymount University (L.M.U.) in Los Angeles, CA. After receiving his B.S. degree in civil engineering in 1964 from Marquette University, Reichenberger spent time with the Los Angeles County Flood Control District while working part time on his M.S. degree in the same field at the University of Southern California, earning the degree in 1967. Between 1967 and 1979, he worked for a large architectural and engineering firm (Daniel, Mann, Johnson & Mendenhall) in Los Angeles, CA. In 1979, Reichenberger joined Engineering-Science, Inc., a large international environmental engineering firm, headquartered in Pasadena, CA, eventually rising to vice president and western regional manager.

During his time in industry, Reichenberger did something that not only made it easier for him to decide to pursue an academic career, but it also impacted his ability to acquire a full-time academic position in competition with younger candidates with doctorate degrees. For most of the last 30 years, he has taught college courses part time at such schools as California State University, Los Angeles, the University of Southern California, and Loyola Marymount University. Says Reichenberger:

> Cal State L.A. was looking for someone to teach hydraulic engineering part time, and I decided to give it a try. I found I liked it, and ended up doing the same thing part time at USC. One thing led to another. Don Anderson, who was chairman of the Civil Engineering Department at L.M.U. in the early 80s, was also a consultant at Engineering-Science, where I was working full time. The latter relationship led to an offer to teach part time at L.M.U. in 1981, eventually reaching the rank of adjunct professor of civil engineering.

Nevertheless, teaching part time at L.M.U. for 12 years did not make Reichenberger an automatic shoe-in for a full-time position when one opened up in 1993.

> The University had to advertise the position nationwide and I had to compete like everyone else. Not having a Ph.D. was also a concern to me and to the school. I had to develop a class lecture outline, trial teach a senior level class, and answer very tough questions on my feet from students. I had a fairly intensive interview with the dean of the college and the academic vice president.
>
> When I began in this business years ago, I wasn't the world's greatest public speaker, and I had to work on it. Teaching part time helped, particularly when students would ask me off-the-wall questions. But what really made the difference was my extensive experience working for a company where I had to make very competitive sales presentations to a wide variety of clients with extensive

question-and-answer sessions. I developed a knack for restating questions, while at the same time mentally forming the answers, and this technique has helped me a great deal.

Why, at age 51, did Reichenberger decide to leave a high-paying industry career for a full-time academic position in which he would still have to earn tenure?

I was at a point where I was really thinking about what I wanted to do with my life. My kids were grown, and my wife had a successful career of her own. I had plenty of offers to help startup and run offices at various engineering companies, but I didn't want to do that. I was making a good salary as a regional manager, but I was burned out. I was tired of all the personnel issues, business plans, and negotiations. What I really liked was working with the younger staff in a mentoring role, but I didn't have time to do this as a vice president and regional manager.

However, for a number of reasons, Reichenberger wanted very much to keep his consulting role with Engineering-Science, Inc. His nine-month salary at L.M.U. was about 50% of his 12-month full-time salary at Engineering-Science. But L.M.U. liked having its faculty spend some time interacting with industry, and its faculty policy specifically allows for outside consulting on a one day a week basis. Such consulting, and related opportunities during the summer, meant that Reichenberger's annual income would not be that different from what it was prior to joining L.M.U.

There were other reasons why Reichenberger wanted to continue with some outside consulting. As he points out, such outside activity is often based on a desire to:

- Improve one's professional expertise
- Stay active in the profession and apply "textbook" techniques to solving real-life problems
- Contribute something back to the profession
- Keep busy because reduced family commitments during midlife years make time available.

Reichenberger believes that L.M.U. clearly benefits from such an arrangement. Through these experiences, faculty are often better able to carry out university obligations such as securing outside assistance to fund or stage a university function, writing proposals to secure grants, and helping with curriculum and restructuring.

However, the real winners, according to Reichenberger, are the students who benefit through the bringing of the real world of the practicing engineer into the classroom to deal with current engineering issues and problems, the career guidance such professors can give students, and

> the bringing in of outside speakers and expertise on projects and topics of interest to students.
> What are the biggest surprises Reichenberger found on returning to academia?
>
> > The biggest shock was the time required on campus. I may have classes two or three days a week, some of which don't start until 11 a.m., but that doesn't mean I'm not here much of the rest of the time. Lots of students stop by for advising, there are seniors looking for career counseling, there is laboratory time, and this year I'm developing two new classes in areas that I have not taught in before.
> >
> > Of course, there are the regular administrative duties, although in my case, they don't appear to be too excessive. My writing skills are pretty good, and so I get called on to help draft department policy statements and proposals, but I don't really mind this.
>
> Reichenberger says he has no regrets about making the switch to academia and thinks the rewards far exceed the costs.

SUMMARY

This chapter began by explaining why, in seeking an academic position, it is essential for you to explore before you search. You need to compare what is available (types of institutions, positions, and locations) with what you need and want (capabilities, interests, and values). Only then will you be in a position to search–apply for specific jobs. We began by looking at how to decide what you want by examining the types of institutions, appointments, and settings of greatest interest to you. Ways of researching what is out there, including background readings, talking to various people, and visiting other institutions, were discussed next. We then looked at the specifics needed to prepare for the job search process that will be discussed in the next chapter. The chapter concluded with a vignette describing a career path to academia taken after a number of years as an engineer in industry.

REFERENCES

[1] M. M. Heiberger and J. M. Vick, *The Academic Job Search Handbook*. Philadelphia, PA: University of Pennsylvania Press, 1992, ch. 3, pp. 24–25.
[2] *Peterson's Graduate Programs in the Biological and Agricultural Sciences—Book 3*. Princeton, NJ: Peterson's, 1995.

[3] *Peterson's Graduate Programs in the Physical Sciences and Mathematics—Book 4.* Princeton, NJ: Peterson's, 1995.
[4] *Peterson's Graduate Programs in the Engineering and Applied Sciences—Book 5.* Princeton, NJ: Peterson's, 1995.
[5] *Engineering Graduate Studies and Research.* Washington, DC: American Society for Engineering Education, 1995.
[6] *Graduate Student Packet.* Washington, DC: American Physical Society and American Institute of Physics, 1995.
[7] F. E. Jandt and M. B. Nemnich, *Using the Internet in Your Research—An Easy Guide to Online Job Seeking and Information.* Indianapolis, IN: JIST Works, Inc., 1995.

CHAPTER 8

Applying for Positions

> We typically get 300–400 applications for every opening. The main reason is that everyone is applying for everything. Only about 10% of the applications are even in the ballpark, although that's still a large number for us. The key is persistence and an ability to convince us that you will be a good *match* with what we need.
>
> Kody Varahramyan, Professor of Electrical Engineering
Louisiana Tech University

> My experience is that it is far better to apply to fewer schools well, tailoring your application to each school, than to blindly apply to a whole bunch of schools. In most places, search committees are looking for a fit with other faculty members in the department, not an overlap, but not too far out either.
>
> Pam Stacks, Professor of Chemistry
San Jose State University

The above statements signal the way to increasing your chances of getting an academic position that is right for you. Because so many people are applying for so many positions, most are bound to do a poor job at most of the places they apply, and as a result, are almost certainly going to be rejected. It is no accident that those who apply to 60 or 70 places often get turned down by all of them. They cannot possibly be a match for that many different positions, and their applications show it. Herein lies the key to increasing your prospects. You have to identify a manageable number of openings consistent with the results of your explorations in the previous chapter, and then do an outstanding job of applying to each of them.

Career counselors often talk about the rifle, the shotgun, and the splatter approaches to applying for jobs. In the splatter approach, you basically throw the same things, your cover letter and curriculum vitae, at every place you can imagine, and see if they stick anywhere. Those who follow this approach tend to believe they will rewarded for sheer volume of effort. For every 100 applications, they will get five interviews, and for every five interviews, they will get one job offer. Well, forget it! In the so-called "good old days," it may have been possible to send out dozens of the same applications to any and every institution, and in effect count on the search committees at these schools to identify a match. No more. Now you have to take a *proactive* approach, and help point out the match you think is there as a result of your background investigations.

In the shotgun approach, you are somewhat more focused, sending a manageable number of applications to schools with which you believe you have some affinity. Your cover letters are tailored to at least some degree to each school, and perhaps you have two or three versions of your curriculum vitae from which you can choose for each application. The problem with this approach is that you still have to compete with the applicants for whom each school to which you are applying is their rifle shot.

Rifle shots are, of course, the most focused of applications. You have researched the school and department to the point where you really believe that you are the right candidate for the job, and your cover letter, curriculum vitae, letters of recommendation, and calls from your advisor show that this is the case. The only drawback to such an approach is that it takes considerable time and effort.

Most experts recommend the rifle approach exclusively, whereas I suggest a rifle approach, and if time permits, a limited shotgun effort. Simply forget the splatter approach; it will only drain your energies and depress you in the process.

Your major effort should go into identifying up to a dozen schools for which you are prepared to do the background research to enable you to present powerful, targeted applications. Beyond that, as time permits, identify another dozen or so schools that fit within the general boundaries of your explorations. Then take the time to prepare a letter and curriculum vitae having at least a reasonable chance of getting the attention of the search committee. Your main time and energy must go into the schools for which you have reason to believe there is a very strong match.

This chapter is a nuts-and-bolts presentation about looking for academic positions. We set the stage by discussing how new positions are established, what departments want in a new faculty member, how to find out what is available in your field, and the typical time frame for filling academic positions. Preparing application materials such as cover letters, curriculum vitae, and letters of recommendation are then explored. Screening interviews at conferences are looked at next. Considerable detail is devoted to a discussion of the campus visit, and in particular, the all-important academic job talk that can make or break your chances of being offered an academic position. The possibility of applying for positions in government and industry is then examined by revisiting the Multiple-Option approach. The chapter concludes with a vignette discussing the impact of increasing job competition on the goals of affirmative action, and the quest for diversity among faculty on college and university campuses.

SETTING THE STAGE

In the last year of my graduate studies, I applied to a number of universities for entry-level assistant professorships. The Physics Department at one school told me in advance that it was not absolutely certain a position was available. Two days after my campus visit, the person who conveyed this information to me suddenly died. On hearing the news, my advisor remarked, somewhat insensitively I thought, "Now you know there's an opening!" Well, not exactly. The position freed up by the demise of my host went to the Chemistry Department!

How New Positions are Established

New faculty slots, or billets, are among the most prized resources in all of academia. As I found out, they are by no means entitlements. Billets for tenure-track appointments that can last a lifetime are dispensed with the utmost care. They can determine the direction and character of a department, school, and institution for years to come.

A common, but by no means universal, approach is one in which new billets are determined by the vice president for academic affairs or the provost who then allocates a certain number to various schools, i.e., engineering, science, humanities, and law. In principle, if a faculty member retires from the Mechanical Engineering Department in the School of Engineering, that billet could be taken away from the School of Engineering and granted to the Economics Department in the School of Humanities. However, such reallocations across schools are quite rare.

More likely is the possibility that a dean will try and use billets to direct the teaching and research mission of the school. He or she may decide that an area such as water resources in the Civil Engineering Department, from which a faculty member just retired, should be deemphasized in favor of the expansion of an area such as telecommunications in the Electrical Engineering Department. However, my case not withstanding, such reallocations across departments are also not the norm.

While billets tend to stay in departments, they certainly do not have to stay in particular fields. It is not at all uncommon for, say, a Biology Department to lose a population biologist, and have him or her replaced by a geneticist. In most cases, the department decides which area to fill after considerable discussion and then a vote of each faculty member. Yet, in some cases, particularly where there are department heads as opposed to department chairs, the former can wield considerable decision-making power and overrule a faculty recommendation.

Given the range of possibilities outlined above, you need to look for ways to find out how things are done in the departments to which you are applying.

What Departments Look for in New Faculty

Once a department obtains a billet, its primary goal is to find someone who can teach certain classes. This need is what really drives new positions, even at research universities. The process of assigning faculty to classes is one of the most important responsibilities of a department chair, and finding someone who is flexible in terms of the type and level of courses they can teach can be quite an advantage.

The degree of flexibility required is often a function of the size of the department. Large departments can afford to have someone who is more specialized than can smaller departments. Kirk Schulz, of the Chemical Engineering Department at Michigan Technical University, puts it this way:

> Faculty from large top-notch research institutions often arrive expecting to teach only in their area of expertise. The reality is that in a department of five or six faculty with both an undergraduate and graduate program, they may have to teach a course significantly out of their discipline.

While most advertisements for new positions specify a particular research or subject matter area, others will list two or three possible areas. Still others can be quite general, i.e., organic chemistry, which has many different subdisciplines. Departments will rarely hire someone because of their specific dissertation topic, but rather because of what such a topic says about their ability to do further scholarship in a particular area.

Finding the right fit is critically important to a department, although "fit" can mean different things to different departments. Most departments are looking for the following:

- Overall promise
- General teaching ability
- Ability to teach courses in need of staffing
- Ability to do research/scholarship in specific areas, a specific research orientation
- Compatibility with department and institution
- Potential for securing external research funding.

The above list is not necessarily in order of priority. What is most important depends on the nature of the department and the institution.

As an example, consider high-energy physics. Virtually all schools have Physics Departments, but very few of these departments are involved in high-energy physics research. One reason is that much of this research is actually done at national laboratories away from the main campuses. Not many schools can support extended absences of faculty who need to be off-site to carry out their research. Even with excellent research and teaching credentials, most Physics Departments, including those at Research I universities, just do not have opportunities for high-energy physicists.

Even schools that do have high-energy physics programs will not be looking for just any kind of high-energy physicist. As noted in Chapter 5, the step from high-energy physics graduate student to high-energy postdoc is a big one. As a postdoc you are expected to write proposals, supervise a research project and associated graduate students, review colleagues' papers, and in various ways contribute to the research around you. The step from high-energy physics postdoc to high-energy physics faculty member is an even bigger one, and even fewer make it. In addition to evidence of teaching and research capabilities, the department will want to see potential for leadership, an ability to direct a high-energy research program, organizational and managerial skills, and a facility for raising funds.

Now, consider theoretical physics. Here, the size of the typical research group and the infrastructure demands are much smaller than for high-energy physics; thus, the requirements may also be less stringent. Many departments, including those at Master's and Baccalaureate schools, will want a theoretical physicist who can teach a variety of courses, will not travel a great deal (as do high-energy physicists), and can conduct research with a much lower overhead.

The point of the above examples is that you need to find out what departments really want. While determining departmental needs is not as difficult as you might first imagine (see "The Application Process" section), it is also true that departments do not always know exactly what they are looking for. As Candice Yano, a professor of operations and industrial engineering who has served on search committees at both the University of Michigan and the University of California, Berkeley, notes, "Sometimes several areas are considered in a search, and the faculty fight it out after they interview a number of candidates. In such cases, it is particularly important that you know something about the faculty who are going to interview you." It should be possible to obtain the names of most of these faculty from the chairman of the search committee. You can then "check them out" by examining their home pages on the World Wide Web.

In almost all cases, you can be sure departments are asking, "Can this person come here, go through the evaluation process, and get tenure in six or seven years?" Some departments are looking for a new extreme point in terms of teaching and other forms of scholarship. Others are looking for more of an overlap. In either case, getting along and being flexible is important. All departments want to know that you are the kind of person who will pull your weight across the board. In small departments where resources are limited, this is absolutely critical.

Finding Out What is Available

It used to be that there were two kinds of academic positions: those advertised in journals and bulletins, and those known to only a few selected faculty on private informal networks. Now, however, all public institutions are required by law to advertise their openings, and most private schools do so as well. This advertising requirement is one, although not the only, reason that there are so many applications for a given position. The existence of that position is much more widely known than it would have been a decade ago.

Although virtually all positions are advertised, just when such advertisements appear can vary widely. Some departments know they are likely to have a position months in advance of their placing an advertisement. They will often *proactively seek candidates* through an informal network prior to putting the word out officially. Hearing about an opening from an advisor or other contact can give you valuable extra time to evaluate the situation and prepare your application material. As my own case illustrated, some schools will advertise a position even if they are not certain it will be available, often going as far as an on-campus interview before knowing if they have an opening. Also, some departments do not know until very late in the academic cycle that a position has materialized. While they are still required to advertise the opening, they may not do so extensively because of time limitations. Being plugged into a network where such information is available can make a real difference.

In some fields, such as chemical engineering, where there are only a dozen or so openings a year and a relatively small applicant pool, word of a position spreads quickly within the graduate student and postdoc community. In fields such as biology and electrical engineering, where both the number of openings and the pool of applicants are much larger, a variety of approaches must be used to identify opportunities.

Most associations publish employment bulletins several times per year in which the great majority of available positions in the field will be listed. In other fields, positions are listed in widely read publications, such as *Science, The Scientists* (life science and medical positions), and *Genetic Engineering News*. Appendix C, "Professional Associations for Academic Job Seekers in Science and Engineering," lists selected professional associations publishing employment notices in their journals or separate employment bulletins.

Search committees, department chairs, department administrators, and university personnel departments often send job announcements to other universities to try to reach candidates. Individual faculty members, affirmative action officers, graduate deans, and career planning and placement centers receive these

Chapter 8 Applying for Positions 189

notices and make them available in various ways. It is important for you to be proactive about searching for such announcements, talking to individual faculty, and availing yourself of career center resources.

When it comes to finding out about academic openings, however, nothing beats the Internet, and in particular, the World Wide Web. While many publications continue to produce hard-copy versions of their magazines and journals (and even this is beginning to change), most are now putting their job announcements on the Internet. The advantages of such a system are obvious: announcements can be updated weekly or even daily, and you can search quickly on a variety of criteria.

There are now dozens of places on the Internet to look for faculty positions in science and engineering, with many more being added all the time. As freelance writer Robert Finn notes [1, p.22]:

> Nearly every major university posts job openings on its gopher server—a system that uses menus to point to various resources. Many scientific societies post openings, as well. Online job ads can be found on Usenet newsgroups, mailing lists, World Wide Web "home pages," and electronic classified sections, not to mention commercial and noncommercial employment services.

An excellent resource for figuring out where to search on the Internet is "Employment Opportunities and Job Resources on the Internet," which contains an annotated list of Usenet groups, telnet services, gopher servers, mailing lists, and Web resources. It is available from the Worcester Polytechnic Institute, 100 Institute Road, Worcester, MA 01609-2280 (http://www.wpi.edu). Another clearinghouse for academic jobs in the United States and Canada can be found at (http://volvo.gslis.utexas.edu/~acadres/geographic.html).

Probably the single best source of electronic information for academic positions can be found on "Academe This Week," the electronic version of the *Chronicle of Higher Education's* classified section. The World Wide Web address is (http://chronicle.merit.edu). This service now gets about 40,000 visits a day, and key word searches pull up 350,000 "hits"—each week [1, p.23]. With that kind of activity, you cannot afford not to be using such a resource.

It is clear that the Internet is a powerful way to find out about jobs. However, it is not necessarily the way to apply for jobs. It may be tempting to knock out a quick cover letter, couple it with your electronic CV, hit the reply button, and send it on its way. But be careful. Quick responses will often only get you quick rejections. You need to take the time for a thoughtful reply just as you would if there were no Internet. Important decisions are being made by search committees who are not simply going to reward the person with the fastest fingers.

You also want to be careful about putting your resume on the Internet. Once it is out there, you lose control over it. Even if you do not care that everyone knows you are applying for positions, you may not be able to retrieve your resume and tailor it to specific situations.

Drawing on Your Network

The old adage that it is not just *what* you know, but *who* you know has never applied more than to the seeking of positions in industry, government, or academia. Certainly, your hard work, accomplishments, and "academic blood line" are the foundations of your candidacy, but if the right people do not know about these things, then they will not matter much when you are looking for a job.

That you should make sure this is the case is one of the tenants of the Multiple-Option/Next-Stage strategy outlined in Part II. The expected outcome of this strategy is that by now, you will have gotten to know a set of people in industry, government, and academia who are *anticipating* the completion of your Ph.D. or postdoc and your subsequent availability for an academic or industry/government position.

Through your interactions with industry, attendance at professional meetings, joint publications, multidisciplinary collaborations, and correspondence with alumni, you should have developed a network of people around North America, even the world, who you can now draw on for further help. These are the people who, in addition to your advisor, can make telephone calls or send e-mail on your behalf to contacts they may know at schools to which you want to apply. Now is the time to call on them for the advice, information, and assistance you need.

The Time Frame for Academic Positions

While some academic positions remain open until filled, most assistant professorships start in the fall with the beginning of the academic year. (See Fig. 8-1.) It is during the prior 12 months that the serious job search activity takes place. Here is a checklist, developed by Kelly Johansen-Trottier [2], formerly of the Stanford University Career Planning and Placement Center, of things you should be doing during this time. Exactly what you do, and when you do it, will depend on your particular circumstances, but this guideline is a helpful way to make sure you are not leaving anything out.

Early fall (August – October)

- Work on a draft of your curriculum vitae (CV) and a basic letter of application, and receive feedback from other students and faculty.
- Consider who you will ask to write you letters of reference, and contact them if they are at other institutions.
- Obtain an employment file packet from the career center at your institution or set up a reference file in your department if they provide this service.
- Produce a final copy of your CV and a general letter of application that can be easily tailored to specific positions.

Figure 8-1 The time frame for academic positions.

The Academic Year

I	II	III	IV
Early Fall August - October	**Late Fall** September - December	**Midyear** November - February	**Late Winter-Summer** March - August
• Develop CV • Obtain letters of reference • Obtain employment packet	• Talk to your advisor • Attend conferences • Apply for positions	• Continue to apply for positions • Practice interviews • Prepare job talk	• Continue to apply for positions • Consider interests and needs • Consider negotiation strategies • Consider Multiple-Option approach

Late fall (September – December)

- Talk to your advisor and other faculty about letters of recommendation or have previous letters updated. Provide references with your CV, teaching portfolio, and abstracts of your research. Discuss with them the types of academic positions that interest you.
- Attend conferences in your field that are publicized through your academic department.
- Apply for positions using listings available in your department, journals in your field, on the World Wide Web, through the *Chronicle of Higher Education*, and any other sources including word of mouth.

During Midyear (November – February)

- Continue to apply for positions.
- Practice interviews with other graduate students and faculty.
- Prepare and practice your job talk. Have others critique you, and if possible, arrange to have at least one practice run videotaped.

Late winter – summer (March – August)

- Continue to apply for positions. Tenure-track and one-year positions continue to be announced during this time period for the upcoming academic year.
- Seriously consider your interests and needs when weighing any offers.

- Take the necessary time to weigh the pros and cons of any offer.
- Discuss negotiation strategies with your advisor or recently appointed faculty in your field.
- If you have not yet found a position, do not despair; discuss strategies with your advisor and recently appointed faculty. (See Chapter 9.) [3]

PREPARING YOUR APPLICATION MATERIALS

The Cover Letter

Most of the cover letters we get are just terrible. They bear no relationship to what goes on in our department or school. You can tell they are word-processed form letters. In almost no time, we can reject half of our applicant pool just by looking at their cover letters. On the other hand, an excellent cover letter may make us consider someone who was not exactly in the research area we were seeking.

Susan Lord, Assistant Professor, Electrical Engineering Department
Bucknell University, Lewisburg, PA

The cover letter is the most important thing in your application. I wrote the best cover letter ever, and it got me the interview. I took the time to tailor it perfectly to the school and department. This took a lot of research, but it paid off.

Emily Allen, Assistant Professor, Materials Engineering Department
San Jose State University, San Jose, CA

In a sense, the first quote is the good news; the second, the bad news. Most applicants for a given position will write poor cover letters. But you can also be sure that at least *some* applicants for a given position will write excellent cover letters. To be in the latter category, you have to take the time to do the job right.

The main thing to remember about cover letters is that they are the first thing the search committee sees. If the committee receives hundreds of applications, it will probably divide them up so that one, or at most two, committee members will look at your application the first time around. These professors, who you are not likely to know, can determine if you are in the pool of the 15–20 applicants who will be moved to the next stage, or if you are out of the running. Based on your cover letter, you want the search committee to look forward to examining the rest of your application material.

To be in the running, you must take the time to find out enough about the school to which you are applying to show the connection between what you have to sell and what the school wants to buy. Pointing out this connection in a one-page letter is no easy task, but everything you do by way of preparation will be helpful if you are then asked to visit the campus for an interview.

Here, according to Johansen-Trottier [3, p.4], are the key points to keep in mind in writing your cover letter:

- Always address each letter to a specific individual, and be sure to use the correct title. (If the advertisement says only "Chairman, Search Committee," try calling the school for the name of the chairperson.)
- Ideally, the cover letter should be one page.
- Individually tailor and address the letters. (More on this below.)
- Content is more important than style, but how you write, as well as what you say, is important.
- Use simple, direct language.

First Paragraph

- Explain how you heard about the position ("Your announcement in…," At the suggestion of…").
- Explain who you are ("I am completing my Ph.D. in…").
- Indicate why you are interested in the position/department ("I am especially interested in this position because of…").

Middle Paragraph(s)

- The second paragraph often provides an overview of your research.
- Subsequent paragraphs highlight your achievements and qualifications, especially those that make you the right person for the position. You want to connect items in your background with the specific needs of the department.
- At smaller colleges or universities, try to point out your interests in the institution as well as the department. At larger colleges or universities, you can concentrate more on your interests in the department.
- You should discuss what you actually accomplished with respect to published articles or research work before describing future activities or interests.

Final Paragraph

- Indicate what you are enclosing along with the letter, such as your vita and teaching portfolio. Offer to provide extra materials or additional information.
- Thank the committee for its consideration, and indicate that you look forward to meeting with them in the near future.

It is particularly important that your letter be error free. Search committees see plenty of letters that are not. The readers will assume that you had all the time you needed to put the letter together, and so are likely to be unforgiving of typographical and spelling errors. Have it proofread by at least one other person. An example of a job announcement and the resulting cover letter that helped get the applicant to the next stage is shown in Figure 8-2a and 8-2b.

Figure 8-2a Sample job announcement.

(Announcing) University Announcement of Position Availability
FACULTY POSITIONS IN MATERIALS ENGINEERING

POSITION:
One tenure track, faculty position in either of the areas: (1) Materials Analysis or (2) Structure/Properties Processing Relationship in Electronic Materials.

RANK:
The department prefers making appointments at the assistant/associate professor levels, but in exceptional cases will consider appointments with the rank of professor.

QUALIFICATIONS:
The candidate must hold an appropriate Engineering Doctorate and must be a U.S. Citizen or a permanent resident. The University is especially interested in hiring faculty members who are aware of and sensitive to the educational goals and requirements of an ethically and culturally diverse student population. Minorities and women are strongly encouraged to apply.

TEACHING ASSIGNMENTS AND RESPONSIBILITIES:
Successful candidates will teach undergraduate and graduate courses in their specialties as well as other courses basic to Materials Engineering, will develop modern laboratories and will conduct externally funded research in their areas of specialties, and will emphasize computer applications in their teaching and research.

SALARY RANGE:
$XX,XXX to $XX,XXX depending upon qualification.

STARTING DATE:
Fall Semester 19XX is preferred; however, the position will stay open until filled.

APPLICATION PROCEDURES:
Candidate should send a resume, proof of citizenship/ permanent residence and names, addresses, and telephone numbers of three references to:

    Name/address/telephone number of search committee chairman

GENERAL INFORMATION:
(Announcing) University is (state's) oldest institution of public higher learning. The School of Engineering has an approximate enrollment of 3,400 students. The school offers both BS and MS degrees. The campus is located on the (city) with a metropolitan population of 3,000,000. (Announcing) University enrolls approximately 30,000 students, a significant percentage of whom are members of minority groups. Many of (state's) most popular natural, recreational and cultural attractions are conveniently close.

The University is committed to increasing the diversity of its faculty so our disciplines, students and the community can benefit from multiple ethnic and gender perspectives. (Announcing) University is an equal opportunity/ affirmative action/Title IX employer.

| Figure 8-2b | Sample cover letter. |

December 10, 19XX

Name of search committee chairman
Address

Dear Professor (name):

I am responding to your advertisement for a faculty position in the Materials Engineering Department at (announcing) University. I am presently a Ph.D. candidate at (applicant's) University in the Department of Materials Science and Engineering. My thesis work is in the area of process modeling for semiconductor fabrication and my specific research topic is the diffusion of dopants in gallium arsenide. My thesis advisor is Professor (name), Director of the Integrated Circuits Laboratory in the Department of Electrical Engineering.

I am particularly interested in a faculty position in a department which values teaching. As you may note from my resume, I have taken every opportunity to teach while at (applicant's) University, and I have also participated in various educational projects outside of the University environment. These included the Computer Literacy Project, which I founded and directed for three years in a predominantly minority middle school in (city), and Expanding Your Horizons at (announcing) University, a workshop for young women interested in pursuing careers in math and science. At (applicant's) University, in addition to being a grader and teaching assistant in several Electrical Engineering courses in semiconductor processing, I helped design and teach a Materials Science and Engineering laboratory course. I also designed and cotaught a new course for undergraduates entitled Electronic Materials Science.

In addition to teaching, I am looking for a research opportunity which would allow me to continue my work in electronic materials processing, structure and properties. At (applicant's) University I have accomplished original research in pursuit of my degree, as well as contributed to the development of a fabrication line for GaAs digital technology. While working at (name) Corporation, I designed and supervised construction of a thin film laboratory research facility and initiated a research project in amorphous semiconductor thin films.

I expect to finish my degree in July 19XX. I have enclosed my resume, including a list of publications, a list of references, and a copy of my passport as proof of U.S. citizenship. Thank you for your consideration. I look forward to hearing from you soon.

Very truly yours,

name/address/telephone number

The Curriculum Vitae

Your curriculum vitae, or CV, is the next part of your application that the search committee sees. As with your cover letter, the visual impact of your CV sends an important message about your thoroughness and attention to detail. You want to present your experiences, accomplishments, and professional qualities in the most positive light. What is the difference between a curriculum vitae and a resume [3, p.2]?

> The curriculum vitae (also referred to as the vita or CV) is a summary of an individual's educational background and experiences. The CV is used when applying for teaching and administrative positions in academia or for a fellowship or grant. In contrast to a CV, a resume is used to summarize an individual's education and experience related to a specific career objective in the public or private sector.

Johansen-Trottier has developed the following list of questions to ask yourself when writing a vita [3, p.1]:

- Is it well designed, organized, and attractive? (well laid-out, and appropriate use of bold and italics).
- Are categories of information clearly labeled? (Education, Teaching, and Research).
- Is it easy to find certain sections of interest to search committee members? (Publications, Postdoc experience, and Professional Associations).
- Has your advisor, and at least one other person, reviewed and critiqued it?
- Have you avoided using acronyms?
- Has it been proofread *several times* to eliminate typographical errors?

Most science and engineering vitae will contain the following [3, pp.2-3]:

Identifying Information

- Name, address, telephone number, and electronic mail address.

Education

- Begin with your most recent or expected degree.
- List degrees, majors, institutions, and dates of completion (or expected date) in reverse chronological order. Also, list minors, subfields, and honors.

Dissertation or Thesis

- Date dissertation/thesis will be finished.
- Dates describing your current status ("Completed coursework, June 19XX," "Passed qualifying exam, March 19XX").
- Provide the title and a brief description of your work, its framework, your conclusions, your advisor (and committee members).

Awards

- Examples include: NSF Fellow, IBM Dissertation Fellowship, and Phi Beta Kappa.

Experience

- Some scientists and engineers like to include their research/thesis/dissertation in this section.
- Include job title, the name of the employer/institution, your responsibilities and accomplishments, dates.
- Use a consistent format. "Experience" works best, but you may want to divide things up by "Research" and "Teaching."
- Stress what you contributed and accomplished using verb/active skills: "Delivered eight class lectures on composite materials and developed five supporting problem sets and a midterm examination" versus "Responsibilities included preparing class lectures, homework assignments, and exams."

Publications/Presentations

- Put these last if you have more than four or five entries.
- Listed in standard bibliographic form, classified by type (journal and conference).
- While it is acceptable to list articles as "submitted" or "in preparation," be careful. You will want to balance these with articles that are either published or in press.

Other possible categories often used are:

- Academic service
- Research interests
- Teaching competencies
- Areas of expertise
- Community service

- Professional associations
- Scholarly presentations
- Foreign study
- Licensure

CVs can be rearranged to fit different situations. Figure 8-3 is a CV designed to appeal to a major research university. Figure 8-4 shows the same person's vita redesigned to appeal to a school with a primary teaching emphasis.

Figure 8-3 Curriculum vitae—research emphasis.

TERRANCE L. JOHNSON

Environmental Science Division
Oak Ridge National Laboratory
Oak Ridge, TN 37831-6056
(615) 574-1234

207 Edinboro Lane,#C32
Oak Ridge, TN 37830
(615) 483-1234

EDUCATION

Stanford University, Stanford, California, 19XX-19XX
Ph.D. in Biological Sciences, 19XX, Area of Specialization: Population Biology
M.S. in Biological Sciences, 19XX

Northwestern University, Evanston, Illinois, 19XX-19XX
B.A. in Biological Sciences, concentration in Ecology and Evolutionary Biology
B.A. in Biochemistry, Molecular Biology and Cell Biology with honors
B.A. in Integrated Science Program, with honors

AWARDS AND HONORS

Hollaender Postdoctoral Fellowship (US D.O.E.), 19XX-present
ARCS Foundation Fellowship, 19XX-19XX
National Science Foundation Graduate Fellowship, 19XX-19XX
Andrew Mellon Foundation Graduate Research Fellowship, 19XX
Phi Beta Kappa, 19XX

RESEARCH EXPERIENCE

Postdoctoral Research: Environmental Science Division, Oak Ridge National Laboratory, 19XX-present (research advisor: Dr. Stephen H. Smith).
- Development of quantitative theory of hierarchical structure in ecological systems.
- Analysis of how ecological communities reflect environmental heterogeneity at different scales.
- Numerical study of foraging behavior with short and long range movement in heterogeneous environments.

Figure 8-3 (Continued)

Doctoral Research: Department of Biological Sciences, Stanford University, 19XX-19XX (research advisor: Dr. James T. Jones).
- Field study of the impact of avian predation on *Anolie* lizards in the eastern Caribbean documents the importance of differences in spatial scale between prey and predators.
- Theoretical analysis of spatial scale and environmental heterogeneity in models of predator-prey communities.
- Analytical and numerical works show how species interactions can sharpen underlying environmental patterns and how heterogeneous environments can stabilize predator and prey populations.

Undergraduate Honors Research: Department of Biochemistry, Molecular Biology, and Cell Biology, Northwestern University, 19XX-19XX (research advisor: Dr. Peter T. Williams)
- Investigation of primary events of bacterial photosynthesis.
- Isolation and spectral analysis of photosynthetic reaction centers.

RESEARCH INTERESTS
- Theoretical and field study of ecological communities.
- The roles that spatial patterns and processes play in shaping communities.
- How populations and processes that act on different spatio-temporal scales affect the behavior of ecological systems.
- Influences of disturbance size and frequency on landscape structure.

TEACHING EXPERIENCE

Instructor: Outdoor Education Program, Stanford University, 19XX-19XX.
- Lectures and weekend outings, emphasis on alpine ecology, animal tracking, and wilderness skills.

Co-Instructor: Biology of Birds, Stanford University, 19XX.
- Lectures and field trips; with Dr. S.T. Phillips.

Teaching Assistant: Systematics and Ecology of Vascular Plants, Stanford University, 19XX.

Teaching Assistant: Core Biology Laboratory, Stanford University, 19XX.
- Ecology laboratory and discussion sections.

Instructor: Chemistry Laboratory, Kendall College, Evanston, IL, 19XX-19XX.
- Sole responsibility for laboratory in biochemistry, general and organic chemistry.

Wilderness Guide: Association of Adirondack Scout Camps, Long Lake, NY, 19XX.

> **Figure 8-3** Curriculum vitae—research emphasis (*Continued*)
>
> - Six-day canoe and hiking trips, with attention to Adirondack natural history.
>
> **UNIVERSITY SERVICES**
>
> *Tour Guide*: Botanical tours of Stanford campus for organizers of Native American students orientation, 19XX.
> • Emphasis on native uses of plants
>
> *Guest Instructor*: Jasper Ridge Biological Preserve Training Program, 19XX.
> • Interpretation of animal tracks and signs.
>
> *Tour Guide*: Ecology laboratory teaching assistant orientation, 19XX.
> • Led natural history tour of field site.
>
> *Student Advisor*: Integrated Science Program, Northwestern University, 19XX-19XX.
>
> *Academic Committee*: College of Community Studies, Northwestern University, 19XX-19XX.
>
> *President and Member*: Northwestern Students for a Better Environment, 19XX-19XX.
>
> **PUBLICATIONS AND PRESENTATIONS**
>
> Jones, J.T. and T. L. Johnson. 19XX. Scrub Jay predation on starlings and swallows: attack and interspecific defense, *Condor* 90:503-505.
>
> Johnson, T.L. and J.T. Jones. 19XX. Avian predation on *Anolis* lizards in the northeastern Caribbean: an Inter-island contrast, *Ecology* 70:617-628.
>
> Johnson, T.L. and J.T. Jones. Predation across spatial scales in heterogeneous environments, *Theoretical Population in Biology* (in press).
>
> Johnson, T.L. and J.T. Jones. Species interaction in space, symposium paper presented at the 19XX meeting of the Ecological Society of America, Snowbird, UT; to appear in R. Ricklefs and D. Schulter, eds., *Historical and Geographical Determinants of Community Diversity*, University of Chicago Press, Chicago.
>
> Johnson, T.L. Species interactions across spatial scales, presented at the November 19XX meeting on Bridging the Gap Between Theoretical and Empirical Ecology, Broaddus, TX.

In Figure 8-3, Johnson makes evident his prior success in acquiring funding, and is also sure to include his scholastic awards, one of which is a substantial postdoctoral fellowship. His teaching experience is secondary in this case, and so is not given the same emphasis. In addition, he may also compose a statement outlining his research interests, as well as a short research proposal to accompany his CV.

Figure 8-4 Curriculum vitae—teaching emphasis.

TERRANCE L. JOHNSON

Environmental Science Division 207 Edinboro Lane,#C32
Oak Ridge National Laboratory Oak Ridge, TN 37830
Oak Ridge, TN 37831-6056 (615) 483-1234
(615) 574-1234

EDUCATION

Stanford University, Stanford, California, 19XX-19XX
Ph.D. in Biological Sciences, 19XX, Area of Specialization:
Population Biology
M.S. in Biological Sciences, 19XX

Northwestern University, Evanston, Illinois, 19XX-19XX
B.A. in Biological Sciences, concentration in Ecology and
Evolutionary Biology
B.A. in Biochemistry, Molecular Biology and Cell Biology with honors
B.A. in Integrated Science Program, with honors

AWARDS AND HONORS

Hollaender Postdoctoral Fellowship (US D.O.E.), 19XX-present
ARCS Foundation Fellowship, 19XX-19XX
National Science Foundation Graduate Fellowship, 19XX-19XX
Andrew Mellon Foundation Graduate Research Fellowship, 19XX
Phi Beta Kappa, 19XX

TEACHING EXPERIENCE

Instructor: Outdoor Education Program, Stanford University, 19XX-19XX.
- Lectures and weekend outings, emphasis on alpine ecology, animal tracking, and wilderness skills.

Co-Instructor: Biology of Birds, Stanford University, 19XX.
- Lectures and field trips; with Dr. S.T. Phillips.

Teaching Assistant: Systematics and Ecology of Vascular Plants, Stanford University, 19XX.
- Laboratory and field trips.

Teaching Assistant: Core Biology Laboratory, Stanford University, 19XX.
- Ecology laboratory and discussion sections.

Instructor: Chemistry Laboratory, Kendall College, Evanston, IL, 19XX-19XX.
- Sole responsibility for laboratory in biochemistry, general and organic chemistry.

Wilderness Guide: Association of Adirondack Scout Camps, Long Lake, NY, 19XX.
- Six-day canoe and hiking trips, with attention to Adirondack natural history.

Figure 8-4 Curriculum vitae—teaching emphasis (*Continued*)

TEACHING INTERESTS
General ecology, community ecology, ornithology, field biology, theoretical ecology, conservation biology, animal tracking, widerness skills, wilderness policy issues.

UNIVERSITY SERVICES

Tour Guide: Botanical tours of Stanford campus for organizers of Native American students orientation, 19XX.
- Emphasis on native uses of plants

Guest Instructor: Jasper Ridge Biological Preserve Training Program, 19XX.
- Interpretation of animal tracks and signs.

Tour Guide: Ecology laboratory teaching assistant orientation, 19XX.
- Led natural history tour of field site.

Student Advisor: Integrated Science Program, Northwestern University, 19XX-19XX.

Academic Committee: College of Community Studies, Northwestern University, 19XX-19XX.

President and Member: Northwestern Students for a Better Environment, 19XX-19XX.

RESEARCH EXPERIENCE

Postdoctoral Research: Environmental Science Division, Oak Ridge National Laboratory, 19XX-present (research advisor: Dr. Stephen H. Smith).
- Development of quantitative theory of hierarchical structure in ecological systems.
- Analysis of how ecological communities reflect environmental heterogeneity at different scales.
- Numerical study of foraging behavior with short and long range movement in heterogeneous environments.

Doctoral Research: Department of Biological Sciences, Stanford University, 19XX-19XX (research advisor: Dr. James T. Jones).
- Field study of the impact of avian predation on *Anolie* lizards in the eastern Caribbean documents the importance of differences in spatial scale between prey and predators.
- Theoretical analysis of spatial scale and environmental heterogeneity in models of predator-prey communities.
- Analytical and numerical works show how species interactions can sharpen underlying environmental patterns and how heterogeneous environments can stabilize predator and prey populations.

Figure 8-4 (Continued)

Undergraduate Honors Research: Department of Biochemistry, Molecular Biology, and Cell Biology, Northwestern University, 19XX-19XX (research advisor: Dr. Peter T. Williams)
- Investigation of primary events of bacterial photosynthesis.
- Isolation and spectral analysis of photosynthetic reaction centers.

RESEARCH INTERESTS
- Theoretical and field study of ecological communities.
- The roles that spatial patterns and processes play in shaping communities.
- How populations and processes that act on different spatio-temporal scales affect the behavior of ecological systems.
- Influences of disturbance size and frequency on landscape structure.

WILDERNESS TRAINING
Animal tracking, avalanche safety, kayaking, mountaineering, outdoor education, rock climbing, wilderness survival.

WILDERNESS LEADERSHIP EXPERIENCE
Back-country skiing, minimum equipment camping, mountaineering, off-trail navigation, snow shoeing, white water canoeing.

RELATED INTERESTS
Back-country baking, basketry, ethnobotany, flintknapping, prehistoric fire-making, shelter design, weaving.

PUBLICATIONS AND PRESENTATIONS
Jones, J.T. and T.L. Johnson. 19XX. Scrub Jay predation on starlings and swallows: attack and interspecific defense, *Condor* 90:503-505.

Johnson, T.L. and J.T. Jones. 19XX. Avian predation on *Anolis* lizards in the northeastern Caribbean: an Inter-island contrast, *Ecology* 70:617-628.

Johnson, T.L. and J.T. Jones. Pattern and stability in predator-prey communities: how diffusion in spatially variable environments affects the Lotak-Volterra model, *Theoretical Population Biology* (in press).

Johnson, T.L. and J.T. Jones. Predation across spatial scales in heterogeneous environments, *Theoretical Population in Biology* (in press).

Johnson, T.L. and J.T. Jones. Species interaction in space, symposium paper presented at the 19XX meeting of the Ecological Society of America, Snowbird, UT; to appear in R. Ricklefs and D. Schulter, eds., *Historical and Geographical Determinants of Community Diversity*, University of Chicago Press, Chicago.

Figure 8-4 is a variation of Johnson's CV designed to impress those institutions that are more oriented toward experiential or applied education. Highlighted here is the classroom and informal teaching experience he has acquired. Also included is his interest and experience in other forms of teaching, for example, outdoor, or nature education. Because he still wants to be active in research, he also provides his research experience and some of his future interests, even though he knows that at these particular institutions, opportunities in these areas might be more limited.

Letters of Recommendation

Mentors write the right kinds of letters when a student or junior faculty colleague is looking for a job, and they call friends in departments to which he/she has applied in order to get the backchannel networks working for the candidate. A mentor writes enthusiastic letters of support, does not damn the subject with faint praise or "attempt" to give a balanced view by articulating the candidate's shortcomings as well as strengths. There are appropriate places for that, but they are not in letters of recommendation [4].

Letters of recommendation are essential to your application for academic positions. Requests for such letters typically look like the following:

Dear Professor Reis:

The Department of Electrical and Computer Engineering is considering Ms. X for a faculty appointment as Assistant Professor. Would you please give your candid opinion of Ms. X's research? How does she compare with others in her field of roughly the same experience? Does she show promise of continued development and professional growth?

Any information you can give us of Ms. X's teaching ability would be of interest. Perhaps you have heard her address classes or seminars, and can report on her style and effectiveness as a lecturer. Any information you can give about her services to professional societies and journals would also be useful.

Thank you for your assistance in this matter.

Sincerely,

John R. Smith, Chairman

As with your cover letter, your letters of recommendation need to be first rate. Of course, the difference between the former and the latter is that you do not get to write the latter. However, you can, and should, have an influence on what appears in such letters.

You want letters from people who will say great things about you. Three–five such letters is usually the right number. Obviously, your primary academic/dissertation advisor will write one of them. So will your supervisor if you are a postdoc. Someone who addresses your teaching interests and capabilities would be another. If you have had any of the teaching experiences discussed in Chapter 6, then supervisors or employers aware of these experiences become a good source.

If, as part of your research, you have engaged in any multidisciplinary or cross-disciplinary collaborations, getting a letter substantiating such collaborations could be quite valuable. Also consider obtaining a letter from someone in industry if you have had substantial interactions with the person during your graduate or postdoc experience.

The timing of your request for these letters can be important. If you have just accepted a postdoc position, now is the time to get letters into your file from the people you knew as a graduate student. Even if they are not used for a few years, they will capture a crucial period in your education. You want to have such letters put in your file while the memory of your recommenders is fresh. You can always go back to them for updates if appropriate. In some cases, you may want your recommenders to write two letters, one for a future academic position and one for an upcoming postdoc or industry position. If you are completing a postdoc then clearly your current supervisor will write a letter about your research experiences, but he/she could also talk about your academic interests.

A key to getting good letters of recommendation is to know your recommenders well enough over time so that they can say substantial things about you, backed up by first-hand experience and a reasonable amount of detail. All letters will have some things in common since they are all about you. This commonalty is good since it reinforces a particular image. However, each letter should also be unique. Specific aspects of your education, character, and capabilities, as seen from the recommender's perspective, should be included.

At least a few months in advance of the need for such letters, you should sit down and talk with potential recommenders about the kind of job you seek. Discuss the balance between teaching and research, and graduate and undergraduate emphasis. Some graduate students and postdocs are reluctant to do this (it is called the Advisor-Avoidance Dilemma) because they fear that what they want, and what their advisor wants for them, may not be the same. In my experience, these concerns are often unfounded, and the discussions, which are an essential prelude to getting good letters, go much better than expected.

Provide each potential recommender with a brief description of the important aspects of your relationship as they relate to your application for an academic, postdoc or industry position. Do not just supply a copy of your curriculum vitae. Also provide one or two pages, perhaps with the main points in bulleted form, about things not in your curriculum vitae that you wish to have expanded in the recommendation letter.

Strike the right balance. Do not appear to tell your recommenders what to say or how to write a letter. Rather, give them needed background (and reminders) about points they will want to write about anyway. I have found that most faculty appreciate this "assistance" if it is presented in the proper way.

THE APPLICATION PROCESS

Conferences

Conferences, as a place for conducting preliminary, or screening, interviews are common in the social sciences, but they also exist in the physical and life sciences, and to a lesser extent in engineering. The annual meetings of the American Mathematical Society, the American Society of Mechanical Engineers, and the American Association of Engineering Education are examples of conferences where such interviews take place. Of course, much can also take place without a formal interview. As noted in Chapter 5, you are going to be scrutinized to a certain degree as soon as you start attending such conferences.

Ann MacLachlan, former academic placement advisor in the Educational Career Services of the Career Planning and Placement Center at the University of California, Berkeley, has written extensively about how to get the most from such conferences. In preparing for a conference, she suggests the following [5, p.2]:

- Write out and rehearse a short statement on your dissertation of about five minutes, no longer.
- Develop a list of courses you have taught, and bring your teaching portfolio with you.
- At a minimum, look over the catalog of the interviewing college or university.

The five-minute talk is something you should be able to give on the spot without notes and overheads. It should also be comprehensible to those not in your specialty area. You will be called upon to give this kind of talk dozens of times over the next year or so. We will discuss it further in the next section.

You also need to be prepared to talk about courses you have taught or would like to teach, as well as to say something about your teaching philosophy. Looking over the school's catalog and World Wide Web displays will help with these questions, and will also give you some additional topics to discuss.

According to MacLachlan [5, p.3]:

Being well organized and prepared reduces anxiety considerably, but attitude also makes a substantial contribution. You should be aware that when you go into an interview that you are making an appearance as a potential colleague. Be agreeable but not servile or overly diffident. Do not be arrogant. Keep your radar finely tuned to the kinds of questions you are being asked. Remember that the purpose of the interview is to get acquainted with you. Avoid speaking at length about unimportant things. Be specific in your answer, but remember that this is not an oral examination. Remember too, that the people interviewing you may be equally fatigued, so respond to them as the individuals that they are and not as "the enemy." Normal courteous expressions of sympathy or convention can break the ice.

Also keep in mind that a certain amount of appropriate humor can go a long way toward setting the right tone.

At the conference, be sure to keep in touch with your faculty and other graduate students and postdocs. Keep them informed about your interviews, exchange information, and do not hesitate to ask your faculty to "talk you up" with colleagues with whom you think you had a good interview.

Finally, be sure to get the names of everyone with whom you interviewed, and follow up quickly with thank you letters. Follow-up is important even if you do not think the interview went well, or if you are not interested in a particular school. Remember, all of these people may be your future colleagues.

The Campus Visit

There is a tendency to be awestruck, to feel as if you're being summoned by God when Princeton calls. Instead, you need to have some self-respect and not act like an over affectionate puppy. This is not a time to lose your professionalism [5, p.4].

Okay, but admittedly, you cannot help but be excited. All the work you have done so far pays off, and you receive a call from a member of the search committee asking if you would be available to visit their campus for an interview. Naturally, you are delighted, and more than a little anxious. Until now, you did not have to tell anyone which schools you had applied to, or how many rejection letters you may have received. Now, everyone will know that you are going for an interview, and ultimately whether or not you get an offer. Adding to your anxiety is the knowledge that there will probably be from three to six other finalists who the school also thinks are very strong, and on whom it is also willing to spend precious time and resources to interview.

There are a few things to keep in mind about your competition. First, they are as nervous and anxious as you. Second, like you, they will probably get asked to more than one interview. Even if one of them is offered the position you are applying for, they may not accept it, and you may end up being a very desirable second choice.

Yet, there is no question that for both you and the school, the campus visit is critical. The search committee has spent a lot of valuable faculty time getting to this stage, and they want it to pay off. So do you. For this reason, you need to prepare carefully. You have already begun this task by writing a cover letter and curriculum vitae that helped get you the interview. Now you need to take additional steps.

If you have not already done so, check out the latest information about the institution and the department on the World Wide Web. This material is likely to be far more current than an annual catalog. Most college and university Web pages have detailed information on faculty in departments of interest, which you will want to print out and review before your arrival. However, do get a copy of the institution's catalog. (You can call the college and ask them to send you a copy if you cannot check one out of your library.) Look over the courses offered in the department, noting the ones you might like to teach. As discussed in Chapter 6, more and more schools are asking candidates to talk about their teaching philosophy and teaching experience. When the department presents your case to the dean, it has to show that you can do the necessary teaching. You are likely to be asked what courses you can teach, and rather than responding with "geomorphology" or "structural mechanics," it will be more impressive if you can point to particular undergraduate and graduate possibilities you have checked off in their catalog. Also, be prepared to briefly describe a course that you would like to develop and teach from scratch. It would be helpful if you had some idea of the texts you would use and the general nature of the topics you would cover.

Be sure that you know what type of institution you are visiting, and see if you can learn something about the composition of the student body. Find out if there are any faculty and graduate students in your current institution who went to the school or who might know someone there. They can be a gold mine of information and suggestions.

There are, according to Johansen-Trottier, other, somewhat more subtle things you should look for. They include [6]:

- Do you feel comfortable with the faculty in the department and the overall feeling at the university regarding your race, gender, sexual preference, religion, and culture?
- If you have children, what will the schools and neighborhoods be like for them? What are the employment opportunities for your spouse at the university or in the community?
- Will you have time to drive around the community in order to get a sense of the quality of life in the academic and surrounding communities?

On your trip, be sure to bring extra copies of your vita, a copy of your dissertation and statement of your research plans, and other relevant material. Whatever you do, *do not* check anything important through to baggage claim. Dress like faculty in your department dress when they are meeting with important people (slacks and sport coat or suit, tailored dress or skirt and jacket).

Find out as much as you can about your schedule in advance. With whom will you be meeting and when? When is your job talk scheduled? Will you have a half hour or so before the talk to prepare and relax? Also, try to find out the backgrounds and dossiers of the people you will meet. Here, again, the best source is their Home Pages on the World Wide Web.

Your academic job talk is usually on your research interests. Make sure this is the case. One candidate I know thought he was going to give a high-powered technical talk, only to find out that he also was expected to give a lecture in an undergraduate course.

If you are not scheduled to meet specifically with students, ask if such a meeting can be arranged. It is something you should want to do, and even if it was not scheduled, the committee will be pleased that you asked.

It will be a good idea if you can remember the names of faculty and other important people you meet as you go along. It will be particularly impressive if you can refer to them by name during the question-and-answer period after your job talk. Doing so is not that difficult if you consciously make the effort.

As noted earlier, being able to talk about your research for about five minutes without notes or other aids is very important. Organization professors John Darley and Mark Zanna call this the "five minute drill." Although specifically referring to the social sciences, their advice applies equally to science and engineering candidates. They say [7]:

> It is useful to be able to give one other sort of presentation. We have labeled it the "five minute drill." Perhaps one faculty member missed your job talk; perhaps for another you want to describe a line (or future line) of research you didn't cover in your colloquium. We suggest that you be prepared to relate the theoretical context of your research, the specific hypothesis you are testing, the general procedures you are using to test them, and the outlines of the results you are getting (or would hope to get) all in five minutes! Your major task is to convey the importance and excitement of the research succinctly so that you can then discuss your work with the person rather than lecture him or her during your time together.

> To convey what you are up to without going into excessive detail is a surprisingly difficult task, and, at first, requires considerable thought and discipline. We suggest that you explicitly think through what you would say and practice it. When practicing, keep in mind that you may be relating your research to a colleague in another area who may need to know a bit more about some aspect (e.g. methods) of the research. Our advice is to think through, in advance, modifications of your presentation as a function of a variety of possible audiences.

Mary M. Heiberger and Julia M. Vick of the University of Pennsylvania have prepared a list of over 30 questions often asked of interviewing candidates. You will almost certainly not be asked all of them, but here are some key ones to keep in mind [8]:

Research

- Describe your dissertation research. How is this a significant contribution to the field?
- Why did you choose your topic?
- If you were to begin it again, are there any changes you would make in your dissertation?
- What are your plans for applying for external funding?

Teaching

- What classes are you now teaching or have taught recently?
- How would you structure teaching a class in your first semester?
- How would you encourage students to major in our field?

Participation in School Activities

- How, and what, can you contribute to our faculty?
- How much are you willing to participate on university committees and extracurricular activities?
- In what ways and in what areas do you see yourself making professional contributions in the next five years?

Self-Image/Career Choices

- Tell us about yourself (memorable but concise, a one- or two-minute summary of background).
- What are your greatest strengths and weaknesses (only a "weakness" that is honest, but would not be something that would be a major negative for performing the job)?
- How do you spend your leisure time?

Of course, you need to be prepared with plenty of questions of your own. Authors Timothy J. Green, Marilyn S. Jones, John G. Casali, and Nancy E. Van Kuren have prepared a list of 120 questions that candidates for engineering faculty positions should ask about the institution they are considering [9]. Most of these questions apply just as well to science positions. Their questions fall into the following ten categories:

- Demographics
- Faculty

- Faculty duties
- Facilities
- Undergraduate program
- Graduate program
- Research
- Promotion and tenure
- Professional development
- Benefits and contractual issues

A list of all 120 questions appears in Appendix D, "Questions to Ask Before Accepting a Faculty Position." You are likely to ask only a fraction of these questions during your visit, and only a fraction more in follow-up conversations. Still, the list is very useful to have, not only as a way of evaluating a school that makes you an offer, but also as a tool to help you determine which schools you want to apply to in the first place.

An interesting question sometimes arises as a result of your campus visit. Should you take advantage of the free travel and the fact that you are in the area to request a visit to another school you are interested in, but which has not (yet) asked you to come for an interview? If you are close to making their short list, or if the school has not yet made up their list, they will sometimes take the time to interview you, particularly if they do not have to pay the extra travel expenses. How much, if anything, to tell the school that is paying for your travel is a matter for you to decide. One approach would be to let them know the situation, and offer to pay part of the airfare yourself, or ask the second school to split the travel costs. I have seen enough cases where this has resulted in an offer from the second school that I would advise you to consider making the effort.

The Academic Job Talk

Let us now take a look at what will probably be the most important talk you have ever given. No, I am not referring to your Ph.D. orals. Your orals were a piece of cake compared to what is called the academic job talk. During your orals, you gave a very technical talk to a generally supportive audience who knew you and who had a strong interest in your doing well. Most likely, your orals were on a date determined by you after considerable preparation and practice. On the outside chance that you were to fail your exam, you would be given an opportunity to take it again. None of the above is likely to be the case with your academic job talk. An excellent talk could get you the job, while a poor talk will almost surely eliminate you from contention.

You have probably given a number of presentations on your research to department seminars, professional meetings, and to industry and government representatives. These talks showcased your technical knowledge and expertise;

that was their primary function. While it is always desirable to pay attention to the broader context into which your research fits, doing so was secondary to demonstrating your technical prowess. If you were applying for a postdoc position, then your technical strengths were a must, and a presentation was one way to demonstrate them. Good communication skills were a nice bonus, but not at the expense of your technical expertise. After all, your postdoc employer viewed your tenure as temporary. He/she wanted to get what they could from you now, and a strong technical knowledge is the essential prerequisite.

Things are different for an academic position. Your ability to communicate well, to show enthusiasm, to place your work in a broader context, and to make good use of various media all correlate highly with good teaching. Your interviewers are going to want to see evidence of these capabilities, as well as your technical expertise. Most faculty in most schools understand that they, and you, are making a decision that could keep all of you together for 30 or more years. Even if a second decision is made six or seven years down the road at tenure time, the cost of making a mistake now is very high. According to Professor Paul Savory [10]:

> Fear of making the wrong decision is their (search committee's) greatest fear. Unlike private industry, once a person is hired, he is there until at least tenure time (five or six years). Many departments would rather not hire anyone than hire someone they are uncomfortable with.

For most department chairs and deans, making the correct hiring decisions is their most important task. It is their legacy. They will be looking for supporting evidence from you for their decision, which must be made in a relatively short period of time. They want to know what kind of researcher, thinker, and teacher you are going to be. They will be trying to analyze your ideas and the contributions you will make over many years. Thus, it is important to place your current work in a broader context, and to tie your interests to the overall issues and problems in the field.

Your presentation needs to be customized to your audience. Do not use overheads showing the name of a recent conference in which you gave a similar talk. Make sure that your cover page contains the name of the institution you are visiting, along with the date of your visit. Consider preparing color transparencies. They are more work, and cost a bit more, but they make a much better impression. Remember, this talk is also one way for you to indicate how you will approach your teaching.

While most of your audience will consist of faculty and students from your host department, do not assume that all of them share your particular technical background. It is also likely that faculty from related departments may be invited to your talk. It is essential that you find out who your target audience will be.

Early in your graduate career, you should start to attend academic job talks in your own and related departments given by candidates for positions at your school. Attending such talks will give you an early look at this important process, and make you less anxious when it comes to your turn to do the same thing.

Michele Marincovich, director of the Stanford University Center for Teaching and Learning, has counseled hundreds of students and postdocs about their academic job talks. She offers the following advice [11]:

General Tips

- Make sure that your talk has a broader context, so that the importance and implications of your work are clear, not merely implied.
- If, when you write your talk, you focus on what you want people to be thinking about as they leave your talk, it will help you concentrate on the essentials.
- Don't wait to prepare your job talk until the last minute—it is more than just a "brain dump" of your dissertation. It's very important to be able to go beyond your dissertation.
- Be prepared enough to allow yourself to be spontaneous; preparation will also help you handle the unexpected.
- Make your talk interesting with good examples, relevant anecdotes, and significant details.
- If speaking to a mixed audience, avoid highly technical or specialized terms.
- Academia is changing and now includes previously underrepresented groups. Use inclusive language—she as well as he, for example—and language that is respectful of all groups.
- The biggest correlates of effective teaching are enthusiasm, organization, and the ability to engage your students.
- Using humor in your job talk can be risky, but if it comes naturally to you, use it. But you don't have to, so don't fake it.
- There will usually be a "Question and Answer" period. There is no way to predict all the questions you might be asked, but you can practice by having friends listen to your talk and and then ask you the hardest questions they can think of.
- Being a good public speaker helps—a well delivered talk will carry your message more effectively.

Practicing and Nervousness

- Practice/do your talk in front of friends who can give feedback.
- Try to view any nervousness in a positive way, as energy or dynamism.

- Few speakers reach everybody all the time—don't focus on unresponsive audience members.
- Some audiences (especially in science and engineering fields) will be serious and unresponsive on purpose to make it more challenging or simply because they're concentrating on the presentation and critiquing it.
- Stay in touch with your audience, but don't try to decide the success or results of the talk during the talk.

If you follow these guidelines and practice several times before your visit, you should do quite well. Indeed, you will probably also find it an enjoyable experience. At the same time, you are certainly going to be glad when your visit is over. There is, however, one important follow-up action for you to take. Write a thank you letter to your host saying how much you enjoyed the visit, mentioning by name the specific people you spent a reasonable time with, and making reference to any matters or observations that you found particularly worthwhile. Conclude by indicating your understanding of the next steps, and be sure to enclose any additional materials you promised to provide. Writing such letters may sound obvious, but you would be surprised how many applicants do not do it—and, therefore, what a difference it makes when you do.

POSITIONS OUTSIDE ACADEMIA

As noted in Chapter 7, you may want to follow through with your Multiple-Option approach by also applying for positions in government or industry. This suggestion is certainly not an exhortation to leave the academic path, but rather a way to expand your options. Knowing that you have choices increases your self-confidence and also strengthens your position in the academic arena. Although the government and industry application process has some elements in common with the academic process outlined above, there are also some important differences.

In her article, "Secrets from the Other Side," Constance Holden, a reporter at *Science* magazine, discusses the industry application process with Denise H. Guthrie, Ph.D., recruiting and placement manager for Dow U.S.A. in Midland, MI [1] [12].

According to Guthrie, the vast majority of Ph.D.s "simply don't know how to prepare for an interview in industry." They just assume that they can answer

---

[1] Adapted with permission from C. Holden, "Secrets from the other side," *Science*, vol. 257, no. 5077, p. 1713, Sept. 18, 1993. Copyright © 1993 by the American Association for the Advancement of Science.

questions as they come, thinking off the top of their head." This assumption, Holden notes, can be dangerous.

With industry positions, there are typically two stages: the on-campus interview and the on-site interview. According to Guthrie, industrial recruiters want to know the answers to four questions:

1. What job does the candidate want?
2. Can the person do the job?
3. Will the person do the job?
4. Will the person be compatible with the existing team?

The answer to the first question is not, as Guthrie notes, "I will do anything." Such a response will be interpreted as either desperation or a lack of thought and preparation. The key is to list the jobs of interest to you in order of priority which, of course, means that you need to know something about the company. As with academic positions, there is simply no substitute for preparation.

In answering the second question, Guthrie advises you to give a concise description of your research by stating: (a) why you did it, (b) how you did it, and (c) the results.

"Will the person do the job?" means demonstrating to the interviewer that you possess leadership and initiative. Search your background to find examples that illustrate this characteristic.

In answering the last question, you need to show you can be a team player, and that you can get along with people from a variety of backgrounds. As noted in Chapter 5, industry has moved away from stand-alone research and toward more team projects, and they now screen and hire candidates who have the ability to lead collaborations and teams in effective ways.

The on-site interview also has some elements in common with the campus visit. The main difference, of course, is that you will be talking to industrial scientists or engineers, and not faculty and students. Your academic job talk becomes your industry seminar, and as with the former, it is all important. As Guthrie puts it, "Based on your seminar you will either be working uphill or downhill the rest of the day." He goes on to say that making a good impression is simple: "Just be exceedingly well organized, concise, clear, confident, professional and enthusiastic as you introduce yourself and explain your research."

Guthrie, according to Holden, also recommends that you leave some minor but semi-obvious question unanswered during the presentation, which will provide a pump primer for later discussion.

As with academic positions, applying for positions in industry takes time and effort. But it may well be worth it. Even if you do not accept an industry position, you will have developed important contacts that will be helpful to you in your new position as a professor.

| Vignette #8 | Diversity Issues in the Hiring of Science and Engineering Faculty: An Illustration from Astronomy |

As competition for academic positions increases, issues having to do with affirmative action and diversity are brought to the fore. In the following vignette, we look at one aspect of this issue as seen from an astronomy perspective.

Geoffrey C. Clayton
University of Colorado, Boulder

Thousands of mostly white male scientists and engineers are currently competing for a far smaller number of permanent faculty positions. What impact is this competition having on the goals of affirmative action, and the quest for diversity among faculty on college and university campuses?

Writing in *Mercury: The Journal of the Astronomical Society of the Pacific,* Geoffrey C. Clayton in his article, "Astronomy in the 90's: Angry, White, and Male?," looks at the situation from an astronomy perspective [13, p.33]. George Musser, *Mercury's* editor, minces no words in introducing Clayton's article by noting, "When jobs are tough to come by, the perceived unfairness of affirmative action sticks in people's craws. In actuality, affirmative action barely hurts the majority, and greatly benefits the minority" [13, p.33]. The article is indicative of where things stand in most of science and engineering.

Clayton is a research associate at the Center for Astrophysics and Space Astronomy at the University of Colorado, Boulder. His research interests are interstellar dust and R Coronae Borealis stars. He was born in Toronto, Ont., and received his bachelor's and Ph.D. from the University of Toronto. Clayton was also a member of the Committee on the Status of Women of the American Astronomical Society. With respect to gender, he notes [13, p.33]:

> When the subject of jobs arises in conversation, as it often does among those of us without secure positions, someone invariably makes a comment about how some woman has been hired for such-and-such position. The clear implication is that women have been hired on the basis of their gender alone and that better qualified men were passed over. These feelings are not backed up by the statistics, which show that women are hired more or less in proportion to their numbers in Ph.D. programs.

Over the years, according to Clayton, gender issues have gained higher visibility in astronomy. He observes [13, pp.33–34]:

> The presence of even one female professor in a department can have a profound effect. Men become used to working with a woman as a professional equal;

students get used to having a professor who is not a man. In astronomy, as in the rest of the country, once people of different genders and races get used to working together, diversity becomes normal.

Although women still make up only about 12% of the total astronomy population, Clayton notes that the number is high enough to form an identifiable group and an active lobby.

The situation is not nearly so encouraging with respect to minorities. According to Clayton, "We are at the point in the awareness of minority issues that we were with women's issues 10 or 20 years ago. It's not something that astronomers give much thought to. But if you look around any American Astronomical Society meeting, the maleness of the crowd is topped by its overwhelming whiteness" [13, p.33].

How can we begin to achieve real diversity when so few women and minorities decide on science and engineering as a career? Short-term attempts at choosing the best qualified candidates will not do the trick when the numbers are so small.

In the long run, Clayton believes [13, p.34]:

If we could increase the diversity of people entering astronomy, then we could make the profession more diverse. But we are left with a chicken-and-egg situation: It's difficult to attract more women and minorities into astronomy when so few are already in the field. For this reason, I still favor promoting diversity in an active way. For all the faults of affirmative action—not to mention the present job market, which makes those who favor affirmative action reluctant to attract *anyone* into astronomy—it is the only game in town.

SUMMARY

This chapter provided a detailed examination of the job search process. It began by looking at how new positions are established, what departments look for in new faculty, how to find out what is available, and the time frame for academic openings in your field. It then discussed the preparation of your application materials including cover letters, curriculum vitae, and letters of recommendation. Conferences, campus visits, and the all-important academic job talk were looked at next. Jobs outside academia and how to accept one of them while keeping your options open for a future academic position were then discussed. The chapter concluded with a vignette examining the impact of increasing job competition on the goals of affirmative action, and the quest for diversity among faculty on college and university campuses.

In Chapter 7, we looked at how to identify the academic possibilities of greatest interest to you. This chapter was about applying for specific academic positions. Chapter 9 is about getting the results you want. In it, we first look at what happens if you receive an offer, or offers, and how then to negotiate arrangements that help ensure the best possible start as a professor. As noted earlier, not everyone will receive a desired academic job offer the first time around. What to do if this is the case is discussed in some detail, with the goal of better positioning you for another look at academic positions.

REFERENCES

[1] R. Finn, "Career-building on the Internet: Hunting for jobs electronically," *The Scientist*, vol. 9, no. 2, Jan. 23, 1995.

[2] Adapted with permission from K. Johansen-Trottier, *The Academic Job Search: A Practical Overview*. Stanford, CA: Career Planning and Placement Center, Stanford University, 1991, p. i.

[3] Adapted with permission from K. Johansen-Trottier, *Curriculum Vitae and Cover Letters*. Stanford, CA: Career Planning and Placement Center, Stanford University, 1991.

[4] I. C. Peden, "Senior friends and cheer leaders," *IEEE Education Society–ASEE Electrical Engineering Division, Newsletter*, p. 9, Apr. 1995.

[5] A. J. MacLachlan, with C. Porter and J. Kettner, "How to get the most out of your professional organizations meetings," Career Planning and Placement Center, University of California, Berkeley, Dec. 1987.

[6] Adapted from K. Johansen-Trottier, *The Academic Job Search: A Practical Overview*. Stanford, CA: Career Planning and Placement Center, Stanford University, 1989, pp. 7–8.

[7] J. M. Darley and M. P. Zanna, "The hiring process in academia," in *The Compleat Academic: A Practical Guide for the Beginning Social Scientist*. New York: Random House, 1987, ch. 1, p. 3.

[8] M.M. Heiberger and J.M. Vick, *The Academic Job Search Handbook*. Philadelphia, PA: University of Pennsylvania Press, 1992, ch. 13, pp. 113–115.

[9] T. Greene, M. Johnes, J. Casali, and N. van Kuren, "Is this an offer you can't refuse?: Criteria for selecting an engineering faculty position," *ASEE Prism*, vol. 4, no. 1, pp. 30–33, Sept. 1994.

[10] P. Savory, "Making the move from student to teacher: Steps in the faculty search process," *Industrial Engineering*, vol. 10, p. 60, Oct. 1994.

[11] M. Marincovich, presentation at Proseminar in Manufacturing Education, Stanford University, Stanford, CA, Feb. 1, 1995. Adapted with permission.

[12] C. Holden, "Secrets from the other side," *Science*, vol. 257, no. 5077, 1713, Sept. 18, 1993.

[13] G. C. Clayton, "Astronomy in the '90s: Angry, white, and male?" *Mercury: The Journal of the Astronomical Society of the Pacific*, vol. 24, no. 3, May/June 1995.

CHAPTER 9

Getting the Results You Want

> Remember, what you are negotiating is the start-up package, not just the start-up salary. How you start out can make all the difference in how well you do, so be sure you get the resources you need at the beginning.
>
> Susan Montgomery, Assistant Professor, Chemical Engineering
> University of Michigan

If you have received one or more job offers that are of interest to you, congratulations; your hard work is paying off! However, your work is not over. While you are feeling the understandable euphoria of a hard-won job offer, you need to stay focused on your ultimate goal: getting the resources you need to enhance your chances of professional and personal success. Getting these resources usually takes some negotiating.

Chapter 9 begins by discussing the approach you need to take to such negotiations. It examines, in some detail, 11 key principles for responding to academic job offers. The special problems faced by dual-career couples in which one, or both, members are seeking faculty positions are examined next, and also in the vignette at the end of the chapter.

In spite of all your hard work and often for reasons beyond your control, you may not receive an offer, or may receive one that is ultimately unacceptable. What to do under such circumstances is looked at in the section, "What to Do if You Do Not Get the Offer You Want." Through the application of the Multiple-Option approach, you will have identified other possibilities such as part-time appointments or positions in government and industry that will enable you to keep open the possibility of a regular faculty appointment in the future.

YOUR NEGOTIATING APPROACH

There is something important you must do before you begin your negotiations. You must go from seeing yourself as an applicant who is *seeking* a job offer to seeing yourself as an applicant who has *received* a job offer. Most job seekers never make this critical shift, and it can cost them dearly. When you were applying for academic positions, your power relative to the search committee, the department chair, and the dean was quite low. It was the committee who could select or reject you. Now it is you who can also select or reject the committee and the administrators to whom they make recommendations. It is the committee who has also put in significant time and energy to get to this stage, and who has a vested interest in bringing its efforts to a successful conclusion. The reputation and judgment of its members are also on the line. This relative shift in power is shown in Figure 9-1.

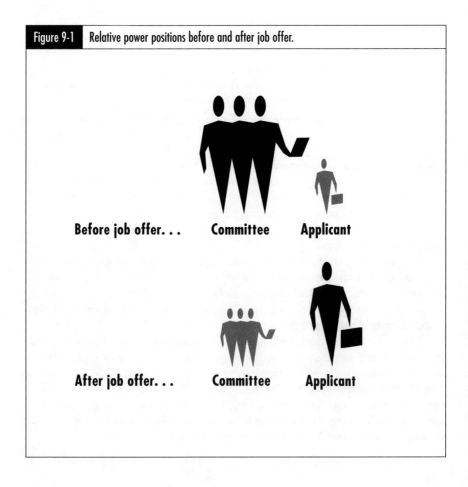

Figure 9-1 Relative power positions before and after job offer.

Before job offer...   Committee   Applicant

After job offer...   Committee   Applicant

Do not misunderstand; this negotiation period is not a time for arrogance. Do not overestimate your power. Yet, in the excitement of having a job offer, you must also think carefully about what it will take to ensure your chances of success in what could be a lifelong employment decision.

Negotiating an academic job offer is not like negotiating the purchase of a house or car, where you are looking for the most for the least up front, and where you do not expect to have an ongoing relationship with the seller. Nor is it like a contract between two parties who will remain in their separate domain (supplier and buyer, union and management), even when there is a continuing relationship.

It is about becoming part of an organization and a group of people with whom you will have a significant relationship. You want to get the things you need to increase your chances of success, while remembering that you are going to work with these people for years to come.

Certain agreements need to be in writing for everyone's protection. Writing also clarifies understandings, and keeps the record straight. On the other hand, you cannot expect to get all contingencies, such as the specific courses you will teach and a long-term guarantee of summer support, locked in ahead of time. You have to decide what you can take as statements of good faith from your future colleagues and what you must have firmly spelled out in writing.

The key question is: What resources do I need to be successful? The department wants to know this as well; after all, consider their investment. Funds to help support research assistants, and specific laboratory equipment, will probably make sense. A reduced teaching commitment in the beginning might also make sense, but may not be as easy to negotiate. At the same time, a corner office, an excessive starting salary, and travel perks will almost certainly not be greeted with enthusiasm.

By all means, do not ignore what it takes to make you and your family happy. But remember the difference between needs and wants. The department will try to meet all your *needs,* but not necessarily all your wants. You may need a computer, but want an office with a view. If you get the office you want, be sure it is not at the expense of the computer you need.

Also, your idea and the department's idea of what you need may differ simply out of misunderstanding or ignorance. You are in the best position to know your needs, but in trying to meet them, be open to approaches that you might not have thought about. The department may not have the computer you need, but may be willing to work with you on obtaining it as a gift from a local technology company.

The key is to try to make it a win–win for both you and the department. Shon Pulley, assistant professor of chemistry at the University of Missouri—Columbia, points to this example: "In explaining to the department why I needed a particular piece of equipment, I also pointed out how other faculty could benefit from its use, and how its acquisition would count as a matching contribution toward additional grant support. "

GENERAL PRINCIPLES FOR RESPONDING TO ACADEMIC JOB OFFERS

Martin Ford, associate dean of the Graduate School of Education at George Mason University in Fairfax, VA, has formulated 11 general principles for responding to academic job offers. They are [1]:

1. Make sure you have an offer.
2. Know what you want—and what you do not want.
3. Clearly communicate what you want—but only to the right people.
4. Always try to use your work quality or productivity as the rationale in your negotiations—align your goals with those of your employer.
5. Make requests in an informational manner rather than in a controlling manner.
6. Negotiate hard on things that are "out of bounds"; negotiate more gently on things that are "in bounds."
7. Learn about the tenure process, but do not get hung up on it.
8. Start as high as you can in institutional prestige.
9. Get as high a starting salary as you can, but be realistic.
10. Create options, and keep as many open for as long as you can.
11. In making a decision, combine logic and emotion.

Ford's advice applies to a wide range of disciplines, not just education. Let us take a closer look at his principles as they apply to negotiating academic positions in science and engineering.

1. Make Sure You Have an Offer

As Ford puts it [1, p.1]:

> If it's not from someone authorized to make an offer (e.g., a dean or department head), it's not an offer. If it's not in writing, it is not an offer. Therefore, the appropriate response to an oral "offer" of a job, salary or fringe benefit (e.g., moving expenses, research space, etc.) is to "put it in writing."

Ford notes that getting such a statement may require you to put the offer into writing, and then ask for written confirmation that what you have written is indeed what is being offered. This advice may sound obvious, but it is sometimes forgotten, usually at the applicant's peril. I recall the following situation:

A geology professor at a midwestern university who wanted to change schools applied for a position on the West Coast. His campus visit and academic job talk went very well in his opinion, and he was quite encouraged. One person on the search committee even called him aside during his visit to say that he was particularly supportive and enthusiastic. Three days after his interview, he received a call from the housing office at the school, asking if he wanted any help in finding living accommodations. Even though he had not heard officially from the search committee or dean, he began to let the word out to his colleagues that he expected an offer, and if it came, he would take it. A couple of weeks went by without a word. Finally, he called the chairman of the search committee, and was told that the position had gone to another candidate from, of all places, the applicant's own department! Whether the school was just keeping him on the hook while it pursued his colleague, or whether there was some genuine confusion (remember the housing office), does not matter. He had spoken too soon, and it cost him dearly with his department colleagues.

2. Know What You Want—And What You Do Not Want

Here, Ford is referring to the kind of exploring we discussed in Chapter 7. Specifically, he suggests the following [1, p.1]:

> Find out as much as you can about what academic jobs are like—salary, working conditions, work activities, work expectations, and lifestyle considerations. Also find out as much as you can about alternative jobs you may consider. Use this information to determine the boundary conditions of what is possible on these dimensions.

In so doing, give some thought to how you will spend your time as a professor, as well as your life outside the institution. Think both in terms of personal goals such as the need to be challenged, freedom of choice, desire to contribute to society, as well as more specific job characteristics such as teaching assignments and summer support. Also, consider what would make the offer, as Ford puts it, "fatally flawed." It may be that location, salary, or research support are simply below your minimum expectations. Use the above information to negotiate the best offer you can consistent with your in-bound needs.

A colleague of mine was offered a nontenure track teaching professorship in the Industrial Engineering and Engineering Management Department at Stanford University. This type of position had some particular advantages for him, such as little research or publication pressure, but it also provided little long-term security. Nevertheless, for my colleague, physical proximity to high-technology, telecommunications, and banking companies as well as some of the "Big Eight" consulting firms was essential because of the consulting opportunities and possible future job prospects they might provide. Stanford was one of the few schools in the country that provided this environment.

3. Clearly Communicate What You Want—But Only to the Right People

This point is important. You do not want people to think you have poor judgment. You also do not want to share confidential information with your future colleagues who might be in a position to resent what you may have achieved. According to Ford [1, p.1]:

> Discussions with potential colleagues and students should be focused primarily on intellectual concerns. Do not discuss salary, or fringe benefits, unless you are talking to the person who will be making the offer (e.g., the dean or department head). One possible exception—often it is appropriate to communicate some of your non-monetary objectives and concerns to your "host" (typically a member of the Search Committee), especially if they involve getting your work done (e.g., space, equipment, research and teaching assistants—but not salary, moving expenses, or housing assistance).

The negotiating process itself can take many forms. A chemistry professor had this experience:

> When I had my final exit interview with the chairman of the department and the chair of the search committee, they asked me what I thought my start-up expenses would be. I had prepared a list prior to my interview trips, and presented it to them. As a result, they already had a pretty good idea what I wanted for a start-up package. When my initial offer arrived, they offered what I had on my list plus some additional equipment. This initial offer made my negotiation very easy. The final offer was amended slightly to include more money for equipment I felt the department should have already but did not, and an increase in the moving expenses. All negotiations were over the phone with the department chairman, and we used a Fax machine to move the counteroffers back and forth. When everything was settled, they Faxed a copy of the offer and sent the original by FedEx for me to sign. I really think my negotiations went very smoothly, mainly because of the initial list I put together. This approach allowed me to avoided haggling over a lot of little things.

Here is the experience of a physics professor:

> All my negotiations took place before the offer letter was actually sent. The dean had insisted on getting all the details about salary and start-up funds straightened out before she would sign off on the offer letter. The process took place almost exclusively by e-mail, between me and the chairman of the department. Although I visited him in June to verify some of the things we had discussed electronically, the letter had already been drafted by then.

It is important that you be forthright in your approach, and not worry that the school is going to withdraw its offer because of it. A professor at a large midwestern research university always assumed that she would work at a community college after graduation, and so when the university made her an offer, she was overwhelmed. As a consequence, she did not ask for much. "I just asked for minimum student support, but later I was told I should have asked for full support for two graduate students for five years," she remarked.

What you ask for is not as important as the *way* you ask for it. Thus, principle #4.

4. Always Try to Use Work Quality or Productivity as the Rationale in Your Negotiations—Align Your Goals with Those of Your Employer

Ford points out that [1, p.2]:

> Employers will respect you even if your requests seem excessive if the underlying goal is to do a better job (e.g., seed grants, RA, computer, and a more manageable initial teaching commitment could significantly enhance productivity; a higher salary, moving expenses, or housing assistance could enable you to focus on your job rather than seek extraneous summer or consulting income).

Let us take your initial teaching assignment as an example. As we will see in Chapter 11, "Insights on Teaching and Learning," getting started on the right foot with respect to teaching is important, not only because you will be doing it on a regular basis, but because if you start out badly, it is hard to recoup and turn things around later on. Thus, you want to be sure to have a manageable assignment in your first year, one that biases you for success in teaching.

Most departments will agree in principle with this goal. A few may even have policies that help you achieve it in one way or another. In most cases, however, you will have to negotiate such arrangements. For example, what if the department says it is sympathetic to your request for a reduced teaching commitment in your first year, but unable to honor it? They say you will be replacing someone who taught a regular number of classes each semester, and so that is what you are going to have to do.

Okay, but do not give up too easily. You might respond by asking if you can borrow ahead with another faculty member who would take one of your classes the first semester in exchange for your teaching one of his/her classes later on.

Other ways of simplifying your initial teaching assignment might include:

- Teaching a course previously taught by someone who is willing to loan you copies of their lecture notes, exams, and homework assignments.
- Reducing the number of different courses you will teach, and thus the number of different preparations you will have.
- Team teaching a section of a class in which you can share the lecture notes, exams, and homework assignments with other professors.

- Teaching a class with fewer students. There is a big difference between a class with 30 students and the same class with 60 students.
- Receiving additional TA help, particularly in laboratory classes.
- Teaching two classes back to back and/or have days without any classes so you can block off time to do other work.

Sometimes, you can negotiate tradeoffs that get you what you need, while at the same time making a very positive impression. David Kazmer is an assistant professor in the Mechanical Engineering Department at the University of Massachusetts. In his first round of negotiations, the department offered to cover his salary for two summers so he could continue to do research. Kazmer countered with a request for salary support for the first summer, but in exchange for the second summer, asked for summer support for two graduate students. He was confident that he could generate his support on his own, and consequently the department was delighted to make the requested trade-off.

Examples of offer letters from various types of institutions appear in Appendix E, "Sample Offer Letters."

5. Make Requests in an Informational Manner Rather than Controlling Manner

Notes Ford [1, p.2]:

> Psychological research clearly indicates that people are much more likely to respond positively to feedback (such as a response to a job offer) if they perceive it to be an honest attempt to inform rather than a manipulative attempt to control behavior or to gain personal resources. This principle is especially applicable to situations involving the negotiation of multiple offers.

Suppose you are not offered the graduate student support you requested. Rather than saying that you must have the support to do your research, and you cannot see any way of accepting the position unless you get it, think about what is behind your request. Try asking a few questions of the department chairman. These might include:

- Is my request out of line or is it a matter of available resources?
- What ways have faculty in my situation gone about obtaining help with their research?
- Are there other ways that I can obtain help?
- Is there a way that I can apply, perhaps with the department's help, for additional resources?

The department should want you to succeed, and by asking questions, you can determine what other approaches might help you do this.

6. Negotiate Hard on Things that are "Out of Bounds," Negotiate More Gently on Things that are "In Bounds"

This is a significant point not always appreciated. Ford refers to it this way [1, p.2]:

> Since a job offer is worthless if there are "fatal flaws" in it that put it "out of bounds," you should stand firm on requests designed to fix these flaws. On the other hand, you can probably afford to compromise (or even give in) on things that are "in bounds" (i.e., satisfactory but not ideal). Some satisfactory elements of a job offer may become "fatal flaws," however, if you are negotiating multiple offers.

If having a contract that gives you time to do consulting is important to you, then bring it up. If the reason is the extra income it brings, then try negotiating on salary. If it is the impact such consulting can have on your various forms of scholarship, then you need to make that case more forcefully since additional salary will not provide this experience.

7. Learn about the Tenure Process, But Do Not Get Hung Up on It

Here, Ford notes [1, p.2]:

> Tenure decisions are too individualized to enable you to use this as a major criterion except in extreme cases. However, make sure you know whether the job being offered is tenure-track, and *get it in writing*. A verbal assurance that a non-tenure-track job will eventually become tenure-track should not be trusted, so get it in writing as well.

I agree with Ford that you cannot expect to know all the details about tenure criteria in advance of a decision that will not be made for five–six years. It is also difficult, particularly in your role as a candidate, to pin faculty down about a process that is often purposely left somewhat vague. (More on this in Chapter 14, "Insights on Tenure"). There are, however, some basic things with respect to tenure that you should explore prior to accepting an offer. It is important to understand what the department sees as its role in promoting, and rewarding the various forms of scholarship introduced by the Carnegie Foundation for the Advancement of Teaching and discussed throughout this book. The degree to which the discovery, integration, application, and dissemination of knowledge are valued by a department in its promotion and tenure decisions are often correlated with the university/college/department's mission, vision, and goals.

Unfortunately, as we saw earlier, not all administrative entities within a given institution share the same mission, vision, and goals. Also, for some institutions, the above are likely to change significantly over the next decade. Yet, it is entirely appropriate for you to at least inquire as to whether your department knows the above factors well enough to quantify your expectations of performance in order to achieve tenure. As Albert Henning, who was an assistant and then associate professor of engineering at Dartmouth College for over eight years, puts it:

> Ideally, the department's tenure expectations should be put forth in writing, but this is rarely the case. Yet, the department should know what role the applicant is expected to fill, and what weight to give each of the four possible areas of scholarship. The candidate should (since tenure decisions are made by the entire tenured faculty) confirm these expectations, at least informally, with tenured members of the search committee.

8. Start as High as You Can in Institutional Prestige

You can probably move down the institutional ladder, but it is almost impossible to move up any significant distance. However, keep in mind that at some schools, the ratings of one department may exceed, by a considerable degree, the ratings for the school as a whole.

In addition to knowing where a school or department is on the prestige scale, you also want to know which way it is heading. As noted in Chapter 1, some schools are clearly making the effort to move up, and they are often willing to hire the very best young faculty by making available the necessary resources.

9. Get as High a Starting Salary as You Can, But be Realistic

Here, Ford makes the point that [1, p.2]:

> A higher starting salary means that future percentage increases will be based on a higher number, thus accelerating your salary at a somewhat faster pace (all else being equal). On the other hand, assistant professor salaries fluctuate only within a very narrow range, so that there's usually not much point in pushing too hard on this component of the job offer.

You do not want to lose a lot of points with the dean by bargaining for an extra $2000 – $3000 in salary. Remember, what you are really negotiating is the start-up compensation package. Academic year salary is only one part of this. Summer income opportunities, consulting time, support for travel, and housing assistance all have an impact on your standard of living. Note: sometimes dissertation committee members and other mentors can provide good information on whether an offer is equitable and reasonable.

10. Create Options and Keep as Many Open as You Can as Long as You Can

Here, Ford notes [1, p.2]:

> Be an active, engaged job seeker—make sure all of the options you would like to have are explored. Be patient and planful—don't make any decisions you don't have to make unless you are certain that other options are closed or less attractive.

In other words, do not assume that you have to say yes right away to any job offer or risk having it withdrawn. Responding to an academic job offer is one of the most important decisions you will ever make so take your time. Unless you are absolutely certain that you do not want a job, do not rush to say no. Remember point 1: if you do not have an offer in writing, do not eliminate other possibilities until you absolutely have to do so. As we will see in the section, "What to Do if You Do Not Get the Offer You Want," an offer that does not look so good right now may look more attractive if other options do not materialize.

11. In Making a Decision, Combine Logic and Emotion

Ford's final point is this [1, p.2]:

> A thorough evaluation of a job offer should combine thoughtful analysis of the degree to which it affords the attainment of desired outcomes AND an appreciation of the fact that emotions are also designed to provide this same kind of evaluative information. If these two kinds of evaluations conflict, you should work hard to try to resolve the discrepancy. In the end you have to trust your gut. If you *feel* really negative about a job, don't take it unless you can resolve why you feel this way.

As noted in Chapter 3, consequential decisions often involve emotional conflict. Psychologists like to talk about "postdecisional regret" followed by periods of "bolstering" when referring to decisions such as whether or not to get married, go to graduate school, or take a particular job. After making such decisions, we often spend a period of time thinking about all the reasons why we should not have done so. We usually do this no matter which way we decided to go. The "postdecisional regret" often lasts in one form or another until we are able to implement the decision. Then we usually shift to "bolstering," in which we tell ourselves, and others, all the positive reasons why we made the decision. Both feelings and actions are appropriate. Postdecisional regret is natural with decisions having significant pros and cons, which is the case with most big decisions. Bolstering is our way of supporting ourselves during the initial period in the new activity. Thus, if you accept an offer in, say, April, but do not actually arrive on the job until the following September, you should expect to feel both anxiety and excitement in the intervening six months.

Remember also that you do not have to do all your negotiating and decision making alone. Talk to your advisors, your student and post-doc colleagues, and others whose opinion you respect. They will have their own perspectives and motives, of course, but taken together, their input can be helpful in coming to the right decision.

There is one more matter of significance. I recently served on a search committee looking for a new assistant professor in the Electrical Engineering Department at Stanford University. Our top candidate was a graduate student from the University of Wisconsin. The committee was so impressed with his visit and job talk that we made him an offer on the spot. This offer was made in late February at a time when the San Francisco Bay Area was experiencing an unusual heat wave, with temperatures in the high 80s. As the candidate and I walked across the campus, noticing that some professors were holding classes outdoors because it was so hot, I asked him if he had any questions. He stopped, turned to me, and said, "Just one. Do I have to return to Wisconsin to finish my degree, or can I just stay here and start right now?"

He did indeed have to finish, and if you are a Ph. D. student, so do you. Almost nothing could be more difficult than trying to begin a faculty assignment, with all the demands it entails, while at the same time trying to complete a dissertation at an institution that you have left. Do not do it. Finish your degree now!

DUAL-CAREER COUPLES

Dual-career couples, those in which both members are professionals, are now quite common. A subset of this group is couples in which one member is a professor, and a still smaller subset is the group of couples in which both members have academic careers. Here are a few examples:

> Jeff Koseff and Thalya Anagnos are married professors of civil engineering, but not at the same institution. Anagnos is professor and chairman of the Civil Engineering Department at San Jose State University, and Koseff is professor and chairman of the Civil Engineering Department at Stanford University, some 25 miles to the north.

> Kirk Schultz and Noel Schultz are married professors in different departments at Michigan Technological University. Kirk is an associate professor of chemical engineering and Noel is an assistant professor of electrical engineering.

> Enid Steinbart and her husband, Lew Lefton, are both professors in the same department (mathematics) at the University of New Orleans.

Finding positions for such couples, or solving the so-called "two-body problem," presents both a difficulty and an opportunity. For couples in which one member is seeking a position outside academia, the probability of success increases with the size of the metropolitan area in which they are looking. This increased probability exists even though the competition is also greater in larger areas. In smaller cities or towns, the number of possibilities in a given field may become vanishingly small. Thus, for couples in which one member is not an academic, larger metropolitan areas afford a greater likelihood of employment, and this fact, of course, has an impact on where the other member looks for academic positions.

If both members of the couple are looking for academic positions, then the size of the geographical area is less important since there is likely to be only one, or at most two, colleges or universities with openings in any city or town.

These couples do face difficulties. As Kirk Shultz points out, "the likelihood that both members of the couple will be the first choice of their respective departments is quite small, given the specialties involved, and the need of a good fit."

Yet, under certain circumstances, if a school makes a special effort to accommodate the employment needs of two-career couples, it can often "land" exceptionally good applicants who, if applying alone, might have gone elsewhere. The University of Nebraska has such a program, and during the last few years, has been able to hire some exceedingly good professors because of it.

The chances that these professors will remain at the institution are also quite good for the same reason; there are just fewer simultaneous opportunities at other universities. A couple at the University of Chicago struggled together toward tenure during their first seven years on the job. They both received it at the same time, after which one of them remarked, only somewhat in jest, "Now we really can't go anywhere."

A question that often arises is: When in the application process should you raise the issue of career opportunities for your spouse? The answer is: sooner rather than later, but not too soon. It is probably not a good idea to bring it up in your initial cover letter, but you should discuss it during your initial campus visit. Most search committees now anticipate that employment of a spouse will be an issue for many applicants, and appreciate the opportunity to consider how they can help.

If both members of the couple are applying for academic positions at the same school, there are special considerations. These are described in some detail in the vignette on Brian Love and Nancy Love at the end of this chapter.

WHAT TO DO IF YOU DO NOT GET THE OFFER YOU WANT

Suppose that, after all your time and hard work, not to mention emotional effort, you end up without a job offer or with an offer that you cannot accept. What do you do now? First, understand that it is certainly normal to feel disappointment, anger, rejection, worry, and even embarrassment. Under such circumstances, we naturally look for someone, or something, to blame, including ourselves.

Perhaps you did not get an offer because someone on the search committee disliked you from the start, you were asked all the wrong questions during your interview, or the winning candidate was known before the search began and the department was just going through the motions. You might imagine that your advisor wrote a less than glowing letter of recommendation, or that he/she is to blame for admitting you into a Ph. D. program in the first place without knowing, or caring, about the poor job market.

If the above does not make you feel better, you can always blame yourself. Should you have researched the department more thoroughly, written a better cover letter, given a better academic job talk, or not asked for so much support? And now what about the future? Has all this work been for nothing, and will you ever get the kind of job that you want?

All of these reactions are normal, and indeed expected. If you do not accept them and deal with them now, it will only take you longer to move on to the next stage. Talking with others is usually a good way to start. You are certainly not alone, and knowing that you are not will help. Most of the time, it has little or nothing to do with you. I have been on a number of search committees and have talked to many others who have also served on such committees, and in most cases, we would have gladly taken our second, or even third, choice if it had come to that. This fact may not make you feel any better, but it should. If it applies to you, then your chances of getting a job the next time around are certainly better, particularly if you learn from your recent experience.

The Decision to Try Again

First, you have to decide if there will be a next time, if you want to try again. I would certainly hope that you do. Pursuing an academic career is too important a matter, and you have too much already invested to stop now. Next year will most likely be a different job market, and you will be a different candidate. You will have finished your dissertation, or published more articles, or have your dissertation revised and accepted for publication. You may have a year or more of postdoc experience. You may have some additional teaching experience. Evidence shows that for most candidates, persistence does pay off [2].

The next step is to consider the possibility that one of the reasons you did not get an offer was, in fact, because of something you did or did not do.

A recent postdoc in high-energy physics applied to a number of schools for beginning assistant professorships in physics. He wrote what he thought were very effective, targeted cover letters, emphasizing his interest in the school and in undergraduate teaching. His CV pointed to the teaching experience he had acquired at night in a local community college. Yet, in spite of all his thought and effort, he received rejection letters from every school to which he applied. Disappointed as he was, he mustered the courage to call each school. After some prodding, he learned that as soon as the search committee saw "high-energy physics" at the top of his application, they put it in the reject pile. As noted in Chapter 8, most schools have no openings for faculty who want to do high-energy physics research because they have no facilities for such work. This applicant was far more interested in just teaching physics, but he never got a chance to make his case. On his second round of applications—which to date have resulted in two job interviews—he rewrote his resume to deemphasize high-energy physics, while emphasizing his physics background in general, and his teaching experience in particular.

Not all changes are this simple, of course. To get the information you need in order to decide what to do, begin by contacting the people to whom you sent applications or with whom you interviewed. Try to learn what you might do in the future to improve your candidacy. You owe it to yourself to obtain this feedback, and you can get it if you ask for it in the right way. (You can also ask a third party, such as your advisor, to make some inquiries on your behalf.) Most people will want to assure you that it was not you, that they simply had a more appropriate candidate, or that they were looking for a particular specialty that you did not have.

Do not settle for such limited feedback. Explain that you will be applying to other schools in the future, and that you would appreciate any specific insights that would enable you to make a better case. Put this way, you are likely to get responses that begin, "I thought your cover letter was pretty good, but a couple of things you could do are…" or "Your job talk went well, except that you spent too much time…" This kind of feedback can make a real difference the next time you apply for positions. It could even get you a new look at the schools that turned you down.

Examining Your Options

A major theme of this book has been the employment of the Multiple-Option approach by which you prepare for possible positions in government and industry at the same time you prepare for a career in academia. There are two reasons for following this approach:

1. Given the relative weakness in the academic job market, such preparation increases your chances of professional employment after graduation or after a period as a postdoc.
2. Such preparation actually enhances your attractiveness as an academic candidate because interactions with industry and government are becoming a necessity for many science and engineering professors.

If you have followed the approach outlined in Chapters 4–6 and the recommendations in Chapters 7 and 8, then hopefully you will have generated some options that fall within one of three categories:

1. Staying in your present graduate student or postdoc position, with some modifications for an additional one or two years.
2. Moving on to another, but still temporary position.
3. Moving on to a more permanent position.

Staying where you are, with some modifications, relieves you of the effort required to meet new people and adjust to new surroundings. This arrangement has its pluses and minuses. If you use the time to concentrate on the additional things you can do to make yourself a stronger applicant, then it can be an advantage. If you use it to just do more of what you have already done, then it will be of limited value.

Rick Vinci had just finished his Ph.D. in materials science at Stanford University. As he approached the end of his studies, he considered applying for positions in both industry and academia. However, his wife, Michelle, was a Ph.D. student in Spanish at Stanford, and two years from finishing her degree. Vinci really needed to stay in the area while she completed her work. Then both of them could look for academic positions. He knew that staying on as a postdoc with the same person who was his dissertation advisor had its drawbacks, so he decided to do so only if he could:

- Have the opportunity to publish additional articles on his thesis topic, an area in which no start-up time was required,
- Identify another area of research, different from that of his dissertation, in which he could establish expertise,
- Generate research proposals in this new area,
- Have full responsibility for teaching a course,
- Mentor both graduate and undergraduate students.

Vinci was able to take advantage of what, and who, he already knew, and at the same time significantly expand his portfolio for the academic job search to follow.

At the same time, moving to another position puts you in a different setting with different people, and this might afford you the opportunity to look for an academic position from a new perspective. One possibility is a full-time teaching position with a fixed term appointment.

Sally Veregge was older than the typical doctoral student when she received her Ph.D. She had had many years of experience in the health care industry when she entered a doctoral program, and could have easily gone back to such a position. As it turned out, San Jose State University in San Jose, CA liked to hire people with experience, at least in biology. Veregge was not offered a full-time tenure-track job (someone else got it), but was offered a full-time temporary position. There was the potential for a tenure-track position when the temporary job ended, but certainly no guarantee. Veregge asked that they evaluate her annually in the same way as they would have for a tenure-track position so if such a position opened up, she would have a good chance of getting it. During her two-year temporary period, she was treated in every way like a regular faculty member. Two years later, she did get the tenure-track position after it was advertised and she applied for it like everyone else.

Another possibility is a temporary position in government or industry.

After completing her Ph.D. at Columbia University in biology, Cynthia Hemmenway decided to join a postdoc program at Monsanto, which has a reputation for its research strengths in plant biology. "I had a really good feeling that I would be able to do well there in terms of publishing, be in an area that is moving quickly, and at the same time learn a little bit more about how companies work," says Hemmenway, now an assistant professor at North Carolina State University [3].

If, on the other hand, you do not want to continue to actively apply for academic positions, then it is time to move on to a more permanent position in government or industry. In doing so, you might want to keep your options open by teaching part time at a local college or university in the evening. Such experiences can provide their own rewards while keeping your "credentials" up for possible academic positions that you might want to apply for in the future.

Vignette #9 The Dual-Career Job Search

Greater attention is now being paid to the problems dual-career couples face in finding positions in the same town, city, or metropolitan area. These problems are particularly acute when husband and wife are seeking academic positions. The following vignette discusses the experiences of a couple in which both members ultimately found positions at the same university.

Brian Love
Nancy Love
Virginia Polytechnic Institute and State University

Husband and wife professionals face enough problems in finding employment in the same city or town. When they are both academics, finding positions at the same institution, and in some cases even the same division or department, can be particularly challenging.

Brian and Nancy Love met as undergraduates at the University of Illinois at Champaign–Urbana. There, Brian obtained a bachelor's degree in chemistry and a master's degree in metallurgy and mining engineering, and

Nancy a bachelor's degree in civil engineering and a master's degree in environmental engineering and civil engineering. Following graduation in 1986, both took jobs in Dallas, TX, and they married a year later. In August 1989, Nancy left Dallas for Clemson University in Clemson, SC to begin work on her Ph.D. In December 1990, Brian acquired his Ph.D. in materials science and engineering from Southern Methodist University in Dallas, after which he took a postdoc at Georgia Institute of Technology in Atlanta, GA. Nancy finished the experimental work for her Ph.D. in civil engineering from Clemson in December 1993. They both wanted to pursue academic careers. Of course, to work in the same geographical area, if not at the same university, was a sine qua non for them.

Brian and Nancy followed a typical job search strategy, sending out vitae and cover letters to those who advertised job openings. Brian decided to mention his wife in his cover letters, but Nancy did not mention Brian at all in her applications. Once a university showed interest, however, Nancy or Brian would directly tell the dean or a member of the search committee that they were looking for jobs together. "We had agreed ahead of time to *not* compromise—we were a package," says Nancy. They were going to work in the same area, even if one of them ended up in industry. "We tried to put the most positive light on everything," recalls Brian. But they were not going to both accept faculty jobs hundreds of miles away from each other. "There was a period when just the same time zone would have been enough," says Nancy, "but not anymore. "

The reaction from the search committees and deans was generally positive. "Nobody withdrew the interview trip, and some places tried to leverage their industrial base," saying that the spouse might be able to work in the local industry, commented Brian. They found that some schools even have official programs for dual-career couples. For example, the University of Nebraska at Lincoln offered Nancy a tenure-track faculty position whose package included a "Faculty Fellowship"—a year's stipend for her husband. However, they found an even better opportunity at the Virginia Polytechnic Institute and State University (VA Tech).

Nancy was invited to give an academic job talk to the Civil Engineering Department at VA Tech. A professor on the search committee took note during her campus visit when Nancy mentioned that she and Brian would come only as a package. This professor also knew that Brian was being considered for a position by the Materials Science and Engineering Department. So when she finished second in the civil engineering search, and Brian made the short list in his search, the professor brought Nancy and Brian's situation to the attention of the dean of engineering and the provost. An agreement was made to consider creating a faculty position in civil engineering for Nancy if Brian ended up number one in the materials science search. He did, and both received job offers from VA Tech. They made the move to Blacksburg, VA later that year. (Nancy completed her Ph.D. from Clemson in June 1994.)"Brian and I are finally living together, after seven and a half years of marriage," exclaims Nancy (they had been apart much of the time while getting their Ph.D.s). Now they are going through the new professor experience together.

Although their schedules vary from week to week, the common denominators are classes, dinner, and working in the evenings. Both are still making it through teaching their first sets of classes. Brian, for example, has taught two new classes and two classes that have been taught before. "Couple this with trying to write proposals," says Nancy, "and we both end up working a great many hours every week." Both have windows in their offices, and Brian comments that: "Sometimes I can look across the way and see her light on—we can almost communicate by semaphores."

Their weekly time together tends to be at dinner, which they have out in local restaurants. How about weekends? "Our weekends tend to vary, depending on who's got what crunch coming up," remarks Nancy. "Another married couple in environmental engineering told us that all you do the first two years is eat, sleep, and work. We don't want this to be a long-term thing."

Both Brian and Nancy are enjoying the experience, although it leaves them worn out at times. As Brian says, "I am happy about the experience, but I never expected so much work. The students are the best part of the job, the fact that they are willing to work so hard. They make the job worth it."

Nancy is trying to fill her pipeline with proposals and teaching. One thing on her mind is what will happen when children come along. She wants to have her lecture notes well organized and prepared, and to have her research going smoothly, so that she may have time for her kids.

Reflecting on the job search, Nancy observes:

> There are a few universities with official programs for dual-career couples. Although Virginia Tech does not have such a formal program, they made an extra effort to work with us. I think some deans recognize that they may lose top quality people by not being flexible.

Clearly, flexibility and the right match are key ingredients in making a dual-career couple academic job search successful. With flexibility and optimism, Nancy and Brian Love diligently sought the right pair of job offers. They succeeded in their search, and are now traveling the dual academic career path together at the same university.

SUMMARY

We began this chapter with a look at the negotiation process, examining in some detail 11 principles to guide your response to academic job offers. We then discussed some of the special problems faced by dual-career couples, in particular those in which both members are seeking faculty positions. We concluded by discussing how to apply the principles of the Multiple-Option approach if you did not receive an academic job offer, or received one that is unacceptable to you.

Your shift from a graduate student or postdoc to beginning professor is going to be exciting and dramatic. At this stage, probably the most valuable thing you could do would be to ask a half dozen professors at the institution to which you are going about what they feel it takes to succeed as a beginning professor. If you chose the right allies, champions, and mentors, the advice you receive could be invaluable. It almost certainly will not all be the same, but out of such discussions there is bound to come a flood of specific ideas that will help you avoid major pitfalls, while concentrating on those critical behaviors that will increase your chances of success.

In the next part, "Looking Ahead to Your First Years on the Job, Advice from the Field," we do the next best thing by capturing key insights for success from science and engineering professors across North America.

REFERENCES

[1] M. Ford, "General principles for responding to academic job offers," unpublished manuscript. Reprinted with permission.

[2] K. Johansen-Trottier, *The Academic Job Search: A Practical Overview.* Stanford, CA: Career Planning and Placement Center, Stanford University, 1991, p. 11.

[3] M. Barinaga, "Checking out industry as a postdoc," *Science*, vol. 257, no. 5077, p. 1721, Sept. 18, 1992.

Looking Ahead to Your First Years on the Job— Advice from the Field

CHANGING GEARS

Your shift from graduate student or postdoc to beginning professor will be dramatic. While your total working hours may remain the same, the *number* of *different* things you will be expected to do will increase significantly. Instead of spending 10–12 hours a day on two or three main tasks, you will spend the same amount of time on at least a half dozen important activities. You will have notable responsibilities with one or more forms of scholarship, i.e., the discovery, integration, application, and teaching of knowledge, as well as undergraduate counseling, professional reading and writing, and in many cases, administration and graduate student supervision. In addition, there will be the need to get to know dozens of faculty and staff and as many as a hundred or more students. Keeping all these balls in the air while also worrying about tenure, finding time for long-term professional growth, family and friends, and even some form of recreation will be a major accomplishment.

One of the most valuable things you could do early in your tenure as a professor would be to ask a half dozen faculty at the institution to which you are going for their advice on what it takes to succeed as a beginning professor in your new environment. Achieving the intimacy and trust necessary to ask the right questions and obtain credible answers will take time. However, not everything has to be done immediately. Yet, the ideas you pick up are bound to help you avoid major pitfalls, while also allowing you to concentrate on those behaviors critical to your success.

The following chapters show you how this might work by capturing insights in five key areas via vignettes with 21 science and engineering professors at schools across North America. The five areas are: time management (Chapter 10), teaching and learning (Chapter 11), research (Chapter 12), professional responsibility (Chapter 13), and tenure (Chapter 14).

Certainly, no one piece of advice will work for everyone in every situation. Only you can evaluate the particular circumstances in which you find yourself, and determine what works for you. Yet, examining what works for particular faculty under particular circumstances will lead to ideas you can adapt to your specific situation.

The five chapters begin with a setting of the stage for the subsequent vignettes on faculty who provide insights on the theme of the chapter. These vignettes are followed by a detailed "In Addition" section describing additional ideas as well as other sources you can turn to for further information and understanding. Each chapter ends with a look at the main conclusions we can draw from both the vignettes and the readings.

CHAPTER 10
Insights on Time Management

> As a graduate student my prevailing thought was, "If I can just find a good problem," or "If I can just find a way to prove this conjecture." As an assistant professor, my prevailing thought was, "If I can just find the time!"
>
> Terri Lindquester, *Associate Professor of Mathematics*
> *Rhodes College* [1, p. 6]

SETTING THE STAGE

There is a common thread running through every conversation I have had with over 70 faculty in all science and engineering fields from all types of colleges and universities. It is the huge number of tasks on everyone's plate, the challenge of figuring out how to do any of them well, and the difficulty of finding a way to have enough time to just sit and think. In late November, I had dinner with a group of professors at an American Chemical Society meeting in San Francisco. A faculty member from the University of Michigan remarked that it was on the four-hour flight from Detroit that she had had the first opportunity since the semester began (in early September) to have a private block of time all to herself. The nods from the others at the table made it clear that she was not alone.

As a new faculty member, you need to find ways to manage your time and tasks efficiently (doing things right), *and* to control what you put on your plate in the first place (doing the right things). If you do not do both, all the effort you put into your professional work will be for naught: you will simply burn out, and in the process endanger your career and your relationships with the people you care most about.

While it may come as a surprise to our colleagues in industry, academics are every bit as caught up in the multitasking, cycle-time reduction, work-anytime-anywhere operating mode with which industry is now so familiar. Beginning faculty in particular tend to place excessive expectations on themselves to do outstanding jobs in their teaching and all other forms of scholarship, while at the same time making sure they are fully available and accommodating to their colleagues, students, and families. Finding ways to manage these conflicting demands can be the key to developing a successful career *and* a rewarding and satisfying personal life.

There is hope. The vignettes that follow are about faculty who have made good progress in managing their time and their lives. They would be the first to admit they do not have the problem solved, not by a long shot. Yet, the insights their experiences offer can be helpful to us all.

In Vignette #10, Alison Bridger, professor of meteorology at San Jose State University (Master's I institution), introduces the concept of "establishing your absence," i.e., of being away from campus on a regular basis to work on projects without interruption.

In the next vignette (#11), Kim Needy, from the University of Pittsburgh (Research I institution), talks about going beyond time management to personal management, which involves focusing on important tasks leading to the achievement of long-range academic goals.

In Vignette #12, Paul Humke, a professor of mathematics at St. Olaf College (Baccalaureate I institution), describes the value of always having something "on the burner" if you want to make the most creative use of your time.

In our last vignette (#13), Thalia Anagnos, chair of the Civil Engineering Department at San Jose State University (Master's I institution), describes the things a department chair can do to help new faculty find more time. (Note: It is a vignette that you might want to show discreetly to your own department chair.)

These vignettes are followed by a look at four books with advice that I and other faculty have found particularly helpful in managing our time and our tasks. We will focus on a contribution from each of them in a specific area. The areas are (1) sources of faculty stress, (2) faculty efficiency, (3) the urgency addiction, and (4) achieving balance in our lives.

Vignette #10	Establish Your Absence

For me the real key came when I realized the value of "establishing my absence." I began by spending every Thursday at NASA, and was amazed at how much I could accomplish. I could collaborate with my colleagues for periods of time that just weren't available to me back on campus.

So says Alison Bridger, a tenured professor of meteorology at San Jose State University in San Jose, CA, a 30,000-student Master's I institution with a strong teaching mission and no Ph.D. students. "At my kind of institution, you have to make research a priority at least some of the time every week; otherwise, everything else will fill all your available time," warns Bridger. She advises faculty in similar situations to:

> Realistically assess your options. Find out what facilities, laboratory space, computing resources, and graduate student support will be available, might be available, or will never be available. If it is never going to be available, then look around your neighborhood; you might be surprised at what you find.

What Bridger found was the NASA/Ames Research Center in Mountain View, CA, about 15 miles northwest of campus. "I began interacting with one person on a project, then I added another and another, to the point where I am now part of a whole research group," she notes. In concert with her group, Bridger has used a computer model to simulate the global flow of air around the planet Mars. Like Earth, Mars has tides in its atmosphere, and Bridger and her colleagues study the relationship between these tides and the huge dust storms appearing on the planet.

Now, Bridger is able to publish papers with herself as the first author and her NASA associates as coauthors. "But this didn't happen overnight," she says. "It took a lot of work and regular meetings to get to this point."

Being away from campus one day a week has yielded other benefits for Bridger.

> I find it really helps to go somewhere else to think and set up. This time away from campus enables me to work on important, nonurgent things I would otherwise ignore. Plus, my students and colleagues are now accustomed to the fact that at certain times during the week, I will not be in my office.

What else does Bridger do to manage her professional life?

> You have to consider all the ways, big and small, that you spend your time. Student advising consumes a lot of time, as does class preparation, all of which takes time away from your research. While I always strive to give my best effort to teaching, it is possible to overdo writing up lecture notes. I was a perfectionist in my teaching, but if you let it, class preparation can take all of your available time.

Bridger also suggests that you look for ways to reduce the complexity associated with your initial teaching assignment. One way is to teach more than one section of the same class. Another is to ask to see the lecture notes, from which you can build your own syllabus, of a senior faculty member who has taught a class you are now teaching for the first time. With this approach, you can still do a good job of teaching while also setting aside time for other things.

How does Bridger handle the pressures and responsibilities of being one of six professors in a small department? "I give my best effort, and then I don't worry myself to death," she says. "The only way you can do everything everybody wants you to do is to overdo it and work all hours of the day." Doing so, Bridger believes, can do great damage to both your professional and your personal life. "In academia, there is a syndrome of workaholics whose marriages break up, and they die early," she admonishes. For Bridger, "It got to the point where I said, 'This is what I can do; take it or leave it.' All you can do is give a sustained, but realistic effort."

Bridger established her pattern of balancing teaching and research early on in her career. By setting this expectation for herself, she also helped set the expectations that others had for her. With some satisfaction, she notes, "The process has made it possible for me to have a successful professional career balanced with time for a satisfying personal life."

Vignette #11   Set Long-Term Goals

One of the biggest transitions you will experience in going from graduate student to professor is the loss of closure. As a student, you knew when a course was over, when your dissertation was finished, and when you got your degree. As a faculty member, things always continue. You teach your course again, you modify your paper one more time, and you attend those committee meetings forever.

For Kim Needy, a tenure-track assistant professor of industrial engineering at the University of Pittsburgh in Pittsburgh, PA (Research I institution), knowing that there are things that will always need doing meant having to be realistic about expectations if she were going to devote her limited resources to what was really important. She remarks:

You can never get it all done, and this is something you need to understand and accept. We used to say just work harder, or longer. Fine, in the short term when you are facing a deadline, but if you make it the rule and not the exception, you are setting yourself up for disaster.

For Needy, reaching a "comfort level" with not being able to do everything was a turning point:

With some things, "good enough" is indeed just that. Time spent doing one thing means time taken away from another, and there are some really important, long-term things I have to pay attention to if I'm going to survive in academia. It does me no good to be the ideal professor, always available to my students, and fully accommodating to my colleagues, if in the end I don't get tenure and am no longer here for anyone.

Needy takes her teaching responsibilities very seriously, and devotes considerable time to them. She notes, however, that teaching preparation time can be more like a gas than a liquid or a solid. In other words, it will fill all the space available to it if you let it, a point made by Bridger in our previous vignette. "You can always add a case study, always improve an overhead, and always revise a handout," she exclaims. "At some point, you have to put a box around it and say, 'enough.' "

The same is true with respect to service. "Women in engineering are frequently asked to serve on committees, counsel women students, and serve as role models," Needy states. "My advice is to really limit your time to only those service activities with the highest payoff. For assistant professors, this usually means focusing on the needs of your department rather than on those of the institution as a whole. Again, if you don't, you won't be here in the long run for the people you care most about helping."

The key, according to Needy, is to follow author Steven Covey's advice and establish long-term goals, and then block off time on a regular basis for tasks that help you achieve those goals. (See the "In Addition" section later in this chapter.) Echoing Bridger's point in the previous conversation, Needy says:

> The only way to do this is to be somewhat selfish with your time. I'm usually here from 7:30 a.m. to 6:30 p.m., and I can never find more than an hour to do any one thing. But one day a week, I stay at home and work on my research, my writing, and my course development.

Such an approach also helps Needy make the switch from traditional time management where we create to-do lists to personal management where we manage ourselves by focusing on relationships and results, not just things and time.

In terms of relationships, Needy has observed some encouraging signs in recent years. She notes:

> In industry, it is common for people to work together in teams. In academia, we teach about the benefits of doing so, and we even have our students working this way on projects. However, for a long time, we as faculty didn't "practice what we teach." I think things are beginning to change. I have observed much more collaboration within and across disciplines, and the result is better papers, better proposals, and more creative ideas. Furthermore, it saves time.

Needy is also a strong believer in physical activity as a way of balancing mental and physical stress. She is able to do so while continuing to develop creative and supportive relationships. As she describes it:

> Every Friday, no matter how busy we are, a bunch of faculty and students in the Industrial Engineering Department drop what we are doing and go for a five-mile run. We do it as a team. We often talk about school but do it in a relaxed way that also does wonders for our mental well-being.

Vignette #12 Keep Something on the Burner

> I remember the transition from graduate student to faculty member; the stunning new demands on my time, the responsibilities for curricular matters I'd never thought about, the energy level necessary to do my job. The foremost question I had about my professional career was whether I would even have one! And I wanted one! [1, p.1]

Although it was over 25 years ago, Paul Humke still recalls those feelings of anxiety and stress as a beginning faculty member. Humke is now a tenured full-professor of mathematics at St. Olaf College, a 3500-student Baccalaureate I institution in Northfield, MN, some 50 miles south of Minneapolis. Since his early days as a professor, Humke has thought about how faculty can best manage their time, particularly when it comes to responsibilities outside the classroom, i.e., research, service, and other professional duties. As he observes, "We hear a lot today about schools wanting to place a greater emphasis on teaching, but at places like St. Olaf, where teaching is the primary activity, those of us who also enjoy research have to look for ways to make it a part of our professional life."

Keeping something on the burner is an approach Humke has found particularly effective. As he puts it:

> If you let your work lie dormant for a month or two, or in some cases even a week or two, your efficiency drops tremendously. When you get back to work, you find yourself spending a great deal of time bringing yourself up to speed with the work you did previously.

As one would expect from a mathematician, Humke stated his point in the form of an equation:

$$WUT = k \exp(TL)$$

or "warm-up time necessary to return to a problem increases exponentially with the time that has lapsed since you last worked on it."

Humke and others have also discovered that the value of "k" increases with chronological age. He told the story of a colleague who remarked [1, p.9]:

Chapter 10 Insights on Time Management

When I was young I'd spend five minutes "warming up" and then I was ready to work. When I was a bit older I had to spend twenty minutes warming up before I was back into my work. Now it seems I spend all my time warming up!

By keeping something on the burner, you minimize your warm-up time. Humke explains it this way [1, p. 10]:

> It is important to have a problem you can work on whenever you have a spare minute. I realize research often takes long periods of concentrated work, but it helps a great deal to have some aspect of your problem to think about when you have a free minute or two, when the party becomes dull, or your lunch date fails to show up.

John Hennessy, dean of the School of Engineering at Stanford University, reinforces Humke's comment when he says:

> You need to keep your creativity cycles free, and the best way to do this is to have something to work on in your head when you are walking across campus, sitting in a dull meeting, and riding in a car. Doing so also keeps you from thinking about a lot of trivial, negative stuff that isn't helpful anyway.

Humke also makes the point about setting aside blocks of time for professional work. It is a theme we have heard over and over again, but Humke managed to state it in a particularly powerful way:

> I treat my research time the way I treat my class time. It's high priority, and I don't cancel my research time unless I would cancel a class for the same reason.

Primary responsibility for your professional development, Humke emphasizes, lies with you. He recognizes that there is an important role to be played by your institution, dean, and department chair, but faculty have to take the initiative. As Humke puts it:

> There are many who believe that if the university doesn't pay for "IT"'or I don't have a grant to pay for "IT," I won't do "IT." "IT" can be attending a conference, visiting a research colleague, or buying books. This is a professionally dangerous attitude, particularly in times of financial famine. It is important to set aside a bit of both time and money specifically for your own professional work. You can hope for help with each of these, and indeed should expect some help in certain areas, but ultimately you have to take the lead in looking out for your long-term development.

One example of the kind of help you can expect from an enlightened institution is described in our last vignette.

Vignette #13   How to Help New Faculty Find the Time—One Department Chair's Approach

I really got dumped on as a new faculty member. I was given all the duties no one else wanted, such as faculty advising for the student chapter of the Society of Women Engineers, serving on an outreach committee visiting various high schools, and a myriad of other committee and administrative assignments. I'm not saying these things aren't valuable in some way; it's that as a young faculty member, they were overwhelming at the time I was trying to get my feet on the ground professionally.

So remembers Thalia Anagnos, currently a tenured professor and chair of the Civil Engineering Department at San Jose State University in San Jose, CA. Now, she tries to protect her young faculty from similar situations. "I don't buy the argument that because we had to suffer, they should as well," she says. According to Anagnos:

Some chairs let their new professors flounder. They bring them on board, assume they can do it all, leave them on their own, and give them no direction. Unless they are lucky enough to have a mentor, they try to figure it out by themselves and, not surprisingly, the results are not as positive as they could be. This is especially true at a metropolitan institution like San Jose State with a large commuter student population, few master's students doing research, no Ph.D. students, and very few (if any) teaching assistants or graders.

Anagnos tries to explain to her older colleagues that the goalposts have moved, and the expectations for young faculty are different from what they were 20 years ago. "Our senior professors need to understand this, and help by picking up more of the administrative and service loads," she remarks. She goes on to say:

We should be asking ourselves what we can do to help young faculty get their scholarship program going, not, "How can I unload lots of busywork on them?" I am not saying young faculty should not get involved; they should. But not to the point where they are under a pile saying, "Help, I can't get out!"

As chair, Anagnos often gets calls from people who want to ask her newer faculty to help out with their committees. Instead, she usually recommends some midcareer people who have established scholarship and would probably be a lot better for the particular committees. "When I explain to them in a nice way why I am saying 'no' [to their request for the new faculty member], people usually understand," comments Anagnos. At times, however, she will help young faculty by recommending that they serve on certain committees where they can gain visibility, learn how the college functions, and make connections.

Chairs can help in other ways with respect to workload. Take, for example, teaching assignments. Anagnos is expected to have her faculty average 17 students per class, four classes per semester. However, if a professor teaches three classes, two with 17 students and the third with 35 students, the commitment can be considered equivalent to four classes of 17 students per class. Or, if the number of students taught by a faculty member in one semester is very high, as chair she can give the faculty member released time the next semester. She uses this system to give young faculty teaching assignments requiring fewer preparations, or to allow them to borrow ahead to get started on research. As she points out, "Young faculty everywhere need to understand that they do not have to take what they are given blindly; *they can ask if there are alternatives.* If they have a good reason, something can usually be worked out."

Much of Anagnos's motivation comes from the unexpected turn of events leading to her appointment as department chair at the age of 36. To reduce costs, the California legislature approved an early retirement plan for faculty. Three senior professors in her department of ten retired in one week, right in the middle of the semester—"It was awful," she recalls. "Suddenly, after being there only eight and a half years, I became the chair. It was difficult, particularly since I was much younger than the other department chairs."

By all accounts, Anagnos has taken well to her new responsibilities, but she remembers her initial experiences well enough to take steps to help others get started on the right foot. And she also has the sympathetic support of her husband, who is a professor of civil engineering at Stanford University. "He was recently named chair of his department," she remarks with some glee. "Now he understands what I have been complaining about for all these years!"

IN ADDITION: SOURCES OF FACULTY STRESS, FACULTY EFFICIENCY, THE URGENCY ADDICTION, AND ACHIEVING BALANCE IN OUR LIVES

In addition to the insights provided in the above conversations, there are resources you can turn to in helping you manage your time and your tasks. Four particularly helpful books are:

- *Coping with Faculty Stress*, by Walter Gmelch [2],
- *Teaching Engineering*, by Philip C. Wankat and Frank S. Oreovicz [3],
- *First Things First*, by Steven Covey, Roger Merrill, and Rebecca Merrill [4],
- *The New Faculty Member: Supporting and Fostering Professional Development*, by Robert Boice [5].

Let us look at the contributions that each of these books has to make in a specific time management area. The areas are

1. Sources of faculty stress (Gmelch)
2. Faculty efficiency (Wankat and Oreovicz)
3. The urgency addiction (Covey et al.)
4. Achieving balance in our lives (Boice).

Sources of Faculty Stress

Walter Gmelch, author of *Coping with Faculty Stress,* interviewed more than 4000 faculty in more than 100 institutions across the United States, and in so doing, identified five major sources of stress [2, pp. 26–27]:

Reward and recognition—Inadequate rewards and insufficient recognition in teaching, research, and service.

Time constraints—Having insufficient time to keep informed of current developments and to prepare for classes, made worse by numerous meetings, interruptions, and other demands on faculty time.

Departmental influence—Attempts to influence departmental and institutional policies and direction, while at the same time recognizing that one lacks authority.

Professional identity—High self-expectations and the realization that professional identity rests upon the extent of scholarship, publications, and presentations.

Student interaction—Conflicts with students over evaluating, advising, and teaching.

Upon further analysis, Gmelch found [2, p.27]:

As faculty received tenure and moved to higher academic ranks of associate and full-professor, not all areas of faculty stress declined…only the stress from time constraints and professional identity declined with age and experience…Married women professors experienced more stress from time constraints and personal identity. Overall, some stress factors are associated with lower rank and untenured status as well as gender, marital status, age, and experience.

Gmelch points out that there is a difference between distress and eustress. Distress is caused by negative stress such as conflicts with students or colleagues, whereas eustress is positive stress whose source is pleasant factors such as receiving tenure or publishing a paper [2, pp. 2–5.] He goes on to list ten most troublesome stress traps for professors. They are [2, p.24]:

1. Imposing excessive, high self-expectations.
2. Securing financial support for scholarship.
3. Having sufficient time to keep abreast of developments in the field.
4. Receiving insufficient salary.
5. Striving to publish one's scholarship.
6. Having too heavy a workload.
7. Job demands interfering with personal activities.
8. Feeling progress in career is not what it could be.
9. Receiving interruptions from telephone and drop-in visitors.
10. Attending too many meetings.

The remainder of Gmelch's book deals with ways of coping with faculty stress. His approaches are based on the following propositions [2, p.28]:

1. The individual is the most important variable; no single coping technique is effective for all faculty at all institutions.
2. Faculty cannot change the world around them, but they can change how they relate to it.
3. Coping techniques must be sensitive to cultural, gender, social, psychological, and environmental differences in individuals and institutions.
4. Faculty who cope best develop a repertoire of techniques to counteract different stressors in different situations.
5. A faculty member's repertoire of techniques should represent a holistic approach to coping, such as exercise, social support, sound dietary practices, self-management skills, personal hobbies, and supportive attitudes.

Coping with Faculty Stress is a small (85-page) book packed with a great deal of practical advice. It should be on the desk of every faculty member and university administrator.

Faculty Efficiency

Wankat and Oreovicz's *Teaching Engineering* is a gold mine of information on all aspects of teaching. It includes an excellent discussion (Chapter 2) on faculty efficiency in which the authors talk about the importance of setting goals, establishing priorities, and preparing to-do lists, as well as how to operate more efficiently while traveling and how to handle stress more effectively [3, pp.14–20]. It also has some excellent suggestions on interacting with others (visitors, students, secretaries, and teaching assistants) in terms of both frequency and style.

I found the comments on work habits and on teaching and research efficiency to be particularly relevant [3, pp.20–24]. Here, the authors reinforce many of the points made earlier in the vignettes. For example, in their discussion of work habits, they talk about the advantages of having a second office or some other place to go on a regular basis to work in an uninterrupted way.

Piggybacking, or using the same work several times, is another important efficiency tool. The most obvious application of this approach is the teaching of the same course several times. Improving on what you have already taught is often easier than starting over from scratch. You can also look for ways to connect your research and teaching. One possibility is to teach a graduate class or seminar in your research area. With this approach, as the authors put it: "Time spent on research will help you present a more up-to-date course, and time spent on the course will help you better understand your research area" [3, p.28]. Having graduate students in your class or seminar participate in literature searches and summaries is also an efficient way to combine a learning experience for them, with an added benefit for you, i.e., using such reviews in research proposals, for example.

The authors also talk about the importance of avoiding perfectionism. As they note, "manuscripts can be revised forever, and the reader will never think they are perfect. At some point you have to let go and put out a less than perfect, but not sloppy, manuscript. This same reasoning is applicable to other work such as lectures" [3, pp.26–27].

Wankat and Oreovicz do caution you to seek a balance between efficiency and effectiveness. Certain activities, such as starting a class period, tutoring, mentoring graduate students, and building consensus within a department, if rushed, can give you more time in the short run, but in the long run are more costly because they will not be done effectively.

The Urgency Addiction

In *First Things First* Steven Covey, Roger Merrill, and Rebecca Merrill, take an approach to time management that goes beyond the familiar reminders and lists, calendars and appointment books, and even planning and prioritizing [4, pp.36–43]. It is the approach Kim Needy (Vignette #11) and others have found so effective. In Chapter 2, "The Urgency Addiction," the authors argue for what is called the "importance paradigm," or the putting of first things first by "doing what's important rather than simply responding to what's urgent." They describe a Time Management Matrix[1] consisting of four quadrants, each containing certain kinds of activities. (See Figure 10-1.) Quadrant I is for important, urgent matters; Quadrant II, important, nonurgent matters; Quadrant III, not important, but urgent matters; and Quadrant IV, not important and nonurgent matters [4, p.37].

[1] Note: Time Management Matrix is a trademark of Covey Leadership Center, Inc.

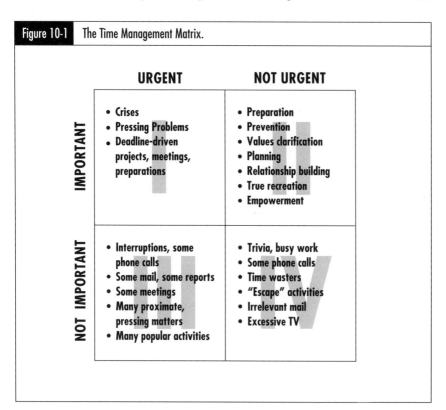

Figure 10-1 The Time Management Matrix.

Source: © 1994, Simon & Schuster, *First Things First*, Steven R. Covey, A. Roger Merrill, and Rebecca R. Merrill. All rights reserved. Used with permission of Covey Leadership Center, Inc., 1-800-331-7716.
Note: Time Management Matrix is a trademark of Covey Leadership Center, Inc.

For faculty, examples of Quadrant I activities include certain meetings, class lectures, student crises, and proposal deadlines. Quadrant II activities include designing a new course, preparing a conference presentation, writing a book, and supervising graduate students and postdocs. Quadrant III activities are such things as some meetings, some phone calls, and some drop-in visitors. Trivial busy work, i.e., junk mail, mindless gossiping, and some phone calls, are examples of time-wasting Quadrant IV activities.

Naturally, we spend time doing the things in Quadrant I. However, most of us spend way too much time in Quadrant III, and not nearly enough time in Quadrant II, which the authors call the Quadrant of Quality.

Writing this book was clearly a Quadrant II activity for me. In my case, it meant staying home on Tuesday mornings and working in the evenings until my wife came home around 9:00 p.m. (We no longer have children living at home.)

Once my wife came home, however, she became my Quadrant II activity, at which time we would often take a walk, talk, and give ourselves the opportunity to listen to each other without the interruption of television.

By increasing the time in Quadrant II, we can reduce the number of interruptions in Quadrant I. As Covey *et al.* put it [4, p.38]:

> Ignoring this quadrant [Quadrant II] feeds and enlarges Quadrant I, creating stress, burnout, and deeper crises for the person consumed by it. On the other hand, investing in the quadrant shrinks Quadrant I. Planning, preparation, and prevention keep many things from becoming urgent.

Also, by understanding that just because something is urgent does not mean that it is important, at least to you, can help you reduce Quadrant III activities and eliminate the need to "escape" to the mindless Quadrant IV activities.

Achieving Balance in Our Lives

In *The New Faculty Member: Supporting and Fostering Professional Development* [5, pp.46–49], Boice studied a number of first-year faculty who had excelled at teaching, research, publishing, and networking. He labeled such successful people "quick starters." Quick starters rated the ability to achieve balance as the single most important thing they had learned to do as new faculty. By balance, they meant three related things: (1) setting realistic limits on lecture preparation so it no longer dominated workweeks, (2) finding time during most workdays to do scholarly writing, typically about four–five hours per week, and (3) generally spending as much time on social networking, on and off campus, as on scholarly writing [5, p.46].

Boice lists four general principles practiced by quick starters: (1) involvement, (2) regimen, (3) self-management, and (4) social networking [5, pp. 46–49].

Quick starters become involved in the campus community early on. They look for ways to collaborate with colleagues on research and proposal writing, and engage colleagues in discussions about teaching. They ask other faculty to guest lecture in their classes, and are willing to do the same for their colleagues if asked.

Quick starters understand the importance of managing their tasks (regimen) in such a way as to take time for the long-term important things, including the involvement described above. They find that they are less rushed and filled with what Boice calls "business displays." In other words, more Quadrant II and fewer Quadrant I activities. As one quick starter described it [5, p.48]:

> When I started feeling less rushed, less behind on everything, I began to do a better, more reflective job of things. Another thing: I noticed that I haven't talked as much about how busy I am. And this, I'm sure is a signal to my colleagues and students that I can be approached.

Being proactive in establishing positive interactions (self-management) with colleagues and students is another characteristic of quick starters. They also take more risks, or as Boice described it, "are willing to communicate with superstars in their disciplinary specialties or go to classes feeling less than perfectly prepared" [5, p.48].

Finally, there is the importance of social networking. Quick starters understand the difference between independence and interdependence. They also, as Boice put it, "Let others do some of the work. [They are able] to relax the usual proud autonomy so often characteristic of new faculty so that colleagues can assist as coteachers, coauthors, and mentors" [5, p.49].

Boice's additional insights into the success of quick starters as teachers will be examined in the next chapter.

CONCLUSIONS

The need to manage time and tasks more effectively exists in all institutions and at all levels, and has received increasing attention during the last few years. Yet, when it comes to actual practice, the results are almost always a function of individuals working alone. Within a given department, some faculty will be quite effective at managing their personal and professional lives, and others will not. Thus, it is *you* who has to take responsibility for striking the right balance, setting long-range goals, determining what you will and will not do, and deciding how to carry out the myriad of tasks on your plate. You simply cannot rely on your department chair, your dean, or others in the administration to do it for you.

Remember, also, you cannot do everything. You cannot please everyone, be available to everyone, and at the same time be the ideal teacher and scholar. There are certain things you must say "no" to, and other things on which it is okay to do less than a stellar job. Doing so will allow you to focus on doing an outstanding job on what is truly important.

An essential point, made over and over again by everyone interviewed, is the need to set aside blocks of time on a weekly basis for long-term important tasks. Making and *keeping* appointments with yourself is the key.

Last fall, a new faculty member joined my department. Immediately, he went about establishing his presence. He felt that it was all right to spend the fall and winter quarters "getting ahead" by always being available to his colleagues and students because he would be getting married in the spring and would then have to cut way back. He did get married in the spring, but of course he did not cut back. By then, he had established the expectation of, and received the positive strokes for, his universal availability, and it was now very hard for him to change. Do not make such a mistake. Establish your presence *and your absence* right from the very start.

REFERENCES

[1] J. Fleron, P. Humke, L. Lefton, T. Lindquester, and M. Murray, "Keeping your research alive," Project NExT Program, Minneapolis, MN, Jan. 9, 1995.
[2] W. H. Gmelch, *Coping with Faculty Stress*. London: Sage Publications, 1993. Copyright © 1993, Sage Publications, Inc. Reprinted with permission of Sage Publications, Inc.
[3] P. C. Wankat and F. S. Oreovicz, *Teaching Engineering*. San Francisco, CA: McGraw-Hill, 1993. Copyright © 1993. Reprinted with permission of the McGraw-Hill Companies.
[4] S. R. Covey, A. R. Merrill, and R. R. Merrill, *First Things First*. New York: Simon & Schuster, 1994. Copyright © 1994, Simon & Schuster. All rights reserved. Adapted with permission of Covey Leadership Center, Inc., 1-800-331-7716.
[5] R. Boice, *The New Faculty Member: Supporting and Fostering Professional Development*. San Francisco, CA: Jossey-Bass, 1992. Copyright © 1992 by Jossey-Bass, Inc., Publishers. All rights reserved. Adapted with permission.

CHAPTER 11

Insights on Teaching and Learning

> Dealing with the education/learning side of our business over the next ten years is going to be a tremendous challenge. The challenge is prompted in large part by the increasing demands for accountability in our teaching. We are going to have to rethink the whole educational enterprise, and in particular how we sell (and package) our intellectual property to what will be a much broader and more varying constituency.
>
> Michael Lightner, *Professor of Electrical Engineering*
> *University of Colorado at Boulder*

SETTING THE STAGE

Teaching is something all professors do. It is, or at least should be, the main reason why you chose to enter the profession. Teaching entails, as one faculty member put it, "an intergenerational obligation," and for many faculty, it represents their professional legacy. Getting started on the right foot with respect to teaching is important in part because, if you start out poorly, it is hard to recoup and turn things around later on.

Is it possible, then, to know what makes a good teacher? In a review of almost 60 studies of student's descriptions of effective teachers, it was found that concern for students, knowledge of subject matter, stimulation of interest, availability, encouragement of discussion, ability to explain clearly, enthusiasm, and preparation were at the top of almost everybody's list [1]. These student assessments correlate highly with the assessments of good teaching from faculty colleagues and administrators [2, p.12].

Teaching, of course, is only one side of the education coin; the other is learning. As Patricia Cross, professor of higher education at the University of California, Berkeley, points out: "It is increasingly clear that effective teachers must [also] have some basic understanding of the learning process. That means that they are able to make connections between what students already know and what we want them to learn" [2, p.9].

As was pointed out in Chapter 3, a revolution is underway in higher education, one component of which has to do with the way professors interact with their students. While many of the current computer and communications technologies provide new ways to carry out such interactions, they are primarily tools supporting a more basic transformation: the one from a traditional emphasis on teachers and teaching, to a more fundamental emphasis on learners and learning.

According to Cross [2, p.12]:

> The ultimate criterion of effective teaching, of course, is effective learning. There is simply no other reason for teaching. But we are beginning to see that learning probably depends more on the behavior of students than on the performance of teachers. Thus, research on teaching is shifting from observing how well the teacher is performing to observing how well students are responding. Good teaching is not so much a performing art as an evocative process. The purpose is to involve students actively in their own learning and to elicit from them their best learning performance.

Two educators who have contributed a great deal to our understanding of teaching and learning are Richard Felder, professor of Chemical Engineering at North Carolina State University, and Linda Silverman, director of the Institute for the Study of Advanced Development in Denver, CO. Felder and Silverman point out that different students learn in different ways, that is, they have different learning styles. (See Figure 11-1.) Different faculty also teach in different ways, that is, they have different teaching styles. According to Felder and Silverman, learning styles can be defined in large part by the answers to five questions [3, p.675]:

1. What type of information does the student preferentially perceive: sensory (external)—sights, sounds, physical sensations, or intuitive (internal)—possibilities, insights, hunches?
2. Through which sensory channel is external information most effectively perceived: visual—pictures, diagrams, graphs, demonstrations, or auditory —words, sounds?[1]
3. With which organization of information is the student most comfortable: inductive—facts and observations are given, underlying principles are

---

[1] Other sensory channels—touch, taste, and smell—are relatively unimportant in most educational environments, and are not considered by Felder and Silverman.

Figure 11-1 Teaching and learning styles.

TEACHING STYLES

1. Concrete or Conceptual
2. Visual or Verbal
3. Inductive or Deductive
4. Active or Passive
5. Step-by-Step or Global

LEARNING STYLES

1. Sensory or Intuitive
2. Visual or Auditory
3. Inductive or Deductive
4. Active or Reflective
5. Sequential or Global

Source: Adapted from R.M. Felder, and L.K. Silverman, "Learning and Teaching Styles in Engineering," *Journal of Engineering Education*, vol. 77, no. 2. p. 675, copyright © 1988, American Society of Engineering Education, reprinted with permission

inferred, or deductive—principles are given, consequences and applications are deduced?

4. How does the student prefer to process information: actively—through engagement in physical activity or discussion, or reflectively—through introspection?
5. How does the student progress toward understanding: sequentially—in continual steps, or globally—in large jumps, holistically?

Teaching styles, according to the authors, may also be defined in terms of the answers to five questions [3, p.675]:

1. What type of information is emphasized by the instructor: concrete—factual, or abstract—conceptual, theoretical?
2. What mode of presentation is stressed: visual—pictures, diagrams, films, demonstrations, or verbal—lectures, readings, discussions?
3. How is the presentation organized: inductively—phenomena leading to principles, or deductively—principles leading to phenomena?
4. What mode of student participation is facilitated by the presentation: active—students talk, move, reflect, or passive—students watch and listen?
5. What type of perspective is provided on the information presented: sequential—step-by-step progression (the trees), or global—context and relevance (the forest)?

The challenge with respect to teaching and learning centers around reconciling the often differing styles employed in each. Problems occur, say Felder and Silverman, because there are often significant mismatches between the learning styles of most college students and the teaching styles of most college professors. As they describe it [3, p.674]:

> Learning styles of most engineering students and teaching styles of most engineering professors are incompatible in several dimensions. Many or most engineering students are visual, sensing, inductive, and active and some of the most creative students are global; most engineering education is auditory, abstract (intuitive), deductive, passive, and sequential. These mismatches lead to poor student performance, professorial frustration, and a loss to society of many potentially excellent engineers.

While acknowledging that the diverse styles with which students learn are numerous, Felder and Silverman believe that the inclusion of a relatively small number of techniques in an instructor's repertoire should be sufficient to meet the needs of most or all of the students in the class. In particular, the authors suggest the following techniques to address various learning types:

- Motivate learning. As much as possible, relate the material being presented to what has come before and what is still to come in the same course, to material in other courses, and particularly to the students' personal experience (inductive/global).
- Provide a balance of concrete information (facts, data, real hypothetical experiments and their results) (sensing) and abstract concepts (principles, theories, mathematical models) (intuitive).
- Balance material that emphasizes practical problem-solving methods (sensing/active) with material that emphasizes fundamental understanding (intuitive/reflective).
- Provide explicit illustrations of intuitive patterns (logical inference, pattern recognition, generalization) and sensing patterns (observation of surroundings, empirical experimentation, attention to detail), and encourage all students to exercise both patterns (sensing/intuitive). Do not expect either group to be able to exercise the other group's processes immediately.
- Follow the scientific method in presenting theoretical material. Provide concrete examples of the phenomena the theory describes or predicts (sensing/inductive); then develop the theory or formulate the model (intuitive/inductive/sequential); show how the theory or model can be validated and deduce its consequences (deductive/sequential); and present applications (sensing/deductive/sequential).
- Use pictures, schematics, graphs, and simple sketches liberally before, during, and after the presentation of verbal material (sensing/visual). Show films (sensing/visual). Provide demonstrations (sensing/visual), hands-on, if possible (active).

- Use computer-assisted instruction—sensors respond very well to it (sensing/active).
- Do not fill every minute of class time lecturing and writing on the board. Provide intervals—however brief—for students to think about what they have been told (reflective).
- Provide opportunities for students to do something active besides transcribing notes. Small-group brainstorming activities that take no more than 5 minutes are extremely effective for this purpose (active).
- Assign some drill exercises to provide practice in the basic methods being taught (sensing/active/sequential), but do not overdo them (intuitive/reflective/global). Also provide some open-ended problems and exercises that call for analysis and synthesis (intuitive/reflective/global).
- Give students the option of cooperating on homework assignments to the greatest possible extent (active). Active learners generally learn best when they interact with others; if they are denied the opportunity to do so, they are being deprived of their most effective learning tool.
- Applaud creative solutions, even incorrect ones (intuitive/global).
- Talk to students about learning styles, both in advising and in classes.

Students are reassured to find their academic difficulties may not all be due to personal inadequacies. Explaining to struggling sensors or active or global learners that how they learn most efficiently may be an important step in helping them reshape their learning experiences so that they can be successful (*all types*) [3, p.680].

Let us now examine the use of various teaching and learning styles through examples presented in four vignettes of professors from four types of institutions. Each professor has a unique story to tell. However, each of their approaches, no doubt with some modifications, could be effective at almost any college or university.

In Vignette #14, we see how Sheri Sheppard, a mechanical engineering professor at Stanford University (Research I institution), develops teaching styles that more closely match the learning styles of her students. Sheppard believes this matching can be done by approaching teaching in the same way as one approaches a design problem, since both involve the task of figuring out how to present a given body of material in a variety of ways. She then discusses five pedagogical elements all new professors should focus on if they want to encourage effective learning.

In the next vignette (#15), Martin Ramirez, a biology professor at Bucknell University (Baccalaureate I institution), points out that there is a lot more to learning science than listening to a professor's lecture. Ramirez believes in the importance of using nonlecture approaches such as reading original scientific papers and using writing-intensive exercises leading to group projects, classroom debates, and oral presentations. Echoing a number of Sheppard's comments, Ramirez describes seven principles he believes will motivate students to become engaged and responsible learners.

In Vignette #16, Rollie Jenison of Iowa State University (Research I institution) describes a unique multidisciplinary design program involving teaching teams made up of professors from many different departments. These teams often include new professors who are able to obtain a cross-disciplinary experience through a highly interactive, yet supportive, environment. The program has an innovative teaching evaluation component involving not only students, but faculty peers. It is modeled after a much more extensive effort involving 12 schools that have joined forces under grants from the William and Flora Hewlett Foundation and the Pew Charitable Trusts to develop new prototypes for peer review of teaching of both tenured and nontenured faculty[2][4].

The last vignette (#17) is about Jo Anne Freeman, chairman of the Industrial and Manufacturing Engineering Department at the California Polytechnic State University—San Luis Obispo (Master's I institution). She describes what her school calls the "upside-down curriculum," which gets students involved early in the work of their chosen discipline. Freeman offers a number of suggestions for faculty who want to adopt elements of such an approach in their own undergraduate teaching.

These vignettes are followed by a look at advice from four additional reading resources, each in a specific area. The areas are (1) characteristics of successful teachers, (2) course planning, (3) technology in teaching and learning, and (4) developing a teaching portfolio. Together with your own efforts to engage knowledgeable professors, they should help you develop the confidence you need to successfully begin the teaching portion of your academic career.

Vignette #14 — Five Elements of Effective Teaching

A teaching challenge can be approached in much the same way as a design challenge. The same body of material can be presented in any number of ways. In design, and in teaching, you just have to keep trying until you find the way that works best.

This is the view of Sheri Sheppard, a tenured associate professor of mechanical engineering at Stanford University (Research I institution) in Stanford, CA.

---

[2] The 12 participating schools are: Indiana University–Purdue University at Indianapolis, Kent State University, Northwestern University, Stanford University, Syracuse University, Temple University, University of California—Santa Cruz, University of Georgia, University of Michigan—Ann Arbor, University of Nebraska, University of North Carolina—Charlotte, and University of Wisconsin.

Sheppard earned a bachelor's degree in engineering mechanics from the University of Wisconsin and a master's degree in mechanical engineering from the University of Michigan, while at the same time working at Chrysler Corporation in Detroit. She then went on to earn a Ph.D. from Michigan, also in mechanical engineering.

At Stanford since 1986, Sheppard teaches classes ranging from large lecture courses for undergraduates to small graduate seminars. Some of her students are "distance learners," who watch classes on television off campus, which creates additional complexities in lecture preparation and execution.

In 1988, Sheppard won the Stanford Tau Beta Pi Award for Excellence in Undergraduate Teaching, and has been similarly honored by the Society of Women Engineers at Stanford. As coprincipal investigator for Stanford's participation in the National Science Foundation Synthesis Coalition, she developed the course "Mechanical Dissection: A Lab-Based Study of Design Context."

Among the elements defining Sheppard's own teaching style are flexibility, a degree of spontaneity, and the willingness to deviate now and again from the curricular "script." "For example, the night before a lecture, I may get the idea that freezing some Bit-o-Honey bars, bringing them in [to class], and pounding them on the table would be a good way to illustrate a talk on brittle fracture at cold temperatures," says Sheppard.

Sheppard notes that, "Excellent teaching is only that if it affects excellent learning." Her advice to starting professors is to focus on five elements: (1) awareness of teaching and (2) learning styles, (3) course infrastructure, (4) creating communities, and (5) continual assessment.

Teachers need to be aware of their own teaching style so they can use it as the base upon which to build and improve. Sheppard did a "self-study" of her own style, and came up with the following:

> Interactive, I respond to questions real-time, high-energy, fast-paced—which is sometimes a problem—I tend to be problem-driven in my motivation; I don't like to do details on the blackboard; I tend to give handouts instead; I like to use a combination of board work and overheads, change the medium...I like to include hardware; I'm spontaneous to a certain extent; I tend to be playful.

Sheppard likes to have students experience problems in three ways: analytically, synthetically, and in a hands-on, real-world manner. "I want them looking around at the physical world, and seeing how what they're doing applies to that world," she says. "I bring in case studies; I bring in real people who are doing real engineering."

Sheppard echoes Felder and Silverman's point about the need for teachers to be aware of, and adapt to, students' varying learning styles. As she notes: "Some students are, for instance, more sensory than cerebral in perceiving and processing information. You need to be sensitive to augmenting your natural style with some other tools that help you to reach out to these students."

Regarding course infrastructure, Sheppard explained that the "scaffolding" around a course usually falls into two categories: the "stuff" category and the "people" category.

"Stuff" essential to classes includes things like textbooks, course readers, and creative use of electronic mail to communicate with groups of students. It also includes the teacher's willingness to be available to students at more than just the standard times.

Sheppard believes that professors and others involved in teaching should "create communities that overlap—a learning community and a teaching community." As she explains, "I'm a very big advocate of encouraging teamwork in class (active learning), in having students see themselves as not just one student in a class of 100, but as colleagues."

"Building a teaching community is also important," notes Sheppard. "In general, teaching is done as this private thing: your colleagues don't ask what you do in there. We don't ask, in part, because we're respecting our colleagues." There is much to be gained from "bouncing ideas off other people," she notes. One way is to coteach a course with another faculty member.

When Sheppard first came to Stanford, she sat in on other courses taught in the design division so that she could "learn about what the students were learning in the classes preceding mine. It was an incredibly valuable experience," she recalls.

Sheppard's final point is about the importance of continual assessment. Teachers, she says, should always be asking themselves: "Is it working? Is there excellent learning going on?"

Instead of waiting until a course is over to get feedback from students, Sheppard promotes involvement of teachers from day one. She likes to poll her students from the start, "to get a sense of what courses [they have] taken, something about their values and motivations."

"You also have to realize that as a professor, you are doing more than professing," she notes, "you are listening to their frustrations." She adds, "Realize that they have a lot more weighing on them than your class, and a lot of the things you do will be put in the proper perspective."

Vignette #15	Developing Engaged and Responsive Learners

Reading original science papers is a learning experience for undergraduates since doing so is a rarity for most of them. Normally, they have a professor telling them how things are in a nice, complete, tidy package, so it is a real revelation to actually read papers that establish particular facts and findings talked about in class.

This tool is just one of the innovative active learning approaches Martin Ramirez has taken throughout his five years of teaching science at liberal arts colleges in California and Pennsylvania. "What students find is that doing science is often 'messy,' " notes Ramirez. "It is fraught with difficulty, complexity, and unexpected challenges in the field and in the laboratory." Consequently, researchers sometimes have to make the best of a less than ideal situation, and as Ramirez points out, "They often make glaring, or not so glaring, mistakes or omissions." He observes:

> The amazing thing for students is to find that some of this material was actually published "as is." "Didn't the journal editors pick up on this mistake?" or "Why do they make this conclusion in their discussion when their data do not support such a conclusion?" Yes, there's a lot more to doing science than you would ever pick up on by just hearing a professor lecture, which is why I think students should wade in deep and get familiar with the literature concerning a particular subject, and then have to make sense for themselves of what it all means, how it all fits together, and what the big picture is.

Martin Ramirez is a nontenured assistant professor of biology at Bucknell University, a Baccalaureate I institution of some 3700 students in Lewisburg, PA. He graduated with a Ph.D. from the University of California at Santa Cruz in 1990, after which he served for two years as a lecturer in the Biology Department at Pomona College in Claremont, CA. He then began his current tenure-track appointment at Bucknell. Ramirez's research and teaching assistant experiences during his undergraduate days at Loyola Marymount University in Los Angeles, CA, convinced him of the benefits of teaching at a small liberal arts college.

At colleges like Bucknell and Pomona, notes Ramirez, the faculty/student ratio is very high, providing opportunities for individual interaction and attention often impossible at larger institutions. Faculty tend to know each other better, and students interact with professors more often, which results in a more intimate environment. Not surprisingly, at such places, undergraduate teaching receives considerable emphasis. Since upper division class sizes tend to be small, on the order of 10–20 students, innovative teaching approaches are possible. For example, in addition to having students read and discuss papers from the original literature, Ramirez's courses at Bucknell and Pomona are writing-intensive, and often culminate in group projects and oral presentations.

Ramirez emphasizes the importance of having students share their knowledge with their peers (and with him) in oral and written form. As he points out:

> There are many things you learn from assembling a body of data for presentation to an audience: the display of information graphically, the ability to sort out that which is really important from that which is not, and so forth. When secondary schools (and even colleges) had public speaking as a normal part of the curriculum, I think students had a much better chance to develop those skills of speaking, writing, synthesizing information, being critical about what one reads, and the value of evidence that are really the keys to success in any

job/profession. In all my courses, I have tried to have students engage in challenging presentations, discussions, and debates; to try to do my part to cultivate that set of skills which old-time public speaking/debate used to nurture.

Ramirez also encourages students to work in groups which, as he notes, matches what they will be doing when they go out into a research setting in academia or industry, or into any job or profession in the increasingly global, networked world. "The sooner we can help students get up to speed on learning to work with their peers to accomplish a common goal, the more quickly such skills will become second nature," says Ramirez.

What Ramirez teaches depends, of course, on the context of related courses and student preparation. But *how* he teaches is based on a set of principles designed to motivate students to become engaged and responsible learners. These principles focus on seven areas:

1. Active learning/involvement (fosters a sense of ownership in the course, via nonteacher-centered class sessions, in which teachers and students have particular roles to play each day).
2. Awareness and understanding of diversity (shows respect for variations in style, culture, points of view, and strategies).
3. High expectations (in a supportive manner, expects a high level of excellence from students).
4. Mutual feedback (provides regular opportunities for student/faculty and faculty/student feedback, and input into the evaluation/structuring of course activities).
5. Student/faculty relations (takes a personal approach with students that is honest and caring).
6. Practice (provides ample opportunities for students to work on or grapple with concepts/problems in such a way that they develop a history of success with the material).
7. Cooperative learning (encourages the formation of small groups to study, present material, and jointly deal with field/laboratory projects).

It is not difficult to see similarities among Ramirez's approach, the teaching and learning styles discussed by Felder and Silverman, and the undergraduate scholarship experiences advocated by Ernest Boyer, late president of the Carnegie Foundation for the Advancement of Science, in the vignette at the end of Chapter 1.

Clearly, for Ramirez, teaching at liberal arts colleges has yielded numerous rewards and considerable satisfaction. "Working at places that subscribe to the teacher/scholar model and that also provide appropriate resources has been a real plus," he states.

Chapter 11 Insights on Teaching and Learning

Vignette #16 — Team Teaching in an Interdisciplinary Program

Imagine a series of multidisciplinary design courses taught by teams of professors assigned to a service unit, and possessing degrees from a variety of engineering departments. Also imagine these professors participating in an evaluation process involving not only their students, but their faculty colleagues.

Rollie Jenison, a tenured professor of general engineering at Iowa State University in Ames, IA, is describing an innovative program with lessons for beginning faculty at many other colleges and universities.

Jenison explains the purpose of the program this way: "We are trying to break down the walls between disciplines in our approach to a group of multidisciplinary lower and upper division engineering courses, and in the process provide a supportive and effective feedback system for faculty."

Consider the Engineering Graphics and Introduction to Design course taken in class sizes of 36 by over 1200 students each year. A significant element of the course is a feature called "mechanical dissection," which has its analog in the more familiar biological dissection courses, and which has many lower division students taking apart and analyzing simple power appliances such as drills, skill saws, and juice blenders, as well as larger objects such as exercise equipment and bicycles. At the junior and senior levels, students dissect power lawn mowers and even full-sized internal combustion engines.

"There is much more to this course than just tearing things apart," notes Jenison. "It involves looking at how basic physical principles are applied to a wide variety of man-made artifacts we find in our everyday environment."

A second approach to accomplishing the division's goals is through the Design Associate program, which brings together upper division students and faculty from different departments who work in design teams to teach and solve specific, multidisciplinary, industry-based problems. One project involved students and faculty from the Departments of Agriculture, Materials Science, and Mechanical Engineering who together worked on the design of a new sulky.

Another project centered on the redesign of a stair-climber so as to make it more lightweight and user-friendly. It involved faculty from the Aerospace, Mechanical, Materials Science, and Industrial Engineering Departments.

An additional project involved students and faculty from the Computer/Electrical and Mechanical Engineering Departments who sought ways to implant sensors in hospital beds to alert nurses when patients got out of bed at unauthorized times.

"That particular project didn't produce a working prototype in spite of many tries," says Jenison. "Not all of them do. But they always provide valuable information and experiences for students, faculty, and industry."

According to Jenison, a common feature of these and the other 12 or so projects developed over the last six years is the way they bring together new and experienced professors from so many different areas. "It really is a diverse group of faculty who get involved, and I think the team-teaching aspect of it expands everyone's knowledge, while at the same time lending support to those participating for the first time."

The program provides a course coordinator who serves as a mentor to new faculty or to experienced faculty who are teaching in the program for the first time. "We have weekly group meetings where we go over the syllabus and look at what is going well and what we want to modify. It's very informal and nonthreatening, and it provides a great support mechanism for the newer teachers," says Jenison.

All of the courses are evaluated by students at the end of the semester, but many faculty ask for intermediate evaluations as well. Even more interesting is the use of three-person faculty teams who evaluate the teaching of all faculty, tenured as well as nontenured. "Tenured faculty are visited once per year, while nontenured faculty, once per semester," says Jenison. "These visits are announced well in advance and with the full cooperation of the faculty member. The team writes a report, and then meets with the faculty to go over the comments. It's actually a very positive approach that can lead to real improvement, and most faculty welcome it."

On occasion, an advanced Ph.D. student or postdoc will teach a section of one of the lower division classes, and of course participate in the course evaluation process. "It's fun to see these people have their first experience teaching freshmen," recalls Jenison. "They come out with a real appreciation of what teaching is going to be like. They also know how to seek feedback and support, not only from students, but from their future colleagues."

Vignette #17 | **The Upside-Down Curriculum**

All my faculty have real-world experience. Most have worked in industry, some have started their own companies, a few have worked overseas, and a number have ongoing consulting arrangements with government and industry. We do not have expensive research facilities so we need to have faculty who are constantly engaged with industry, and who then find ways to bring their experiences and issues into the classroom.

So says Jo Anne Freeman, chair of the Industrial and Manufacturing Engineering Department at California Polytechnic State University (Cal Poly), a primarily undergraduate teaching institution of 16,000 students in San Luis Obispo, CA.

Freeman advises new faculty to "find ways to establish or maintain your corporate contacts, and then use the corporate problems you encounter in your teaching." As she points out, "For us, it is not just a way of spicing up a lecture or two; it is our whole way of teaching." This way of teaching derives from Cal Poly's use of something called the "upside-down curriculum." "It means we turn everything inside out," says Freeman. "You see it in the small classes: the active learning, the hands-on polytechnic tradition, and the way we get students involved early in the work of their disciplines."

At Cal Poly, all students are required to take at least one course in their major every quarter starting from the first quarter they arrive as freshmen. While they have to declare a major early, i.e., when they apply as high school seniors, it also gives them an early exposure to the field, something Freeman believes makes a big difference. "We don't have to wait for co-ops in the junior year to build interest," she remarks. She went on to point out that students can change majors fairly easily if they find that they have made a mistake.

Classes at Cal Poly are lecture, activity, and laboratory-based. While you would expect activity classes (programming courses, for example) and laboratory classes to be small, this is also true for lecture classes, many of which average 20–25 students. "I think we have only one very large lecture hall on campus," notes Freeman, "and it is used mainly for music recitals." Smaller classes allow for greater student–faculty interaction, as well as the opportunity for students to work in small groups on a variety of practical projects.

Theory and practice are covered concurrently in the classroom, which is part of the school's polytechnic tradition. For example, one group of students might work on a plant location problem provided by General Motors Corporation, another a quality control problem presented by Applied Materials Corporation, and a third a database design problem presented by the U.S. Navy. In addition, all students have senior projects, for example: designing and implementing an inventory system for a local mountain bike manufacturer, evaluating and reviewing safety and hazardous materials methods for the San Francisco Airport, or designing a machine to reduce carpal tunnel syndrome in operators at a magnetic tape manufacturing plant through assisting in their repetitive work in testing tapes.

In Freeman's words:

> These design problems are messy, iterative problems, often with significant political and societal overtones. To tackle them effectively requires exceptional teachers, and here is where real-world experiences can make a difference. Advising these senior projects requires a balance between good solutions offered by the advisor and restraint in which the student is encouraged to invent and implement solutions based on his or her own experience and insight. Since our main goal is to produce undergraduates with excellent professional engineering skills and knowledge, this is very important to us. Professional practice through these senior projects and through the student's cooperative experiences

provide much needed self-confidence and polish, as well as exposure to many real-world constraints that help students to understand the real, engineering world.

Freeman herself is an excellent example of the type of teacher needed at such an institution. After receiving her bachelor's degree in industrial engineering from the Georgia Institute of Technology, she worked in the garment industry in Southern California while studying for her master's degree in industrial engineering from the University of Southern California. She also consulted for many companies in the Los Angeles area, specializing in consumer products and in the application of industrial engineering techniques to those firms. She began teaching at Cal Poly in the early 1970s, but after a few years realized that a Ph.D. was a necessity, and therefore went on to earn one from Stanford University in 1982. Freeman returned to Cal Poly for a few years, but then served a one-year stint at the General Motors Technology Center in Warren, MI, where she studied advanced manufacturing techniques, including artificial intelligence, robotics, and cell design. This experience led to a position on the faculty of industrial engineering and operations research at Virginia Tech (VPI&SU) in Blacksburg, VA. Freeman then took a one-year teaching assignment at Stanford University and a concurrent consulting contract with the Hewlett-Packard (HP) Company in Palo Alto, CA, where she worked on manufacturing education and training with HP's Headquarters' staff. She returned to Cal Poly as a tenured department chair in 1991.

In her engineering economics course, Freeman uses her experience with financing and economics to bring in real-world examples gleaned from industrial and commercial users of credit. For example, leasing versus buying equipment may be related to car buying for individuals. Students are most engaged when they understand the good sense of particular techniques that may be applied to themselves as well. In addition, individuals attend the orientation course in which the new students (both freshmen and transfer students) learn about their new major and about their new university. In these presentations, the students have a chance to learn about such things as the devices they might be designing and building, consulting possibilities for large garment manufacturers, "smart" products that need integration of mechanical, electronic, and computer science knowledge, or opportunities to work in the health-care industries.

Cal Poly's approach to teaching, with its small classes and individual attention, does not come free. The faculty workload in the California State College Systems is based on a 15-unit per quarter commitment of which three units may be used for such things as student advising, committee assignments, and other service activities. The remaining 12 units must be assigned to actual teaching, but on a weighted basis. Lecture classes count 1.0 unit per in-class hour, activity classes, 1.3 units per in-class hour, and laboratory classes, 2.0 units per in-class hour. "This means," said Freeman, "since a laboratory class is only about half the size of a typical lecture class, it actually costs four times as much to teach." Because Cal Poly is committed to small, hands-on classes, many of which fall into the activity and laboratory category, the institution is the most expensive on a per-student basis of all the 21 campuses of the California State College System.

"But we definitely think it is worth the cost," argues Freeman. "Our graduates are highly sought, our faculty are actively engaged in practical research, our students and graduates make enormous contributions early and long term, and we have fun! What more could you ask for?"

IN ADDITION: CHARACTERISTICS OF SUCCESSFUL TEACHERS, COURSE PLANNING, TEACHING, AND LEARNING WITH TECHNOLOGY, AND DEVELOPING A TEACHING PORTFOLIO

Beyond the advice found in the above conversations, there are several other sources you can utilize to help you get started in your teaching. As noted in Chapter 6, most colleges and universities have centers specifically designed to help faculty with their teaching. These centers are not just for professors with poor teaching records. They are also for those who want to improve on what is already a good thing. An early visit to the center at your new school should be high on your priority list.

Among other things, there are a number of books designed to help faculty with teaching, student learning, and curriculum development. Four that I have found particularly helpful are

- *The New Faculty Member: Supporting and Fostering Professional Development*, by Robert Boice [5],
- *The New Professor's Handbook: A Guide to Teaching and Research in Engineering and Science*, by Cliff I. Davidson and Susan A. Ambrose [6],
- *Teaching Engineering*, by Philip C. Wankat and Frank S. Oreovicz [8],
- *The Teaching Portfolio: Capturing the Scholarship in Teaching*, by Russell Edgerton, Patricia Hutchings, and Kathleen Quinlan [9].

Let us look at a contribution from each of these books in a specific area. The areas are (1) characteristics of successful teachers (Boice), (2) course planning (Davidson and Ambrose), (3) teaching and learning with technology (Wankat and Oreovicz), and (4) developing a teaching portfolio (Edgerton, Hutchings, and Quinlan).

Characteristics of Successful Teachers

We mentioned Boice's book briefly in the previous chapter in connection with time management and his study of "quick starters," those faculty who had excelled at teaching, research, publishing, and networking.

In Chapter 3 of his book, "Establishing Teaching Styles," and Chapter 6, "Establishing Basic Teaching Skills," Boice continues his study of quick starters with a focus on what makes them successful teachers [5, pp. 30, 75–76, 130]. In examining beginning faculty at both comprehensive and research universities, Boice found that virtually all of them had experienced a great deal of stress during their first two years. With respect to teaching, he noted [5, p.30]:

> Typically they over-prepared lectures and presented too much material too rapidly, they taught defensively so as to avoid public criticism, and they had few plans to improve their teaching beyond improving the content of their lectures.

Quick starters, on the other hand, adopted a number of behaviors to help ameliorate the stresses associated with teaching [5, pp. 75–76]. They:

- Had a positive attitude about students at their institutions,
- Gave lectures that were paced in a relaxed style so as to provide opportunities for student comprehension and involvement,
- Had low levels of complaints about their campuses, including collegial support,
- Showed evidence of seeking advice about teaching (especially the mechanics of specific courses), often from a colleague in the role of guide or mentor,
- Made a ready transition to moderate levels of lecture preparation (less than 1.5 hours per classroom hour), usually by the third semester,
- Made a generally moderate but meaningful investment in time spent on scholarly and grant writing (mean = 3.3 hours per work week),
- Showed a greater readiness to become involved in campus support programs.

Boice believes faculty need the most help with first-order basics such as classroom comfort and rapport, balancing teaching and other academic activities, and obtaining appropriate rewards from teaching. Without attention to this foundation, he fears new faculty will postpone involvement in higher order teaching skills, perhaps indefinitely [5, p.130].

Course Planning

Cliff Davidson and Susan Ambrose are both at Carnegie-Mellon University in Pittsburgh, PA. Davidson is a tenured professor of civil engineering and engineering and public policy, and Ambrose is the director of the University Teaching Center. Their book, *The New Professor's Handbook: A Guide to Teaching and Research in Engineering and Science*, is a "nuts-and-bolts" document for beginning faculty focused on science and engineering.[3] The first half of the book (Chapters 1–6) looks at teaching in a variety of settings, not just the classroom.

Of particular interest to new professors is Chapter 2, dealing with how to plan an undergraduate course in science or engineering. Whether you are teaching a course taught before by others, or developing a new course that is entirely your own, you need to do the appropriate planning.

The authors begin by setting forth eight principles for undergraduate education helpful in any curriculum and instruction effort. These principles, which are similar to those discussed by Ramirez in Vignette #15, are (1) encourage active learning, (2) design effective learning experiences, (3) provide prompt feedback, (4) emphasize the importance of time and effort spent learning, (5) encourage student–faculty contact, (6) encourage cooperation among students, (7) communicate high expectations, and (8) respect diverse talents and ways of learning [6, pp.20–23].

Davidson and Ambrose discuss six steps, based on the above principles, to be followed in planning a course. They are

- Assessing the backgrounds and interests of your students,
- Choosing the course objectives,
- Choosing the scope and content of the course,
- Developing the learning experiences within the course,
- Planning feedback and evaluation of student learning,
- Preparing a syllabus for the course.

Assessing the backgrounds and interests of your students helps you develop appropriate learning strategies. Some of this assessment can be done in advance by talking with faculty who are familiar with the types of students you are likely to have in the course. However, you may also need to hand out questionnaires and a pretest during the first day or two of class to obtain more specific understandings.

Developing clear course objectives helps you set high expectations of your students, and at the same time lets them know what to expect of you. The authors stress that objectives should indicate the expected knowledge and competence of students at the end of the course to enable you to observe and measure the extent to which they have satisfied these objectives.

Choosing the scope and content of the course is particularly important, although not particularly easy. The tendency of most faculty is to attempt to cover too much. Davidson and Ambrose [6, p.28] suggest that you begin by listing topic areas that are candidates for the course, and then rank them in order of

---

[3] The following material is adapted from [6].

what you consider most important in the field and what you believe will interest the students. To accommodate the variability in backgrounds, interests, and abilities among students, they suggest that you:

> Consider three categories of subject matter to include in the course: *basic* material that should be mastered by every student who passes the course, *recommended* material that should be mastered by those students seeking a thorough knowledge of the subject, and *optional* material that is intended only for those students with special interests who desire to learn more than what is offered in the course.

Developing the learning experiences within the course involves such things as: (1) choosing the right textbooks and other reading material; (2) examining the reading material critically to determine where to place your emphasis and the logical order of coverage; (3) determining the types of activities that will provide the students with opportunities to practice newly acquired skills, apply new information in different contexts, and ultimately achieve the course objectives; (4) creating a course schedule based on the university calendar and on the amount of time needed to cover each topic adequately; and (5) examining all the components of the course to make sure that they are consistent and complement each other in the manner intended [6, pp.28–30].

When it comes to planning feedback and evaluation of student learning, the authors emphasize the importance of tailoring your methods to the goals of the particular topic area, and not just to your personal preferences, philosophical beliefs, time available, and habit. They note the following [6, p.30]:

> Testing students on their creative problem-solving abilities may require you to develop open-ended problems. On the other hand, closed-form mathematical problems with unique solutions may be more appropriate for testing analytical skills. Written assignments may be best suited for testing synthesis and evaluation skills.

Finally, preparing a course syllabus, a document that explains the rationale, purpose, content, and procedures of the course, is essential to enable students to understand the purpose of the course and what is expected of them. Davidson and Ambrose list the following items for inclusion in most course syllabi [6, pp. 31–32]:

- The name and number of the course, number of credits, the name of the university, the date by semester and year, the classroom meeting place, and a list of prerequisites.
- The names, office locations, office hours, phone numbers, and electronic mail addresses of the professor and teaching assistants.
- A brief course description that provides an overview of the subject matter and a brief explanation of why students might want to take the course. The

syllabus might also contain a brief explanation of how the parts of the course fit together.
- A list of course objectives, stated in terms of what the students should be able to do by the end of the course.
- Information about the learning experiences in the course, including in-class activities such as lectures and discussions, and out-of-class activities such as readings and homework assignments.
- Information about policies established for the course, including attendance, late work, and make-up work.
- A course calendar that includes (to the extent possible) a list of dates for homework assignments, readings, quizzes, tests, papers, projects, and other work.
- Information about how grades will be determined, including the percentage of the grade for each major element of the course.
- A caveat that indicates that parts of the course are subject to change to meet the needs of students in the course. This qualifier allows instructors to slow down or speed up the pace of the course if students show a need.

Teaching and Learning with Technology

The use of electronic and computer technologies in teaching is, of course, not new; however, the nature of these technologies and how they are used have clearly changed over the years. The overhead projector is a case in point. It has been in use in education for decades, although it did take 20 years to move from the bowling alley to the classroom. What is new is the linking of the overhead projector (actually a very high-powered version of the original projector) with the laptop computer in a way that enables teachers to project on-line interactive computer displays to the entire class.

The plethora of new software and hardware technologies linked to the Internet has resulted in new opportunities to improve student learning and teacher productivity. Such opportunities raise important questions for new faculty to consider as they seek a good start in their teaching. How will you decide what technologies, if any, to adopt in your classes? How do you know that what you adopt is better, or worse, than traditional methods of teaching and learning? How do you choose approaches that will have a real impact on student interest and student learning, while at the same time reduce unnecessary risks, and not require an excessive amount of your time? Finally, how will you know if your efforts in these areas will be rewarded when it comes to retention, promotion, and tenure?

There is no question that a number of faculty, young and old, are reluctant to adopt many of the new teaching technologies. A recent study by the University of Southern California revealed that less than 5% of college and university faculty currently use computing to aid in classroom instruction [7, p.47]. As Robert DeSieno, a professor of computer science at Skidmore College, notes with respect to digital technology, one form of the new information technologies [7, p.47]:

> In the context of their courses, many faculty view digital technology as the latest collection of gadgets unable to deliver the educational merit promised by proponents. For these faculty, digital technology requires too much time and effort, supplies too many distractions, and yields too little value for the investment. Worse, this technology delivers only what the marketplace provides, and too often those provisions do not meet the local curricula needs of the faculty and students.

The situation with respect to quality software and courseware is improving, however, and as DeSieno goes on to observe [7, p.47]:

> ...marketplace pressures, student interests, the influence of professional organizations, and the example of colleagues are persuading faculty to employ digital technology in their teaching and scholarship.

Many of the issues related to the use of technology in the classroom are explored by Philip C. Wankat, a tenured professor of chemical engineering at Purdue University, and Frank S. Oreovicz, a communications and educational specialist in the School of Chemical Engineering at the same university, in Chapter 8 ("Teaching with Technology") of their book, *Teaching Engineering* [8]. Although specifically written for engineering professors, much of the material is applicable to science, and indeed to most other disciplines.

The authors begin by echoing DeSieno's point when they note that: "Over the years, the introduction of new technology for education has generated initial high excitement, but that has been followed by disillusionment, although eventually most technologies find a niche in the educational system" [8, p.143]. Wankat and Oreovicz offer the following guidelines for the successful use of any new technology in education [8, p.144]:

1. Plan use for a specific audience.
2. Define objectives which are relevant to the audience.
3. Pick a technological medium and a teaching method which are appropriate to the topic.
4. Pick educators interested in using the technology.
5. Plan for personal interaction, particularly among students.
6. Monitor the course and change materials and methods as appropriate. [22]

In their chapter, the authors discuss a number of technological innovations, two of which, television (and video) and computers, we will briefly examine here.

According to Wankat and Oreovicz, television and video have three advantages over other means of delivering instruction:

1. They can provide instruction at remote sites (extensively used for continuing education and graduate programs).
2. They can be used to break huge classes into much smaller sections.
3. They can be used to make "electronic" field trips to observe technology (great for showing visuals not available in any other way).

Many schools, in particular community colleges, have become heavily involved in distance learning via television. The biggest provider by far, however, is the National Technological University (NTU) whose satellite broadcasts vary from "single two-to-three hour programs for continuing education, to three-credit courses, to ninety-hour certificate programs, to master's programs for engineers" [8, p.145].

In the typical format of television teaching, the professor lectures to a live class, with his/her lectures also being broadcast to remote sites with students who can interact via two-way audio links and/or computer keyboard connections. Not only can students ask questions, but they can take objectively formatted diagnostic examinations and quizzes, the results of which can be displayed instantaneously to the instructor and the entire class.

However, television and video are not without their problems. As Wankat and Oreovicz note: "The major instructional difficulties with live television are the lack of contact between the students and the professor, and the cost and difficulty of doing anything other than 'straight' lecturing" [8, p.146]. As we all know, television encourages passivity. Electronic mail exchanges, phone office hours, and visits by the professor to various sites can all help, but, of course, they all take additional time. As with large lecture classes taught in the traditional way, the time needed to interact with off-site students should be taken into consideration when teaching assignments are made.

Wankat and Oreovicz's discussion of computers and teaching is divided into three parts: computer tools, computer-aided instruction, and interactive laser videodisks.

Computer tools include such things as spreadsheets, equation solvers, and symbolic algebra programs. These tools have gained increased acceptance in recent years to the point where they are now quite robust, inexpensive, and relatively easy to use. As Wankat and Oreovicz put it: "In many applications computer tools are a significant advance over both hand calculations and programming. Because of this advantage, computer tools, particularly spreadsheets, have been widely adopted" [8, p.153].

With computer-aided instruction (CAI), the computer either supplements or replaces traditional forms of instruction. CAI can be used in three modes: (1) drill and practice, (2) tutorial, and (3) simulation [8, p.157].

In the drill and practice mode, the computer presents questions to the student, the student responds, and the computer provides feedback on the responses. The authors liken this to textbook homework assignments with feedback from a teaching assistant or grader, except that the feedback is instantaneous and private.

In the tutorial mode, the computer provides instructional material, and may actually replace traditional delivery methods such as lectures and textbooks. According to Wankat and Oreovicz [8, p.157]:

> In addition to content material, the tutorial should contain example problems and figures, include questions and problems, and have a richer feedback than typical drill-and-practice programs. Tutorials can guide a student to different lesson parts depending on his or her response...The tutorial can guide the student through problem solving with prompts and then gradually reduces the number of prompts until he or she is solving difficult problems without help.

Simulation tools involve application software programs, many of which are computer-aided design (CAD) programs found in use in industry. They can be extremely powerful and realistic, although not always user-friendly. They are most commonly used in advanced undergraduate and graduate level courses.

The major downside to CAI is the time needed to author CAI programs. Wankat and Oreovicz compare it to the effort required to write a textbook. As noted above, more commercial publishers are moving in the direction of developing their own programs, yet other barriers to the adoption of CAI exist. The authors believe the main ones are: computer incompatibility, the "not invented here syndrome," cost to the students, lack of rewards to faculty for development effort, and too close an identification with television and its weak image of scholastic rigor [8, pp.158–159]. Wankat and Oreovicz believe CAI will be limited to courses with large enrollments across North America such as calculus, physics, chemistry, and certain lower division engineering classes [8, p.159].

Interactive laser videodisks (ILV) store a tremendous amount of information that can be randomly accessed in fractions of a second. ILV systems can incorporate a number of different media such as [8, p.159]:

- Photographs and text material
- Overhead transparencies
- Slides
- Motion pictures
- Videotape
- Computer text and graphics

Wankat and Oreovicz see three major advantages of ILV: (1) the ability to store as much as an entire course on one disk, (2) the ability to consolidate a variety of media into a single package, and (3) the ability to incorporate extensive visual material [8, p.160]. Uses for ILV include: operation at remote sites, individual self-study, student tutorials, in-class use, laboratory simulations, and outreach programs [8, pp.159–160]. The authors believe that interactive video laserdisks will develop a unique but small niche in places where it is clearly the best delivery system.

Many colleges and universities are investing heavily in new information technology, and such technologies offer significant opportunties for improvement in teaching and learning. However, faculty, particularly those just starting out on the tenure track, are not going to invest the time and energy required to effectively use such technologies unless the administration provides the appropriate rewards. As DeSieno comments [7, p.47]:

> Colleges should encourage educational uses of digital technology that can be assessed for their effectiveness, and that testify, at promotion and tenure time, to one valued aspect of a faculty member's work. These incentives, thoughtfully applied, will increase informational reach of students across the curriculum and equip them to explore digital forms of information after they graduate.

Developing a Teaching Portfolio

In Chapter 6, we talked about the value of establishing a teaching portfolio while at the graduate student or postdoc level. We also described some of the items you might include in such a document. If you have not started developing your portfolio, now is the time to do so. *The Teaching Portfolio: Capturing the Scholarship of Teaching*, by Russell Edgerton, Patricia Hutchings, and Kathleen Quinlan [9], is an excellent resource to help you, and perhaps others in your department, get started.

The book begins by discussing the concept, content, and format of portfolios in some detail, and follows with eight samples from a variety of schools across the United States. While only one of the samples is from science and engineering (Computer Conferencing About Computer Languages), all provide useful ideas for any portfolio preparation.

Of particular interest is Chapter 4, "Portfolios on Your Campus: Getting Started." It offers guidelines for those faculty wishing to establish the practice where it does not yet exist. The authors offer suggestions [9, pp.49–53] on how to:

- Find the right purposes and occasions to introduce portfolios
- Involve the right people in their preparation
- Learn from the development process
- Encourage collaborations
- Evaluate the products
- Use portfolio evaluation as an occasion for standard setting
- Establish a spirit of experimentation via "pilot" projects that are more acceptable and less threatening to existing practice.

Finding the right purpose and occasion to introduce portfolios can be tricky. At some schools, such as Syracuse University (Research I) in Syracuse, NY, such portfolios are an explicit part of the effort to shift reward structures more toward teaching [9, p.49]. However, Edgerton *et al.* caution against pushing such high-stakes use too early at schools where the groundwork for portfolios is yet to be laid. As they put it [9, p.49]:

> Such high-stakes context can in fact work against some of their [portfolios] strengths, washing out rich detail and variety for the sake of a more uniform and efficient decision making. On the other hand, where nothing at all is at stake, portfolios are likely to become an empty exercise.

Involving the right people in the initial development of portfolios is also important. Here, the authors argue for starting with faculty who are admired *teachers* and who are also prominent *researchers,* particularly at research universities. Since the status and prestige of such faculty is already established, they can give credence to the effort to promote the scholarship of teaching.

The authors also suggest that the *process* of preparing a teaching portfolio itself may be more valuable than the specific product or data that result from such an effort. For beginning faculty, an emphasis on portfolio preparation rather than portfolio review might provide the greatest benefit [9, p.50].

As an example, the authors cite biology professors participating in the Stanford University Teacher Assessment Project, who indicated that the process of preparing teaching portfolios was valuable because: (1) someone was very interested and concerned about their teaching, (2) the portfolio captured evidence that *looked like* their teaching, and (3) selecting evidence and writing captions and reflections had impelled them to clarify their intentions and beliefs about students, about biology, and about teaching" [10].

The discussion on collaboration is particularly relevant for beginning professors. As educator Kenneth Wolf puts it [9, p.51]:

> One of the drawbacks of portfolios is the difficulty of ensuring that the work presented is entirely that of the person whose name is on the folder. But this potential stumbling block can be turned into a stepping stone.
>
> Instead of treating authorship as a problem, treat collaboration as a virtue. In this view, teachers would be expected to seek out the assistance of others in their teaching and in constructing their portfolios.

Following up Wolf's comments, Edgerton *et al.* [9, p.51] suggest that:

> [A] powerful arrangement would be one based on mentoring, where the more experienced faculty member works with a younger colleague to organize and present

appropriate work samples and reflections. Indeed, one of the reviewers of this monograph argues that developing a teaching portfolio is an ideal way to establish long-term, positive working relationships between senior and new faculty members.

Of course, in some departments, mentors of this kind do not yet exist. In such situations, you might start by looking for possible mentors in related departments. Also, check with your institution's teaching and learning center.

Who should evaluate them and what criteria they should use are among the most frequently asked questions about teaching portfolios. The Stanford University Teacher Assessment Project found that: (1) looking at only a few entries per category was sufficient to adequately assess a portfolio, and (2) a holistic evaluation can be more helpful than a fine-grained, analytical scoring system.

Edgerton *et al.* point out that the purpose of portfolio evaluation should be to drive practice. Thus, if you want to encourage conversation about teaching, then a group of faculty from across departments might be helpful. On the other hand, if your goal is a content-specific evaluation of teaching, then evaluations from your department would be more appropriate. If individual improvement is the primary goal, then a small group of department colleagues might be the best approach. Finally, if portfolios are to be used in such high-stakes areas as tenure and promotion, then the school might seek to develop a small group of highly trained evaluators [9, p.53].

The idea of using teaching portfolios as a means of setting teaching standards is a powerful one. Of course, answering the question of what we mean by "excellent teaching" is not easy. One approach, suggested by the authors, is the inductive one in which portfolios of outstanding teachers are reviewed for their common elements and themes. This approach could be particularly effective if done in a group setting such as a department retreat [9, p.53].

While teaching portfolios are beginning to catch on at a number of schools across North America, they are still in the early stages of development and acceptance. We should not deceive ourselves into believing that an expanded use of such portfolios will come easily. Until there is a greater recognition of the various forms of scholarship discussed throughout this book, and an understanding that peer review of such scholarship is central to such recognition, advances will be limited. Edgerton *et al.* put it quite directly when they say [9, p.56]:

When and if the scholarly communities apply peer review to teaching and service as they now do to research, then and only then, will they have finally said, "we respect not only research but teaching and service as well."

It is my hope that we will see a stepwise process in which the pilot use of portfolios leads to a greater acceptance of peer evaluation of the various forms of scholarship, which in turn leads to a greater use of all types of "scholarship portfolios."

CONCLUSIONS

The four vignettes and follow-on book topic discussions in this chapter only hint at the amount of activity underway in teaching, learning, and course development in science and engineering. The use of student-led design teams, industry-based case studies, original source materials, special writing and speaking assignments, as well as the explosion in World Wide Web and other interactive distance learning tools are all reflections of the innovations taking place in schools across North America.

Not all of this innovation is taking place in traditional colleges and universities. A considerable amount is coming from community colleges looking for better ways to serve their constituencies and at the same time increase income, "private learning" companies, corporate "universities," and commercial publishing houses. One interesting example is The Teaching Company of Springfield, VA, which has registered the term "SuperStar Teachers," and in advertisements promises "Audio and video recordings of brilliant college lectures by SuperStar Teachers ranked highest by students at top universities" [11].

Most of these developments, however, have come from independent, and often solitary undertakings, particularly when done at four-year colleges and universities. Collaborations are much more common in the research realm. It is in research, not teaching, where peer review is the norm. We generally do not take the time to leverage and learn from each other when it comes to teaching. New professors often start from scratch, and/or teach what, and how, they were taught. This approach need not be the norm. There is so much to be gained in effectiveness and creativity, not to mention time and effort, from sharing with others what we are doing in the classroom. Remember Boice's quick starters; they were the ones who were not afraid to seek advice and materials from experienced teachers and other institutional resources, and then build on what they had acquired to improve their own teaching. By doing the same, you will increase your chances of success while freeing some time for other important activities.

REFERENCES

[1] K. Feldman, "The superior college teacher from the students' view," *Research in Higher Education*, vol. 5, pp. 43–48, 1976.

[2] K. P. Cross, "On college teaching," *Journal of Engineering Education*, vol. 82, no.1, Jan. 1993.

[3] R. M. Felder and L. K. Silverman, "Learning and teaching styles in engineering education," *Journal of Engineering Education*, vol. 77, no. 2, Apr. 1988. Copyright © 1988 by the American Society of Engineering Education. Reprinted with permission.

[4] S. Sheppard and L. Leifer, "Overview of the ME-PEER Project at Stanford," unpublished report, p. 1, Mar. 12, 1996.

[5] R. Boice, *The New Faculty Member: Supporting and Fostering Professional Development*. San Francisco, CA: Jossey-Bass, 1992. Copyright © 1992 by Jossey-Bass, Inc., Publishers. All rights reserved. Adapted with permission.

[6] C. I. Davidson and S. A. Ambrose, *The New Professor's Handbook: A Guide to Teaching and Research in Engineering and Science*. Bolton, MA: Anker Publishing, 1994. Copyright © 1994 by Anker Publishing Company, Inc. Reprinted with permission.

[7] R. DeSieno, "The faculty and digital technology," *Educom Review*, July/Aug. 1996.

[8] P. C. Wankat and F. S. Oreovicz, *Teaching Engineering*. New York: McGraw-Hill, 1993. Copyright © 1993. Reprinted with permission of the McGraw-Hill Companies.

[9] R. Edgerton, P. Hutchings, and K. Quinlan, *The Teaching Portfolio: Capturing the Scholarship in Teaching*. Washington, DC: American Association for Higher Education, 1991.

[10] A. Collins, "Portfolios for biology teacher assessment," *Journal of School Personnel Evaluation in Education*, vol. 5, no. 2, pp. 147–167, 1991.

[11] Advertisement for, "The Teaching Company," *The New York Times Book Review*, p. 21, May 12, 1996.

CHAPTER 12

Insights on Research

> You can't imagine what it's like as a junior faculty member. You get a research grant, and that's great! But no one ever tells you how to budget, how to plan, how to manage people in your laboratory. As a grad student, you are just doing your research. But now you are on the other side of the fence. Now you have to supervise the graduate students in the laboratory, and get them to do the experiments properly. And all the while they are scrutinizing your every word and action.
>
> *Elizabeth Komives, Professor of Chemistry*
> *University of California, San Diego*

SETTING THE STAGE

As a graduate student and a postdoc you spent most of your time doing research. Now, as a professor, you will not only "conduct research," but also "direct" research, and the change will be significant. Indeed, you may find it an even greater shift than the one from student to teacher. Your role as a teacher is much more structured and defined than is your role as a researcher. Many, if not all, of the classes you teach will have been taught before, and some kind of assistance with syllabi, homework, examinations, and even class notes will be available to you. As a teacher, you will know on a regular basis the kind of progress you are making. You will have taught a certain number of sessions, given a certain number of homework assignments, and graded a certain number of examinations. In most cases, the resources you need in order to do a good job will be available to you. You are the one in charge in the classroom, and it is relatively easy to assign credit (or blame) for your efforts. At the end of the semester, there will be closure, and you can move on to your next teaching assignment.

Such is rarely the case with research. Your research is not public in the same way as is your teaching. You must make time for research; it is not automatically set aside for you. You have to decide what areas will allow you, even at schools with large teaching responsibilities, to make a worthwhile contribution. Indeed, you will soon discover that sustaining an academic career is as much problem *finding* as it is problem *solving*. You also have to determine how and when to collaborate with colleagues, while at the same time distinguishing yourself from them. Classroom teaching and "teaching" as a research supervisor are not the same thing, and your role and authority in each are quite different. And then there are all those matters having to do with budgets and record keeping.

The biggest difference, of course, between being a student, and in most cases a postdoc, and being a professor is that now *you* are responsible for obtaining the funding you need to carry out your research and support your graduate students. This requirement means that very early on, you will have to start writing research proposals, which in almost every case will face considerable competition. Overall, considering both government and private sources, approximately 80% of all first-time proposers will have their proposals rejected [1, p.2]. On the other hand, and this is certainly something you should keep in mind, resubmissions have a much greater chance of acceptance than do original submissions. For example, the Geography and Regional Science Program at the National Science Foundation recently funded only 23% of standard first-time submissions, but 43% of proposals that had been resubmitted [2, p.1]. (Of course, not all proposals rejected on first submission were asked to make resubmissions.) Given the above statistics, it would be worthwhile for us to set the stage for our vignettes on faculty research by first looking more closely at the proposal development process.

As we discussed in Chapter 5, "Research as a Graduate Student and Postdoc", you need to propose compelling research that seeks answers to specific questions designed to excite those who will be reviewing your proposals. As Ronald F. Abler of Pennsylvania State University and Thomas J. Baerwald of the National Science Foundation put it [2, p.1]:

> A proposal that begins, "The question I wish to answer in this project is..." starts out well. You will strengthen your case if your question is rooted in general ideas. Questions rooted in theory are compelling. Questions specific to particular times and places are only interesting. In the competition for scarce funds, compelling questions garner awards, merely interesting ones do not.

In preparing a research proposal, you first need to know the various types of research support available to faculty. Roger V. Smith whose book, *Graduate Research*, was referred to in Chapter 5, lists six common sources of support [3, p.228]:

1. Free gift or grant-in-aid
2. Grant

3. Cooperative agreement
4. Contract
5. Fellowship
6. Scholarship

Free gifts or grants-in-aid money are often the most desirable because they come with the fewest constraints. Such gifts are usually small institution awards in the $10,000–$50,000 range. Although small, they often bear little or no overhead, and can be used in a general way to "support the research interests of the recipient." In other words, they can be used in a discretionary way for such things as salaries, student support, equipment purchases, and travel.

A grant, on the other hand, is often more substantial, and frequently requires a higher degree of accountability. A grant, in the words of William Dando of Indiana State University [1, p.2],

> is any form of financial assistance or support to carry on intellectual activities. The word "grant" means to entrust and connotes the implications of faith and obligation. Every grant agreement includes the understanding that the recipient (grantee) is entrusted with funds and has a commitment to perform some activity or fulfill some expectation of the benefactor (grantor, agency, or foundation).

Different granting agencies operate in different ways with respect to oversight and accountability. For example, the National Science Foundation (NSF) has an extensive peer review process prior to the awarding of a grant, yet is relatively hands-off once the award is made. The Department of Defense (DoD), on the other hand, does no external peer review of proposals (program officers make recommendations to DoD panels), yet is fairly hands-on after the award. The U.S. Department of Agriculture (USDA) and the Environmental Protection Agency (EPA) are further examples of agencies that provide a fair amount of direction during the course of your research. Smith refers to grants with such agencies as "cooperative agreements" [3, p.230].

Contracts, unlike grants, are awarded for research that requires fairly specific timetables and "deliverables." Government contracts are awarded in response to publicized RFPs or Requests for Proposals that appear periodically in the *Commerce Business Daily* [3, p.231]. Private contracts, usually with industry, are often negotiated individually by faculty who may do such work in the form of a consulting agreement. (See also Vignette #23, Chapter 13.)

Fellowships provide support for graduate students and postdoctoral researchers, while scholarships are usually designated for undergraduates. In some cases, students will come to you having already been awarded a fellowship or scholarship. In other cases, you will need to encourage students to apply, often with your support, for such awards. While students who have financial support are clearly desirable, remember that fellowships and scholarships usually only cover the students' tuition and stipend (and in some cases, not all of that), and not other costs of doing research such as computers, laboratory equipment, and travel.

Once you have an idea of the research you want to do and the types of grants you want to go after, you need to identify potential sponsors. Obviously, you should begin by looking at your own institution. Some departments, schools, and academic research centers provide support, often in the form of seed or start-up grants, for beginning faculty. While competition for these resources is often keen, the chances of success are usually greater than they are with outside support, and the length and complexity of the required proposals are not as high. Keep in mind, however, that seed or start-up grants are just that. Sooner or later, you are likely to have to seek external funding to sustain your research.

One of the best sources of information about potential sponsors is IRIS (Illinois Research Information System), a national database of funding opportunities for researchers and scholars in the sciences, arts, and humanities. The IRIS World Wide Web address is

http://www.grainger.uiuc.edu/iris/

IRIS contains federal and private agency information, sponsor descriptions, programs, and deadlines. It allows you to sort by keywords or phrases, deadlines, and sponsor/agency type (federal or nonfederal). IRIS also provides links to other agencies such as the National Aeronautics and Space Administration (NASA), the National Institutes of Health (NIH), and the National Science Foundation (NSF).

Prior to developing your proposal, you need to find out as much as you can about the agency to which you are applying and their degree of interest in your particular idea. Often, this means contacting the agencies' program managers or program officers whose responsibilities lie in your general area of expertise. While many beginning faculty feel uncomfortable with this approach, you should know that: (1) it is essential (over 75% of NSF awards are made to faculty who had prior contact with their program manager), and (2) the program manager expects, and even welcomes it [4].

Do your homework before making contact. Check things out on the World Wide Web, and write to the agency for a program description that will tell you what kind of work they are interested in and what kinds of research they have funded in the past.

Program managers and officers are often hard to reach on the phone. It is better to first send a brief e-mail message outlining your ideas in no more than three pages of text. Then follow up with a scheduled phone call and/or visit. Visits, which may require you to go to Washington, DC, can be important. As one successful researcher put it, "It is good to have the full bandwidth that such meetings can provide." On such a visit, be prepared for anything from formal stand-up presentations to informal sit-down discussions.

Keep in mind that we are talking about a relationship that will build over time. Even if you are not successful the first time out, you will need to go back to these individuals in the future, so think long term. Also, think about the possibility of starting your research under the tutelage of an experienced faculty member who has a contract/relationship with his/her funding agency. This kind of relationship is not always possible, or even desirable, but it is an approach worth considering.

When it comes to the actual preparation of your proposal, it is important to follow the agencies' guidelines to the letter. As Susan M. Fitzpatrick, program officer for the James S. McDonnell Foundation, comments [5]:

> We have received applications requiring us to retrieve the mailing label out of the trash in an attempt to identify who the proposal was from or try to obtain contact information (although we provide a proposal cover sheet). We have had to contact investigators to clarify whether their proposed budget was for one, two, or three years. Furthermore, I have read through proposals searching for the experimental plan, trying to figure out what it is the researcher intends to do. About 50 percent of the proposals we receive exceed the recommended 5,000 word limit, and almost as many are missing a requested support document.
>
> Applicants should try putting themselves in the shoes of the program officers and external peer reviewers. We will be reading 50 to 100 proposals, with very tight turn-around schedules. It doesn't take a genius to realize that the well-organized and thoughtful proposals will have a competitive edge.

To further assist you in your proposal preparation, I have included in Appendix F a list of the elements found in most successful proposals. The list was developed by Rebecca Claycamp, assistant chairman of the Chemistry Department, University of Pittsburgh. In addition, Appendix G contains Robert Smith's list of the common shortcomings found in most grant proposals.

Keep in mind that you do not have to do all of the work by yourself. Virtually all research universities, and now many master's and baccalaureate schools have in-house research project offices that can help you in identifying funding sources, developing budgets, viewing successful proposals, as well as the actual writing of your proposal. If your university does not have such an office, try teaming up with a colleague at a university that does.

Now, let us turn to specific advice from four faculty with respect to conducting and managing research. In Vignette #18, Lew Lefton, professor of mathematics at the University of New Orleans (Doctoral II institution), looks at how young faculty can keep their research alive in the midst of large teaching responsibilities and service obligations. He urges beginning professors to develop contacts with experts in areas adjacent to their particular research specialty, to seek institutional colleagues with whom to collaborate, and to make research one of your priorities by setting aside time for it on a weekly basis.

Our second vignette (#19), with Mark Hopkins of the Rochester Institute of Technology (Master's I institution), describes a program involving Hopkins, his institution, and a local research company. The program helped him get started with his research in a way that also contributed to his effectiveness as a classroom teacher.

In Vignette #20, Greg Petsko, director of the Rosentiel Basic Medical Sciences Research Center at Brandeis University (Research II institution), tells of an operation ideally suited for young faculty who want to work at the boundaries of disciplinary research. He discusses how his center's approach works, and the elements necessary to enable young, untenured faculty to be successful in such an environment.

Our last vignette (#21) features Nino Masnari, director of the Center for Advanced Electronic Materials Processing, headquartered at North Carolina State University (Research I institution). Masnari picks up on Petsko's theme, and extends it to cross-university collaborations. Here, the potential benefits for young faculty are even greater, but so are the difficulties. Masnari outlines some of the keys to surviving, and thriving, in such an environment.

These vignettes also provide illustrations of the various nonteaching forms of scholarship discussed throughout this book. While more than one form of scholarship is often represented in a given situation, generally speaking, we see the discovery of new knowledge in Lefton's efforts, the application of knowledge in Hopkins's work, and the integration of knowledge in the work of the centers described by Petsko and Masnari. The vignettes also demonstrate how the various forms of scholarship can take place at different types of institutions, in this case Doctoral II, Master's I, and Research I and II schools.

These vignettes are followed with a look at the other end of the research continuum, the writing of papers reporting on the results of your research work. In this discussion, we draw on specific contributions from two books: *A Ph.D. is Not Enough: A Guide to Survival in Science*, by Peter J. Feibelman, and *The New Professor's Handbook: A Guide to Teaching and Research in Engineering and Science*, by Cliff I. Davidson and Susan A. Ambrose.

Vignette #18	Keeping Your Research Alive

Most new Ph.D.s are too cautious when it comes to moving forward. You are no longer in a protected lagoon, and you need to venture out into the ocean. You do not have to try to cross the ocean, but you do need to determine a general direction and set sail for it.

Lew Lefton, a tenured professor of mathematics at the University of New Orleans (Doctoral II institution) in New Orleans, LA, feels that it is very important for new Ph.D.s to look beyond the narrow focus of their

Chapter 12 Insights on Research 295

dissertations if they want to keep their research alive. "Taking this broader perspective is what enables you to do interesting work other people care about," says Lefton. "At the research level, things change so quickly that you could easily be out of date if you don't have more than one problem to work on."

Lefton has three suggestions for faculty who want to keep their research going:

- Connect with experts in areas adjacent to your current research specialty.
- Seek out colleagues within your own institution.
- Make research a priority (in a way consistent with your institution's culture).

With respect to the first point, Lefton suggests setting aside an hour each week to browse through the library to see what other people are doing and to get a sense of what is hot in your field. Today, the "library" is more likely to be the World Wide Web, which is just fine because it also allows you to exchange e-mail messages with people who have posted their work on the Internet.

Of equal importance, says Lefton, is to start going to conferences. The size of the conferences you attend will be determined in part by your institution's research emphasis. At Research I universities, for example, it may be important to attend major national, and even international meetings. Yet, much can also be gained by attending smaller conferences, particularly if your travel budget is limited. As Lefton puts it [6]:

> I recommend the smaller regional conferences as opposed to the large annual meetings. Not that the annual meetings are bad; in fact, if there's a special session in your area it can be quite productive. But smaller conferences often have fewer distractions and more opportunities to interact with people who have similar interests. Even if you don't know anyone, go to listen and ask questions and learn. Once you attend a few such conferences, you will start to see several familiar faces and you will have begun the important task of establishing professional contacts in your area. Don't be intimidated; most active research groups are happy to welcome new people and they may well suggest some interesting open questions for you to work on.

Lefton stresses the importance of not only attending conferences, but of *participating* in them. "By giving talks and getting feedback, you learn if your work is interesting and relevant, or if it is just too esoteric or bizarre," he says. "Conferences are also a place where you can find the magic that got you interested in your field in the first place as an undergraduate, and which may have been lost while a Ph.D. student." Finally,

Lefton reminds us that: "You will need references for upcoming tenure and promotion reviews, and these are the people who can give them to you. You can, after all, rely on your graduate school faculty for just so long."

A second way to keep your research alive is to find colleagues at your institution with whom you can collaborate on a regular basis. Lefton meets regularly once or twice a week with colleagues to work on problems of mutual interest. As he describes it:

> We think aloud, go to the board, write down equations, and bounce lots of ideas off of each other. We are not judgmental, and neither of us is embarrassed to ask simple questions. It takes awhile to build up this trust, but when we do, it works wonders.

Where possible, Lefton urges new faculty to pair up with older, more experienced colleagues, even if this may seem intimidating at first. "You need to work with someone who will really challenge you and stretch you as much as possible," he notes. "I did this as a postdoc with a leader in my field, and it really made a difference. Indeed, our relationship continues to this day."

Finally, Lefton stresses the importance of making your research a priority, particularly at institutions where teaching commitments are rather high. His advice in this respect is quite familiar:

> Set aside particular times of the week and find a quiet place to go where you'll be free from interruptions. This step is really important, and I know it can be difficult to do. It may help to try saying "no" when you are asked to do additional time-consuming activities. There are many worthwhile projects, but you can only do one thing at a time. In the early years of your career, you need to establish a research record, and saying "no" to certain other things may be necessary.

Good advice, although Lefton notes that it is also important to do things you enjoy outside of your professional life. When he is not working out new proofs or trying to explain them to his students, he works as a stand-up comic in and around New Orleans.

Vignette #19   A High-Leverage Approach to Industry–University Collaboration

The Rochester Institute of Technology (R.I.T.) is on a year-round quarter system, and for seven years, I spent every other quarter doing full-time research at the Xerox Wilson Center for Research and Technology in nearby Webster. The other two quarters were spent as a full-time professor at R.I.T., while also spending one day per week at Xerox. Throughout the year, I averaged 60% time at Xerox and 40% time at R.I.T.

Getting started well with your research not only depends on your own interests and initiative, but also on the resources, modes of operation, and expectations of the department and institution to which you belong. According to Mark Hopkins, an associate professor of electrical engineering at R.I.T. (Master's I institution) in Rochester, NY: "Finding the combination that works for you is the key, and this arrangement happened to be what made sense for me." During the seven-year period with Xerox and R.I.T., Hopkins received outstanding teaching evaluations, published papers based on his research at Xerox, acquired two patents, with one pending, and perhaps most importantly, obtained academic tenure.

"Originally, some of my colleagues were skeptical and thought I had not spent enough time at R.I.T. to be awarded tenure in the normal time period," comments Hopkins. "But this arrangement was part of my original appointment, and when they looked at the entire effort, including my teaching record, they chose to award me tenure."

The relationship with Xerox was the brainchild of Professor David Perlman, and was set up prior to Hopkins's appointment. Students at R.I.T. are required to spend considerable time in co-op programs, and the institution wants faculty who understand industry, who can carry out research of interest to industry, and who can bring industry problems and solutions into the classroom. Perlman arranged for a contract to be signed between Xerox and R.I.T. creating a special position having the above characteristics and in which Xerox would pay R.I.T. 58% of the professor's salary, plus benefits. Notes Hopkins:

> When I applied to R.I.T. as an assistant professor, I had just finished going straight through from my bachelor's degree (Southern Illinois University at Carbondale) to my Ph.D. (Virginia Polytechnic Institute and State University) without acquiring any industrial experience. I very much realized such experience would be important in my teaching and research, particularly at a school like R.I.T. When they suggested I apply for the (Xerox) joint position, I jumped at it.

Hopkins's research at Xerox was done with a group of about six people who worked for five years on developing a way to stabilize the color xerographic process. "This problem was much more difficult than anyone had first realized, involving various novel control architectures," comments Hopkins. "The problem is complicated by the fact that: (1) the process itself is highly nonlinear, time-varying, and noisy; (2) measurements of the process are infrequent, low-bandwidth, and not necessarily representative of the process; and (3) the process is subject to uncontrollable, and unmeasured (or unmeasurable) disturbances."

The research team developed a strategy and an architecture to solve the problem, and as Hopkins notes with some satisfaction, "was disbanded after five years, having achieved our mission."

The main payoff of this experience in terms of teaching was Hopkins's understanding of how a large technical company works, and of the "tools of the trade" used in industry R&D operations. As Hopkins puts it:

The main thing is that I'm now able to speak to students about what it is like to work in a big company. I also now understand much better the use of tools such as data acquisition systems, graphical interfaces, and various software products, and this knowledge helps tremendously with the material I introduce in my course, as well as in my discussions with students. It also improves my credibility.

The on-again off-again teaching and service at R.I.T. was not much of a problem given the way classes begin and end each quarter. But Hopkins admitted it was a little more difficult at Xerox. "It sometimes became tricky to wrap things up by the end of the quarter," he says. "But continuing on one day a week during the time I was at R.I.T. helped a lot."

Hopkins feels that the arrangement was particularly valuable during his first few years at R.I.T., but now he has decided to move to a regular full-time professorship on the campus. "I want to establish myself as a professor at R.I.T. who does his own research," he says. "When I was at Xerox, I was part of their research program, which was fine at the time. I didn't have to worry about going through the hassle of finding research support and the like, which really helped me as a new professor. But now I'm at a point where I want to pick what I do and study the stuff I'm particularly interested in."

Hopkins is currently working in the area of predictive diagnostics, the goal of which is to find feasible means of extracting information (from the color xerographic environment) and using it for automatic failure prediction, as well as automatic diagnosis. "Clearly, this work derives from my earlier studies at Xerox," says Hopkins, "and this is important for any professor who wants to build a research reputation. Another faculty member has taken my place at Xerox, and I anticipate that he will have an equally successful experience."

Vignette #20 Multidisciplinary Research and the Untenured Professor

My approach is to bet on people, not projects. During my tenure as director, I want to bring in outstanding, nonsenior faculty. It makes things much more exciting from a research point of view, and besides, these are the faculty who make the best teachers.

Greg Petsko, who has appointments in both the Chemistry and Biochemistry Departments at Brandeis University in Waltham, MA, is talking about the Brandeis Rosentiel Basic Medical Sciences Research Center, a multidisciplinary facility ideally suited to beginning faculty who want to work at the boundaries of disciplinary research.

Established in 1973, the Rosentiel Center is housed in a single building designed for extensive interaction. Faculty from the Physics, Biology, Biochemistry, Chemistry, and Biophysics Departments are all represented here. The physical attributes of the center, which play an important part in its overall philosophy, are as follows [7]:

> The first floor contains a classroom, administrative office, and an electron microscopy facility. The second floor houses shared services such as fermentation, glassware washing, electronics and machine shops, supply stores, and peptide and DNA synthesis facilities. Floors three through six house the research laboratories of the faculty members who belong to the center, and who are grouped, four per floor, according to broad research interests. The third floor is developmental and cell biology; floors four and six house the structural biology laboratory, and the fifth floor is devoted largely to immunology. The seventh floor is a penthouse/cafeteria where all members of the center can mingle for lunch and Friday tea.
>
> The laboratory space is located in a totally open architecture with no physical barriers between research groups. Students from different groups are lumped together in offices and labs around the periphery of each floor, and common equipment and other facilities occupy the central region, where people naturally interact as they work. Faculty offices are located in the laboratory areas, so that faculty are always accessible and in touch with what is going on.

The result of this physical arrangement, according to Petsko, is a multidisciplinary environment unlike any other.

> There are yeast geneticists sharing space and equipment with physicists building X-ray area detectors, who in turn rub elbows with cellular immunologists, who in turn use the same facilities as protein and virus X-ray crystallographers, and so on. You can find, as close as the next office, a person to answer questions on unfamiliar fields, to teach your students a new technique, or to collaborate in a broadly based research endeavor.

How well does this work for beginning faculty who may want to become involved in multidisciplinary research, but who also want to develop an individual research identity and an affiliation with a particular department?

"Every faculty member at Brandeis has an appointment in a regular department," notes Petsko. "If we then want to locate them in the center, we go about convincing them it is to their advantage to do so." Petsko feels there are a number of reasons why such an arrangement would appeal to a young faculty member.

> First, there are the common set of facilities, equipment, and tools. For a new faculty member, in particular, having this infrastructure available and supported by a critical mass of people can be very helpful. We charge faculty from their

grants for such support, but the center also has funds to help faculty through "dry" periods between grants. We also have the administrative support to help faculty with the preparation and tracking of proposals and grants, which can make a huge difference for beginning professors.

The center also aggressively promotes its younger faculty for awards and positions on editorial and review boards. Given the above, it is not surprising that most faculty consider it an advantage to be asked to locate their office and laboratory in the center. Department chairs support such a move for their young faculty because it helps defray costs and increases the likelihood that faculty members will be successful.

How does all this interdisciplinary activity play out when it comes to tenure? In virtually all schools, as Petsko points out, tenure is granted by the host department. However, the center does offer its input based on the recommendations of its committee. The key is the way Brandeis looks at multidisciplinary research in tenure decisions. According to Petsko:

> Here, collaborations of an interdisciplinary nature are not only *not* seen as a problem at tenure time, they are seen as a *giant plus*. We regard collaboration and interaction as the right thing to do, and consequently we look for people from the start who intend to operate this way. In the life sciences, most young people want to do interdisciplinary work because the field is so "systems-focused." This systems emphasis is less common in physics and chemistry, but here we are certainly making progress.

Petsko, who also teaches an undergraduate course in detective fiction at Brandeis, offered an interesting metaphor to sum up how his university in general, and the Rosentiel Center in particular, supports its young faculty.

> At some schools, the faculty and administration stand on the deck and throw their young professors anchors, and wait and see if any of them can stay afloat. At still other schools, they just stand there and watch you thrash about. At Brandeis, we help you learn how to swim, and if it's necessary, we will throw you a life preserver.

Vignette #21 Cross-University Collaborations

> Few schools can stand alone anymore. Joint experiments with other universities are going to be the norm. We will need a new paradigm, one that values and recognizes team research where each person draws upon, and contributes to, the work of others. Yet, we may need a generation to pass through before such collaborations are really valued and completely accepted.

Chapter 12 Insights on Research

So says Nino Masnari, professor of electrical engineering and director of the Center for Advanced Electronic Materials Processing (AEMP) at North Carolina State University (N.C.S.U.) in Raleigh, NC. Masnari should know. His center, established by the National Science Foundation in 1988 and headquartered in the College of Engineering at N.C.S.U., is one of the oldest examples of multiuniversity research collaboration. It is also a likely model for other universities in the coming decades.

According to Masnari:

> Part of our original proposal was to have projects ongoing within different departments and at the various universities and institutes. It was understood these would not be stand-alone efforts, but ones that were integrated into the overall mission of the center. This approach is one of the reasons why there is such close collaboration among faculty in so many different locations.

In addition to the research at N.C.S.U., work under the auspices of AEMP takes place at the University of North Carolina—Chapel Hill, the University of North Carolina—Charlotte, Duke University, the North Carolina Agricultural and Technical State University, the Research Triangle Institute, and the Microelectronics Center of North Carolina.

As Masnari describes it, AEMP:

> is developing technologies for *in situ*, single-wafer processing (cleaning, deposition, etching). The center deals with automation and control of the individual processes and their integration into single-wafer processing module clusters.
>
> Our efforts involve collaboration among chemists, physicists, materials scientists, and electrical, mechanical, chemical, and computer engineers, and involves graduate and undergraduate students in various disciplines, as well as advisors from industry.

What does such a center mean for beginning faculty who do not yet have tenure, who are looking for ways to obtain resources, who may want to collaborate across institutions, but who must also make a name for themselves in research? These are similar to the questions raised by Greg Petsko in the previous vignette, except that now there is the added dimension of collaboration outside the four walls of a particular building at a single institution.

"The National Science Foundation was concerned about this situation from the very beginning," says Masnari. "With young faculty working across disciplines and across universities, how were they going to do what was necessary to meet the traditional requirements for promotion and tenure?"

Since its inception in 1988, eight of the nine young faculty associated with AEMP have gotten tenure and/or have been promoted. "These include professors in electrical engineering, chemical engineering, and physics, plus one or two other fields," says Masnari. This kind of success only takes place if both the center and the department take ownership of the faculty member. "There has to be an understanding up front that there will be a clearly defined path toward research and tenure; otherwise, it just won't work."

That is the good news. Masnari went on to note, however, that:

> There are still some faculty who hold to the old classical faculty scholar model where a professor works alone with a few graduate students and takes sole responsibility for obtaining the resources he needs. Some of these faculty look askance at the younger professors who have a leg-up on getting resources through the center. They feel these faculty haven't had to work as hard to obtain resources and/or support, so the senior faculty tend to devalue their work.

This situation is changing, but as Masnari noted earlier, it may take a while before everyone sees value in this kind of research. "Not everyone is cut out for operating in this team-oriented way," says Masnari. "But for those who are, it can be a wonderful way to make significant contributions." Of course, the real test will come when Ph.D.s based on dissertations involving such collaborations are accepted by the broader academic community. AEMP is certainly doing its part in this regard. Since its founding in 1988, the Center has had an impact on the work of over 85 Ph.D. and 75 master's degree students. Approximately half of the Ph.D. dissertations were based on work involving teams from departments such as electrical engineering, materials science, chemistry, and physics.

While the major focus of AEMP is on advanced research, Masnari is also proud of the center's educational efforts. One example is AEMP's summer program for undergraduates involving students from North Carolina Agricultural and Technical State University and other historically black colleges and universities. At N.C.S.U., the students design, build, and test integrated circuits, give presentations on their work, and meet with faculty mentors from a number of institutions. Masnari notes proudly, "Sixty-five to seventy percent of these students end up going on to graduate school."

IN ADDITION: WRITING RESEARCH PAPERS

In the first section of this chapter, we discussed the writing of research proposals. Now, let us look at the other end of the research continuum, the writing of papers reporting on the results of your research work. We examined this topic briefly in Chapter 5. Here, we will take a more in-depth look at what two books have to say about such an important activity. The books are: *A Ph.D. is Not Enough: A Guide to Survival in Science*, by Peter J. Feibelman [8], and *The New Professor's Handbook: A Guide to Teaching and Research in Engineering and Science*, by

Cliff I. Davidson and Susan A. Ambrose [9]. We referred briefly to parts of Feibelman's book in earlier chapters, and discussed the contribution of Davidson and Ambrose to course preparation in Chapter 11. The discussion on the writing of research papers appears in Feibelman's Chapter 4, "Writing Papers: Publishing Without Perishing,"[1] and Davidson and Ambrose's Chapter 9, "Writing Research Papers."[2]

Both chapters begin with a discussion on the importance of writing and publishing high-quality research papers. The authors admonish you not to view publishing as a "requirement" imposed by the system, but rather as a way for you to gain acceptance of your ideas while also telling the world of the results of your hard work. Publishing can also be, as Feibelman points out, "a timeless advertisement for yourself" [8, p.39]. That said, you also need to remember that you cannot take back a publication. While excellent papers will serve as a permanent public record of your accomplishments, poor papers will damage your reputation [9, p.127]. And poor papers do get published. One measure of the quality of a paper is the degree to which it is cited by others in the field. While the norms for citing papers differ among fields, approximately 55% of all articles published in top journals worldwide are never cited by anyone [9, p.127].[3]

Davidson and Ambrose follow their discussion of the importance of publishing with a look at four types of research papers: (1) peer-reviewed papers in journals, (2) conference papers, (3) research reports, and (4) books or book chapters [9, pp.128–129]. While there are exceptions, generally peer-reviewed papers are of a higher quality than are those that appear in other forums. These are the papers that count most when reviewing a publication record for promotion and tenure. Conference papers, which cover in more detail the topic you are going to present orally at a conference, may or may not be peer reviewed. Research reports are those usually required by funding agencies, are more detailed than the journal or conference papers, and generally have a more limited audience. Of course, they are often the basis for the first two types of papers. Books or book chapters give you even more space to tell your research story, although the level of detail you present depends on the audience [9, pp.128–129].

Both Feibelman and Davidson and Ambrose spend most of their respective chapters discussing the writing of peer-reviewed journal papers. In this connection, is it better to publish shorter, more frequent papers, or longer, less frequent papers? As was pointed out in Chapter 5, some fields, such as microbiology, lend themselves to shorter investigations that can be reported more frequently, whereas

---

[1] Adapted from [8].

[2] Adapted from [9].

[3] Given this statistic, most authors would be happy to have their papers cited just once. However, if you want a standard to shoot for, look no further than physicist Edwin Witten, whose 96 papers on superstring theory published between 1981 and 1990 were cited by other scientists a total of 12,105 times [10]!

other fields, such as population biology, may take longer to produce publishable results. While all three authors argue for high quality, Feibelman takes the view that, where possible, shorter, more frequent papers that later can be combined into a significant review paper are a good approach for beginning researchers. Such papers can: (1) provide evidence to your supporters of the results of your work, (2) keep your name in the academic spotlight, and (3) reduce the chances of your being scooped by your competitors [8, pp.40–43].

Feibelman suggests that you organize a research paper somewhat as you would a newspaper article, that is, tell the same story several times by going into increasing levels of depth and difficulty [8, pp.43–44]. The title itself (equivalent to the newspaper headline) is very important since it will determine if busy readers will go further. It needs to be, says Feibelman, "concise, accurate and compelling" [8, p.44].

In both of their chapters, the authors emphasize the importance of the abstract, which follows the title, and which is usually from 50 to at most 300 words. It is often circulated much more widely than the article itself, so pay close attention to it. Davidson and Ambrose describe two types of abstracts: *descriptive*, which lists the contents of the paper, and *informative*, which describes the most important results and their significance [9, p.132] Research papers should use the latter.

The abstract is followed by the introduction. According to Davidson and Ambrose, the introduction serves several purposes [9, p.133]:

> It describes the general topic area of the paper, lists the specific problems of interest, and presents the motivation for the work. Unless the manuscript is very short, the introduction should also include statements about the organization of the paper: listing the major sections helps the reader understand the flow of ideas that follow.

Feibelman has an interesting way of overcoming the difficulty most of us feel when writing an introduction to capture what new knowledge has been produced by our research. He says [8, pp.45–46]:

> In my introduction I want to let my reader know what this new information is, in a nutshell, and why it is worth reading about. Sitting at the word processor, I imagine that I am on the phone with a scientist friend whom I haven't spoken to in some time. He asks me what I have been doing recently. I write down my imagined response. If, when you try this, you feel an attack of writer's block coming on, turn on a tape recorder and actually call a friend. It works.

Next comes the literature review where citing the work of others is essential. Obviously, you want to show the connection between their work and yours. Not only does your professional integrity demand this, to do otherwise would be a case of fraud. It also certainly does not hurt to give credit to colleagues, some of whom will be reviewing your paper for publication.

Davidson and Ambrose then discuss the stating of your research objectives and how to describe your methods of achieving them. With respect to the latter, they suggest that you describe the equipment and experimental procedures for laboratory or field work, and the mathematical relations and solution techniques for a theoretical study [9, p.134].

In the results section, which usually follows the section on objectives, you should refrain in most cases from including raw data, but rather, you should present the results themselves with some explanation of how you got to them [9, p.134].

The discussion section, which usually comes next, explains the significance of your results. According to Davidson and Ambrose, a poor job here is the main reason for rejection of most journal papers. As they put it [9, p.135]:

> In some cases, there is a fatal flaw: the research results are simply not significant enough to warrant publication. In other cases, the findings are interesting and worth publishing, but the discussion is inappropriate. For example, the author may be afraid to make a bold statement even when it is supported by the data (perhaps the true significance of the work is not recognized), or conversely the author may make unsubstantiated, sweeping claims when in fact only modest claims are warranted.

Some papers follow the discussion section with a section on future work. If this is the case, such statements should be limited to broad overviews of the directions you see your research taking, and not the kind of detail that would fit into a forthcoming proposal [9, p.136].

Next to the title and abstract, the section most likely to be read by most readers is the one with the conclusion or summary statement. A conclusion states the outcome of your work, whereas a summary is a brief statement covering the main points of the paper [9, p.136]. Either, or both, may be found in a research paper.

While the two books discussed above, plus standard writing references such as *The Elements of Style* by Strunk and White [11], can be extremely helpful in the development of your research papers, do not forget what may be your most important resource: your research colleagues. Asking colleagues to give you critical feedback on your drafts before submission to a journal can save you tremendous time and considerable grief, and will go a long way to increasing your chances of eventually having your paper accepted for publication.

CONCLUSIONS

What messages can you take away from the vignettes of this chapter as you undertake the research component of your academic career? There are several. First is the stress placed on going beyond your dissertation or postdoc research. You can certainly build on this research, but now is the time to broaden your perspective, and as Hopkins says, "set sail beyond the lagoon." Recall the example of Brian Love, professor of materials science at the Virginia Polytechnic Institute and State University. Within the more fundamental field of surface science, Love specialized in the bonding of semiconductor packaging materials. However, as time went on, he expanded his "applications portfolio" to include dental crown bondings and bondings for implanted prosthesis devices. Staying within the surface sciences theme, but expanding the areas to which it could be applied, allowed Love to, if you pardon the expression, broaden his contacts, while also increasing his funding options.

The second message of these vignettes is about the importance of setting aside time to work on your research, that is, to place it in Steve Covey's "Quadrant of Quality." (See Chapter 10.) Doing so may seem easier at Research I and II universities, where research expectations are higher and teaching commitments are lower, but, in fact, having even more unstructured time may make it more difficult. At all institutions and at all levels, it is incumbent upon you to manage your time and set aside uninterrupted periods for all forms of scholarship.

The third message of the vignettes is about the value of collaboration. As we have seen, research collaborations are now the norm, be they Lew Lefton meeting with a fellow mathematics professor down the hall, Mark Hopkins working with his research team at Xerox, the physicists, chemists, and biologists working on cancer research at the Rosentiel Center, or the electrical engineering and materials scientists working across universities at AEMP. To be sure, some faculty still prefer to work alone or with just a few graduate students, but this is less and less the case as topics become more complex and the advantages of collaboration become ever more obvious.

Keep in mind that collaboration can take many forms. Most efforts are discipline-focused, with two or more people working in the same area, each contributing ideas, special expertise, and unique resources. Increasingly, however, such collaborations are multidisciplinary, as we saw in the Brandeis example. In all situations, an important issue, particularly for beginning faculty, is how to make significant contributions to a team effort while at the same time establishing your credentials as a creative, independent researcher. We will look more closely at this challenge in the next two chapters.

The other message of this chapter is that no matter what type of research you choose to engage in, you need to pay attention to what happens before you actually do the research (proposal writing), and what happens after you complete the research (report writing and publishing). Several suggestions with respect to both of these tasks were offered, based on material found in three very worthwhile books. Remember, however, that you do not need to undertake these tasks alone. Follow the approach of the "quick starters" of Chapters 10 and 11. As is the case with teaching, you have numerous resources at your institution to which you can turn, the most important of which are your successful faculty colleagues.

REFERENCES

[1] W. A. Dando, "Grantsmanship," *The Professional Geographer*, a journal of the Association of American Geographers, vol. 41, no. 1, Feb. 1989.

[2] R. F. Abler and T. J. Barewald, "How to plunge into the research funding pool," *The Professional Geographer*, a journal of the Association of American Geographers, vol. 41, no. 1, Feb. 1989.

[3] R. V. Smith, *Graduate Research: A Guide for Students in the Sciences*. New York: Plenum, 1990.

[4] B. Hale, Stanford University Sponsored Projects Office, unpublished paper, Apr. 1996.

[5] S. F. Fitzpatrick, "Survival in today's tight funding climate depends on following agencies' rules," *The Scientist*, vol. 10, no. 9, p.11, Apr. 29, 1996.

[6] J. Fleron, P. D. Humke, L. Lefton, T. Lindquester, and M. Murray, "Keeping your research alive," Project NExT Program, Minneapolis, MN, pp.3–4, Jan. 1995.

[7] From the Rosentiel Center at Brandeis University, World Wide Web page (http://www.brandeis.edu/), 1995.

[8] P. J. Feibelman *A Ph.D. is Not Enough: A Guide to Survival in Science*. Reading, MA: Addison-Wesley, 1993, p.39. Copyright © 1993 by Peter J. Feibelman. Reprinted with permission of Addison-Wesley Publishing Company, Inc.

[9] C. I. Davidson and S. A. Ambrose, *The New Professor's Handbook: A Guide to Teaching and Research in Engineering and Science*. Bolton, MA: Anker Publishing, 1994. Copyright © 1994 by Anker Publishing Company, Inc. Reprinted with permission.

[10] J. Horgan, *The End of Science: Facing the Limits of Knowledge in the Twilight of the Scientific Age*. Reading MA: Addison-Wesley, 1996, p.68.

[11] W. Strunk and E. B. White, *The Elements of Style*, 3rd ed. New York: Macmillan, 1979.

CHAPTER 13

Insights on Professional Responsibility

> For 2,000 years the biblical injunction, "No one can serve two masters," has been considered sound advice on the issue of conflict of interest. Faced with seemingly contradictory mandates from essentially three masters—industry, society, and their own academic tradition, many engineering departments and their faculty can appreciate such advice. In fact, some are beginning to wonder if serving industry's needs while at the same time appeasing the public's demands, and maintaining institutional loyalty is becoming an order too tall to fill.
>
> Vincent Ercolano,
ASEE Prism [1, p.20]

SETTING THE STAGE

Teaching and other forms of scholarship are the primary duties of all new faculty. There are, however, other activities that will command your attention. Departmental service, participation in professional societies, and consulting and other relationships with government and industry are just a few possibilities. These latter activities not only have intrinsic value in themselves, if chosen carefully, they are capable of adding significantly to your primary responsibilities.

In all of your work, you will want to follow the highest ethical standards, and in most instances, doing so will not be a problem. Yet, there will be times, particularly in your teaching and research, when knowing and doing the "right things" are not as simple as they sound. For this reason, it is important to look at how you can find guidance in making the "right calls" in ethically problematic situations.

Robert E. McGinn, author of *Science, Technology, and Society*, has taught a number of courses on technology and society and on ethical issues in science and engineering. He has generated a list (see below) of 15 "ethically problematic behaviors in science." The list focuses on research-related conduct, and as you can see, with the exception of a few items (1, 2, 5, and 8, for example), these are not simple black and white matters with easily prescribed courses of action. The list includes the following behaviors [2, pp.1–2]:

1. Falsifying (e.g., "cooking" or "trimming") data obtained from a genuine experiment.
2. Fabricating experiments to "obtain" or "generate" data.
3. Misrepresentation in funding requests (e.g., hyperbole regarding previous accomplishments or future value of research).
4. Giving undue credit or failing to give due credit to someone regarding authorship of research work.
5. Deliberately misleading research competitors to "throw them off the trail" in order to improve one's chances of "getting there first."
6. Failure to secure bona fide "informed consent" from experimental subjects (for example, the Tuskegee experiment involving subjects with syphilis, or recent Department of Energy revelations regarding testing of civilians with radioactive substances).
7. Failure to take steps to insure "fair play" in one's laboratory (e.g., discrimination against or sabotage of the work by one or another party or group).
8. Plagiarism.
9. Demeaning a competitor's work to boost one's own.
10. Allowing one's research findings to be used in a misleading or potentially harmful way for personal or group political or economic gain.
11. Publishing one's work in LPUs (Least Publishable Units) to increase the number of one's publications.
12. Failure to "blow the whistle" on someone whose work is known to be defective where failure to do so may endanger the public interest or put a private party at risk of incurring unjustifiable harm.
13. Failure to conduct a fair-minded and scrupulous review of a scientific paper for which one is a referee.
14. Providing a biased or facile evaluation of a proposal for research funding for which one is a reviewer.
15. Influencing scientific research projects of one's subordinates (e.g., graduate students) in order to advance research in which one has a vested economic interest (e.g., because of owning stock in a company which stands to benefit from the skewed research).

The first step in avoiding many of these behaviors is to acknowledge their existence, and by so doing, bring them out into the open for discussion. Many of your more experienced colleagues will have encountered one or more of these behaviors in their careers, and may be able to share experiences that can help if you find yourself in similar situations. In discussing these matters, it helps to be aware of the pressures leading some faculty, in spite of their best intentions to the contrary, to engage in such conduct. McGinn has looked at this issue in some detail, and has postulated a dozen "factors conducive to misconduct in contemporary science." They are [2, p.2–3]:

1. The institutionalization of contemporary science (with all that this implies regarding the indispensability of obtaining substantial, ongoing funding).
2. The concept of an obsession with "success" in U.S. society, something which translates into great value being placed on obtaining desired results and which tends to devalue the importance and integrity of the process by which the results are obtained.
3. The difficulties that stand in the way of replicating previous experiments (e.g., difficulty of obtaining funding to replicate someone else's experiment).
4. The time that must be spent writing and marketing proposals to obtain funding for one's laboratory or institution, resulting in less time being available for transmitting "integrity values" to one's students "at the bench."
5. Fear of being hit with a lawsuit if one blows the whistle on a colleague or superior.
6. Fear of ostracism by colleagues if one blows the whistle.
7. The highly competitive nature of contemporary science regarding obtaining funding, being first in print, and obtaining one's own laboratory or a coveted endowed chair.
8. The high prestige attached by institutions and departments to having colleagues who publish prolifically and the related reward system.
9. The unprecedented degree of specialization in contemporary science (resulting in the prevalence of "a vulgar quantitative mentality" regarding publications).
10. The huge (about 40,000) number of scientific journals extant (resulting in the publication of much work of dubious scientific value and the difficulty of detecting fraud).
11. The lack of will and absence of an effective mechanism in science to root out fraud.
12. The pressure on young scientists to obtain significant funding and publish a lot to get tenure.

Perhaps you recognize a number of these pressure factors from our discussion of forces for change in teaching and research in Chapter 3.

We will now take a look at four faculty vignettes, each with ethical implications. The first two describe activities that have the potential to enrich your professional life, while contributing significantly to your teaching and research. The third and fourth vignettes explore ethical issues in teaching and research likely to be encountered by most science and engineering faculty. (See Figure 13-1.)

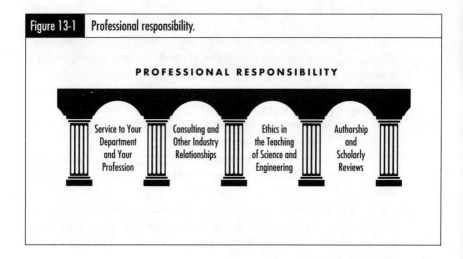

Figure 13-1   Professional responsibility.

In Vignette #22, Mary Anne Carroll, of the University of Michigan (Research I institution), urges young faculty to stay away from university-wide committee assignments, but points out that there are good reasons to become involved with the work of your department and one or more of your professional societies. Not only does departmental service demonstrate your willingness to be a good citizen, in some cases it can lead to direct benefits in your teaching and other forms of scholarship. While professional society service can make a difference when it comes to obtaining support from faculty outside your institution, at tenure time, for example, it can also help you identify a research thrust that has broad support within your discipline.

Our second vignette (#23) focuses on how the right kinds of consulting and other relationships with industry can enrich your teaching, research, and professional development. Professor Hau Lee, of the Industrial Engineering and Engineering Management Department at Stanford University (Research I institution), looks at various types of consulting arrangements, their strengths and weaknesses, and how you might go about preparing for the ones that are right for you. He then offers specific guidelines designed to help you get the most from your efforts while maintaining the desired ethical standards.

Vignette #24 looks at teaching and learning standards, an area not often discussed by either beginning and experienced faculty. Norm Whitley, professor of mechanical engineering at the University of New Orleans (Doctoral II institution), raises important questions about what it means to be a responsible teacher and learner. He then discusses the impact unacceptable faculty behavior can have on both students and the engineering profession. Whitley concludes by proposing five ethical principles that he feels should be adopted by all teachers and students.

In our final vignette (#25), Donald Kennedy, professor of biology and president emeritus of Stanford University, takes a look at professional responsibility and academic duty, particularly in the context of academic authorship and the review of scholarly papers, research proposals, and grants.

After examining these vignettes, we will take a further look at faculty ethics in our "In Addition" section by discussing three specific ethical dilemmas involving: (1) the appropriation of another person's ideas, (2) conflict of interest, and (3) the questionable use of another investigators data. We conclude with a suggestion on how new faculty can find help in dealing with these difficult and challenging "gray areas."

Vignette #22 Service to Your Department and Your Profession

> Stay away from university service until you get tenure. Departmental service makes sense for a lot of reasons, but only if done within strict limits. Active participation in your professional societies, on the other hand, can yield real benefits.

This statement sums up the view of Mary Anne Carroll, a tenured associate professor of atmospheric, oceanic, and space sciences at the University of Michigan in Ann Arbor, MI. Carroll echoes the views of many experienced professors when she recommends staying away from university-wide service in your years prior to tenure. True, such service can be a way to get to know how your institution functions, to be "seen" by higher ups in the administration, and possibly meet faculty in other departments with whom you may wish to collaborate. Yet, the time involved in such activity is often great, and there are other ways of accomplishing many of the same objectives. After you receive tenure, there will be many opportunities for institution-wide work.

Departmental committees are another matter. While service is never as important as teaching and research when it comes to tenure decisions, there are some good reasons for you to be on such committees early in your career. As Carroll points out:

> Serving on departmental committees demonstrates your good citizenship and your willingness to pull your weight. It also shows you are someone the rest of the department wants to have around. In addition, it gives people a chance to

see and hear you in action, to get a read on how you think, and to measure your reliability and responsibility. Remember, these are people you are going to work with for many years to come, and you can't simply wait until you get tenure to start being part of the department.

Many departments really want and need input from younger faculty who often have a different and refreshing perspective. As one professor put it, "We need these young Turks to push us in new directions, faculty who don't know all the history and all the reasons why something can't be done."

By serving on a limited number of departmental committees, you will also come to see how others in the department think and feel, and also identify the people with the most influence and the best ideas.

Of course, in some cases, you may not have a choice, as in small departments where it is essential to spread work out among a limited number of faculty. Under such circumstances, you want to serve on committees where you can have a real impact in areas directly related to your teaching and/or research.

Carroll mentions a case of a colleague in another department who agreed to serve on his department's graduate student admissions committee, with the understanding that he would only need to look at the folders of candidates who he thought were promising for his particular field.

Another area of service is the one within your professional community beyond your college or university. Every beginning faculty member should engage in such service at some level. As Carroll explains it:

> It is not enough to know the few people in your particular specialty. You need to become known to the broader community of which your specialty is a part. These are the people who you will need to go to for your tenure support letters. But there is more to it than that. By making contacts outside your specialty, you broaden your research horizons, and this broadening is critical for someone trying to establish a future research direction.

Professor James Sweeney, of the Engineering Economics Systems and Operations Research Department at Stanford University, backs up Carroll when he urges young faculty to:

> Start becoming a colleague with people beyond your institution. Start networking now, even in other countries. And don't confine yourself to your current discipline. Broaden your perspective. After you become a full professor, you will want to move into a leadership position; therefore, it is important now to make contacts, volunteer to serve on professional committees, and organize sessions. This involvement is all very important for your professional development.

There are many ways to get known in your research community. Attending conferences in your field, asking questions after a presentation, *and* giving your own presentations on a consistent basis are the key.

Chapter 13 Insights on Professional Responsibility   315

According to Carroll:

> Once you start submitting papers, you will be asked to review other papers as well as grant applications. A certain amount of this activity is *very* important. For now, stay away from reviewing books or writing chapters, but do get involved in some peer review of articles and research proposals. It gets you known by the granting agencies and the editors of journals, and will prove very helpful with your research and with those tenure letters.

Carroll points to an interesting program in the atmospheric sciences designed to help young faculty become established in their professional community. Called ACCESS (Atmospheric Chemistry Colloquium for Emerging Senior Scientists), it is intended specifically for new and recent Ph.D.s. ACCESS holds its meetings every other year in connection with what are known as Gordon Conferences, week-long retreats held at a remote college or university campus. About 25 new and recent Ph.D.s typically attend the ACCESS meetings. From 50 to 100 researchers attend a single Gordon Conference. In Carroll's words:

> The ACCESS meetings are a chance for new and recent Ph.D.s to give presentations on their dissertations or other recent research, meet funders from various agencies, and talk with journal editors. Such meetings are a way to begin to be identified with your research community.

The Gordon Conferences focus on a particular theme, with invited speakers, panel presentations, and discussions. However, the participants also take time to hike and play games together as an additional way to get to know each other. The ACCESS meetings take place two days prior to the Gordon Conference. All ACCESS participants are invited to attend the Gordon Conference, with their expenses covered by the National Science Foundation and other supporting agencies.

While not unique in science and engineering, ACCESS-like programs are not common. More such efforts would go a long way toward helping new faculty get started on the right foot in relationship to their research community.

Vignette #23 | **Consulting and Other Industry Relationships**

Engaging in consulting and other industry projects can be important for a successful professorial career in engineering, and even some sciences. In my field, there is a trend toward placing more and more value on such activities, sometimes even in tenure decisions. The key is to do it well, leverage such activity toward research and teaching goals, learn from your mistakes, develop the right principles, and maintain the discipline that enables you to use these activities to further your teaching and research.

All good advice from Hau Lee, a tenured professor of industrial engineering and engineering management in the School of Engineering at Stanford University. Lee feels there are a number of reasons for faculty, both beginning and experienced, to develop working relationships with industry. Engineering is, after all, an applied field and industry; in addition to being a good reality check, it is an excellent source of ideas, data, and problems. Lee has found "tremendous wisdom and experience out there." Relationships with industry enrich his research, teaching, and professional development. In addition, consulting and other collaborations with industry can provide research support, additional income, and help with the placement of students. Lee explains it this way:

> Such experiences will enable you to say to students and colleagues that you have seen the work of government and industry. They give you a more seasoned credibility while providing you with an important window on what is truly relevant.

With respect to new faculty and their research, Lee makes the following point:

> At the Ph.D. stage, you have been guided mostly by your advisor in terms of what research you do. As a new professor, you are responsible for selecting your own topics and your own direction. You have to find your way, and do what is relevant to you and not to your advisor. Working with industry can provide you with insights to help you determine your own direction.

With respect to consulting, however, Lee cautions young faculty not to move too fast with arrangements outside the university structure. "Don't do this only for financial gain," he advises. "You simply don't have the time." Lee urges young faculty to stay away from expert witness or pure service assignments that can cause considerable stress while doing nothing for your teaching and other forms of scholarship.If you do choose to do consulting, Lee recommends that you:

- Abide by university regulations, and make sure the work does not interfere with teaching and research;
- Choose subjects within your areas of expertise and interests;
- Set up rules for pricing (e.g., travel time, court time, teaching, initial meetings...),but be prepared to be somewhat flexible if needed later;
- Always look for teaching and other scholarship opportunities through such engagements;

- Spell out clearly the terms of confidentiality and publication rights;
- Identify and work with individuals and managers who have strong interests in the success of the engagement.

Lee points out that most young professors are not likely to be hired as consultants. As he puts it:

> Senior professors are more well known, and are hired for their expertise as consultants. Young professors are less well known, and are more likely to be engaged for their research capabilities in a particular area, usually under a research contract or industrial gift arrangement. It is often better for new professors to bring work into the university through research contracts or gifts. With this approach, everything is above board, which is particularly important if you are using students where you don't want there to be even the appearance of a conflict of interest or commitment.

Lee lists other examples of faculty–industry relationships such as:

- Company gift funds, usually for less defined research activity and without specific deliverables
- Company partnerships with university research centers or affiliate programs
- Foundation-funded research studies
- Company–government–university research partnerships.

When engaging in such arrangements, Lee advises that you:

- Make sure there is an explicit delineation between sponsored research and gift funds,
- Seek a clear understanding of expectations and deliverables, including the intent to publish parts of the work,
- Specify the participation and involvement of company personnel,
- Recognize the importance of periodic site visits and management briefings,
- Demonstrate your willingness to listen, observe, and change focus when necessary,
- Actively solicit coauthorship with industry participants,
- Do not "nickel and dime" everything; think of the long-term relationship,
- Spell out clearly confidentiality terms and publication rights,
- Involve students as much as possible.

There are a number of ways for you to get started with consulting and other industry projects. Former advisors, and even former student colleagues, can be good sources. As time goes on, your own former students and postdocs will provide contacts, as will liaisons from industry. Lee points out that word-of-mouth after a certain number of successes, visibility from publications and presentations, referrals by colleagues, and even cold calls into the university have all worked for him.

Lee obviously believes that these relationships have real value to him as a professor. He sums it up this way:

> I try to develop a teaching case or find other ways to integrate the material into my courses. Often, I am able to write an application paper from the work I do, sometimes with a coauthorship from someone in industry. In addition, I can usually find ways to extend the work I did with industry and the data they provide me with by stimulating doctoral students to work on such problems in their dissertation research. With all this kind of leveraging, it's hard to see why I wouldn't want to develop such collaborations.

Vignette #24 Teaching and Learning Standards

> Today, more attention is being given to the teaching of ethics, and this is good. But what about the ethics of teaching? As a professor, you have to constantly ask, "Am I doing the right thing?" If I teach statics by just reading out of the book, and then fail 50% of my students, or pass all of them with As and Bs, is this ethical? At what point have I stopped teaching them? Professors need to look long and hard at these questions, but most of them don't.

These are strong words from Norm Whitley, a tenured professor of mechanical engineering at the University of New Orleans in New Orleans, LA. Whitley is clearly concerned about a very important, but little discussed, area of professional responsibility. As he puts it:

> Too many faculty believe if students don't learn, then it is entirely the students' fault. There needs to be a concern for the bottom line here, particularly when you are preparing students to go out and serve the public. Some faculty are quite comfortable giving tests in which the high grade is 30/100 and passing students with something like 15/100. As a professor, you have to take some responsibility. There are professors who believe the marketplace will take care of these problems. Sure, a poorly trained graduate may eventually get fired, but in the meantime, he or she could make a lot of mistakes and do a lot of harm.

As an example, Whitley refers to the rush to use computers in the classroom. He is not against such use, but wonders if we have the cart before the horse. Students can now run programs, use spreadsheets, and use software that does the mathematics for them. "But what about understanding the fundamentals behind the spreadsheets and the simulations?," says Whitley. "If they don't have this understanding, how do they (and we) know what they are doing is correct?" As Whitley notes, many students, even in engineering, lack hands-on experience with "real things," and this hampers their ability to make appropriate judgments. "A lot of engineering is intuition," he says, "and you don't build intuition by looking at a computer screen."

Whitley also raises the issue of whether or not teachers have an obligation to cover the material described in the course catalog. As he explains it:

> At my institution, faculty are not asked on a timely basis if they are following and completing the syllabus. Everyone agrees we have a responsibility not to send students on to the next course if they don't know the material, yet we do it all the time. In so many science and engineering courses, what you learn at a higher level depends on what you learn at a lower level. If we don't cover the material, or if we pass students who don't know what's going on, we are putting our students, and our faculty colleagues, at a terrible disadvantage. It only increases the likelihood we will have to go over old material, which then puts us further behind in completing the syllabus. Eventually, we end up with students who are ready to graduate, but are clearly not prepared.

Recognizing the lack of guidance for faculty in these and other ethical areas, Whitley, and some of his colleagues, have been working on a code of ethics for the College of Engineering at the University of New Orleans. The code, which is not yet in final form, is meant as a complement to the existing university-wide list of ethical principles found in the faculty handbook. The college of engineering code, which applies to both *students and faculty*, would have as its basis five ethical principles, wherein each individual member shall:

1. Hold foremost that only after the acquisition of necessary and sufficient knowledge shall an individual be an acceptable candidate for the Profession.
2. Have responsibility for the acquisition of knowledge crucial to a successful educational experience.
3. Promote an atmosphere for the free pursuit of knowledge and intellectual fulfillment for each member of the Community.
4. Have responsibility to uphold the integrity of the College, the University, and the Profession.
5. Represent work, ideas, or intellectual property as the efforts and results of their true owners.

Among other things, the code, which in most respects could apply equally well to science, will also list examples of unacceptable faculty and student behavior.

As we discussed earlier, in the decades to come, some colleges and universities are not going to survive because of various demographic pressures and demands for accountability from students and their tuition and tax-paying parents. Schools succeeding in the marketplace will, among other things, have to pay much closer attention to teaching and learning. And, says Whitley, "An important component of this teaching and learning will have to be attention to the ethical issues we have just discussed."

Vignette #25 Professional Responsibility and Academic Duty

We barely attend to such matters in academia, and it's certainly not part of the curriculum. We talk about ethics in law, medicine, and to some extent in engineering practice, but we don't do so for the one group (Ph.D.s) that will replace us. It may have to do with the notion of academic freedom, the idea that somehow if we have academic freedom we can't have academic duty or responsibility, that the two are in some way in conflict. Well, academic freedom doesn't mean you are "free" from constantly asking yourself if you are doing the right thing.

So says Donald Kennedy, a tenured professor of biology and president emeritus of Stanford University. Kennedy teaches a seminar, Professional Responsibility and Academic Duty, for Stanford Ph.D. candidates who plan to undertake academic careers. Topics covered in the course include: teaching responsibilities, ethical challenges to teaching, university governance and faculty service, research as a competitive venture, intellectual property, outside activity—conflict of commitment and conflict of interest, the university and the faculty, and faculty duties and obligations. In this vignette, Kennedy focuses on two themes: (1) academic authorship and the review of scholarly papers, and (2) research proposals and grants.

The most important issues, Kennedy notes, are not the ones that make the headlines. "Faking data or fabricating experiments are in a sense easy to deal with because they are so obviously wrong." However, as he goes on to point out, "In many other areas, there are huge variances in what is commonly regarded as the 'right thing,' and this creates 'zones of difficulty' not always easy to resolve."

Consider the subject of authorship. As we discussed in Chapter 2, coauthorship of scholarly papers is now the norm in all areas of science and engineering. Such authorship raises important questions about the allocation of credit since, as Kennedy puts it, "Determining from 'where our ideas come' is not always easy." He goes on to observe:

> The consequence of people working together is that "ideas are in the air." To a certain extent, we all "steal" from each other, and figuring out who thought what and when, who gets credit, and in what order is a nontrivial problem.

Conflicts often occur between faculty and graduate students, but they can also take place among faculty. Different credit norms exist within institutions as well as among various disciplines. Kennedy describes it this way:

> In some cases, the laboratory director's name goes on every paper. In genetics and microbiology, for example, things tend to be "shared," and the director of the laboratory is almost always on the list of authors even if he or she did no direct work on the project. In population biology, it is the people who actually do the work, usually graduate students, who are the only names on the paper.

Kennedy believes "complementary authorships," which also result when a student's name is put on a paper as a career boost even if he or she did little or no work, is a form of fraud. It is, he says, "a case of authorship being awarded, not earned."

For Kennedy, there is a simple test: "Can every one of the coauthors give a talk on the paper at a scientific meeting and defend it publicly in a question and answer session?" If not, then some attribution other than coauthorship should be found. He suggests that in many cases, "with technical assistance of" will suffice.

Of course, the other side of credit is blame, and as Kennedy notes, this fact puts a different perspective on coauthorship."If everybody's name is on the paper, and if someone cooks the data, who should be responsible? How should blame be allocated? Should it be the same as credit?"

Kennedy urges faculty to: "Make clear to your students and postdocs the norms of your culture and what is expected in your laboratory. Make these expectations clear right from the start, and you'll save yourself a lot of problems later on."

Another area providing a great many ethical challenges is the reviewing of scholarly papers, proposals, and grants. Here, Kennedy echoes a point made in Chapter 3, by Leon M. Lederman of the American Association for the Advancement of Science. As Kennedy puts it:

> Research/scholarship is a very competitive activity, and there are lots of pressures on those who do it, evaluate it, support it, and measure it. These pressures raise a number of issues around the role of reviewers and editors, how we recognize the contributions of others, how we assign priority to various works, and how we deal with moral restraints on the use of information.

Kennedy points out that criticism of most scholarship, published and submitted, is negative and unfriendly. "The idea," he says, "is to be 'cleverly critical.' " This behavior leads to disputes among authors and reviewers. One example has to do with possible conflict of interest in the evaluation of grant applications. Comments Kennedy:

> Originality and creativity are highly prized in academia, but in the review process, we put our ideas in the hands of others prior to publication. Under such circumstances, it is natural to ask if some of our competitors will "appropriate

our ideas." In some grant areas, it is now possible for authors to list persons they want *excluded* from their reviews.

A related matter has to do with who signs the reviews of papers, proposals, and grant applications. At one level it makes sense for a professor to ask his/her graduate students and postdocs to review such material since this kind of activity can be an important part of their education. But is this approach a substitute for the faculty member also reviewing the application? Who signs the review sent back to the editor or granting agency, the professor of the graduate student? Also, what about confidentiality when the paper or application is shared with others? Under such circumstances, Kennedy notes, "The assumption of confidentiality is needed, but perhaps not provided."

These are just a few of the many ethical issues Kennedy and other thoughtful faculty are concerned about. Let us take a closer look at some additional ones in the section below.

IN ADDITION: APPROPRIATING THE IDEAS OF OTHERS, CONFLICT OF INTEREST, AND FREEDOM OF INFORMATION?

One way to gain a greater appreciation of dilemmas you are likely to encounter in teaching and research is by looking at specific examples of ethically challenging cases. Let us briefly examine three such cases, the first having to do with the appropriation a student's experimental ideas, the second with a conflict of interest over the use of industrial funding for research, and the third with a questionable use of another investigator's data.

Appropriating the Ideas of Others

In, *The Ethics of Teaching—A Casebook*, authors Patricia Keith-Spiegel, Arno F. Wittig, David V. Perkins, Deborah Ware Balogh, and Bernard E. Whitley, Jr. use a case presentation approach to highlight ethical dilemmas in such areas as: class issues, lessons and evaluations, activity outside the classroom, relationships in academia, and responsibilities to students and colleagues [3, pp.v–x].

These brief 220–300 word commentaries (there is a total of 165) are quite thoughtful, and manage to focus on the gray areas while staying away from the more obvious egregious breaches of ethical conduct. Here is one example, reprinted in its entirety, to illustrate the approach. (Note: The authors' use of unlikely fictitious names relating to the themes of the cases is not meant to be irreverent, but rather to assist in recall and reduce the incidence of names of real people.)

Case 72. Swiping a Student's Idea [3, pp.64–65]

Richie Bright handed in an outstanding research proposal as his senior honors thesis. After the student graduated, Loot conducted the study that Bright had designed, making very few modifications. When a colleague who had consulted with Bright mentioned the similarities, Professor Loot explained that he had supervised the student from the beginning so the idea was partially his. Furthermore, Bright had gone on to study in a field unrelated to this project. Finally, Loot noted that Bright could not have conducted the research or have written a publishable manuscript himself because he lacked the skill and experience.

The story is beginning to blur...because Loot was involved from the beginning and conducted the work that brought the idea to fruition. Furthermore, the originator of the idea was long gone. However, Loot still profited from the original effort of another, and without crediting his benefactor.

This matter could probably have been easily resolved had Loot consulted Bright as soon as he decided that he wanted to implement Bright's design. Bright should have been offered direct involvement in the conduct of the research. In this case, it appears that Bright would have declined the invitation because his life had turned in a different direction. However, it is likely that some agreement about credit for Bright's original contribution could have been reached (e.g., a junior authorship or a footnote credit, depending on considerations such as the innovation of the design or a new technique that Bright created to test his hypotheses).

That Bright did not possess the competencies to continue on his own is an irrelevant defense in and of itself because Bright might have resumed study of the same subject in graduate school, caught the interest of the faculty, and run the study as his master's or doctoral thesis. Using the senior thesis for such purposes is not unusual.

Students at any level can inspire us in ways that direct our own scholarly work. When that contribution is deemed to have had sufficient influence, professors must objectively assess the question of the student's rights to credit. If a student offhandedly presents an idea during a class discussion that triggers a chain of events that eventually leads to a project, acknowledgment may not be warranted. When in doubt, consultation with colleagues may assist in resolving the matter.

Conflict of Interest

Free-lance writer Vincent Ercolano, in his article, "Ethical Dilemmas: When Faculty Responsibilities Conflict," talks about faculty conflict of interest, and suggests that it be looked at in terms of limits, i.e., "How much is too much?" [1, p.22]. Here is one of Ercolano's examples, along with his commentary on how some universities deal with the potential difficulty [1, p.22]:

> *If a private corporation arranges to sponsor a project at your university, say, to develop a new material, would it be a conflict of interest for you to participate in that project if you have a substantial financial interest in that company?*
>
> Some universities base their response to this question on how such an arrangement would appear to—in the words of one institution's policy—"reasonable people." Other universities are more specific, defining substantial financial interest as ownership of a specific minimum percentage of a company's equity or ownership of stock that has a specified minimum value.

If your financial involvement exceeds that threshold, the dean of your school or the university's research dean might, in addition to requiring full disclosure of your holdings in the sponsoring company, assemble an oversight committee to monitor things such as your use of students' time on the project and your adherence to requirements for prompt dissemination of the results of your research. These thresholds vary from university to university. One university considers holdings of more than one percent of the company or more than $500,000 in value to be sufficiently substantial to raise conflict of interest concerns. Another sets the threshold of substantial interest at 0.5 percent or $100,000.

Freedom of Information?

A few years ago, in order to stimulate a more active and reasoned discussion of various ethical issues in science, *Science*, the magazine of the American Association for the Advancement of Science, initiated an interactive Internet project based on its special "Conduct of Science" section printed in its June 23, 1995 issue [4]. This special 14-page report focused on "gray areas" of behavior, such as allocating credit for research work, assigning authorship of research papers, and sharing materials. Among other things, five case scenarios were presented to challenge readers to examine what they would do in each situation and to respond with their own examples. The scenarios were (1) Whose Data are They, Anyway?, (2) Freedom of Information?, (3) A Grant Reviewer's Quandary, (4) A Suggestion of Fraud, and (5) Who is the Culprit?

Here is the "Freedom on Information?" scenario for you to consider [5]. It was developed by John C. Bailar III, chair of the Department of Epidemiology and Biostatistics, Faculty of Medicine at McGill University in Montreal. After reading it, ask yourself how you feel about the courses of action proposed by each investigator, and why.

> *A large research study was supported for many years by a series of government contracts. The study was completed, and the resulting data were submitted to the contracting agency three years ago. However, the principal investigator and her colleagues have not yet published any of the results. An investigator not affiliated with any of those who did the research requested and was sent (as required by the United States Freedom of Information Act) a copy of the data, research protocols, and other materials. He has written a paper based on the data and submitted it to a journal; a footnote accurately explains the circumstances. In response to the journal editor's request for comment, the principal investigator replies that she has had a paper almost finished for two years, that she will immediately get a colleague to finish it, and that she will sue the journal if her competitor's paper is published. The author of the paper in hand says that the data were generated with public money and are therefore public property, that he has used the Freedom of Information Act as Congress intended it to be used, and that three years is long enough to wait for publication of important results.*

For a period of time, it was possible to use the on-line component of the *Science* effort to interact with a panel of experts on scientific conduct and initiate a dialogue on the subject of the scenarios with other readers. While the interactive aspects of the project are no longer in operation, the scenarios, and comments concerning them, are still accessible in the "Beyond the Printed Page" section of the *Science* home page at: (http://www.aaas.org/science/science.html).

CONCLUSIONS

Given the issues raised in this chapter, how do you as a new faculty member proceed? As noted in the introduction, you want to engage in activities having value in themselves, but that also contribute to your primary mission of teaching and other forms of scholarship. And you want to do everything in a way that is consistent with your personal values and your high level of integrity.

With respect to the latter, it is encouraging to see the increased attention paid to ethical issues in courses, books, journals discussions at professional society meetings, and on the Internet. Such attention also makes it easier for young faculty to seek out additional advice and guidance. One helpful approach is to identify senior faculty who share your values, and who have experience in the particularly difficult and challenging "gray areas." These people can be found in your department, your institution, at other colleges and universities, and in your professional societies. I think you will find that the Internet is a particularly effective forum for locating and promoting useful dialogues with individuals beyond your immediate physical environment.

REFERENCES

[1] V. Ercolano, "Ethical dilemmas: When faculty responsibilities conflict," *ASEE Prism*, vol. 4, no. 2, Oct. 1994.

[2] R. E. McGinn, "Ethical issues in science: A topology and list of contributory causal factors," unpublished manuscript, 1996.

[3] p.Keith-Spiegel, A. F. Wittig, D. V. Perkins, D. W. Balogh, and B. E. Whitley, Jr., *The Ethics of Teaching—A Casebook*. Muncie, IN: Ball State University, 1993. Copyright © 1993 by Ball State University. Reprinted with permission.

[4] J. Benditt, "Conduct in science," *Science*, vol. 268, pp.1705–1718, June 23, 1995.

[5] Taken from J. Bailar *et al.*, Eds., *Ethics and Policy in Scientific Publication*. Chicago, IL: Council of Biology Editors, 1990. Appears in J. Benditt, "Conduct in science," *Science*, vol. 268, pp.1705–1718, June 23, 1995.

CHAPTER 14

Insights on Tenure

SETTING THE STAGE

> Tenure is a powerful force. A desire for tenure at times grips people as strongly as a desire for romantic love. The outcome of both tenure and love shape and mold one's self-image and self-esteem. Success is cause for great joy and celebration. Failure implies dejection and defeat. The prospect of earning tenure, like love, provides a roller coaster of emotions. As prospects brighten, emotions soar. As prospects dampen, gloom and doom set in. At times, both tenure and love force people to pursue activities past the point of fatigue. Both tenure and love make some people sensitive to others. Both make some people crazy.
>
> *From* Getting Tenure
> by Marcia Whicker, Jennie Kronenfeld, and Ruth Strickland [1, p.1]

As we pointed out in Chapter 1, tenure is unique to academia, and in spite of occasional stories to the contrary, it is going to be with us for some time to come. If you want a permanent faculty position at a college or university, you almost certainly need to become tenured.

I believe that significant changes are called for in some aspects of the tenure-granting process, and these are discussed in Chapter 15, "Insights on Academia: Needed Changes." In this chapter, we take a look at the tenure system as it currently exists, and what you need to do to succeed within it.

We begin with a look at the various paths typically taken toward, and away from, tenure. This discussion is followed by vignettes featuring five individuals. Two of these individuals received tenure in a fairly standard way, one received tenure at her second school after being denied it at her first school, and two did not receive tenure, and have, at least for the time being, left academia. In our "In

Addition" section, we offer some further insights on tenure, and also look at how one university is making a special effort to help faculty understand the tenure and promotion process.

Paths toward—and away from—tenure[1]

Clearly, not everyone follows the same path toward tenure. What works for one individual at one type of institution might not work for that same, or another individual, at another institution. What follows is a brief look at the ten most common paths toward, or in some cases away from, tenure. (See Figure 14-1.) Keep in mind that many faculty will follow a path that is a combination of two or three of the ones presented. The ten paths are:

1. The traditional path
2. The accelerated path
3. The delayed entry path
4. The late practitioner path
5. The late career child-bearing path
6. The from-one-school-to-another-school path
7. The fail to get tenure—try again path
8. The fail to get tenure—other career path
9. The walk-away-from-tenure path
10. The never-try-for-tenure path

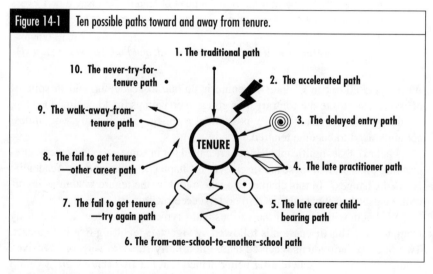

Figure 14-1 Ten possible paths toward and away from tenure.

[1]Adapted in part from [1].

1. The traditional path

The traditional, or on-time path, is the one most commonly taken by full-time assistant professors who start out on the tenure track. On this path, tenure is typically granted at the end of the sixth year of service. At most schools, a starting assistant professor receives a three-year initial appointment which is then renewed for a second three-year period at the same rank. At the end of the sixth year, if the professor is awarded tenure, he or she is usually promoted to associate professor. Some schools, however, promote faculty to the associate professor level prior to tenure.

For all full-time tenure track professors, tenure decisions must be made no later than the sixth year according to the rules set forth by the American Association of University Professors (AAUP). Since the process of making the tenure decision can take up to a year, you basically have five years of full-time service in which to put together a tenurable record [1, p.39]. The AAUP rules also specify that those faculty who are not granted tenure must be given a years notice. Thus, a professor who is not granted tenure could remain on the tenure track for up to seven years before having to leave the institution. However, your "tenure clock" can be stopped for a period of usually no more than two years for special circumstances, such as child rearing, administrative or sick leave, and research or service leave, as will be discussed below.

The on-time path is the one currently being taken by Shon Pulley (Vignette #5), Brian Love and Nancy Love (Vignette # 9), and Kim Needy (Vignette #11). It was the path successfully taken by Alison Bridger (Vignette #10), Paul Humke (Vignette #12), Sheri Sheppard (Vignette #14), Rollie Jenison (Vignette #16), Lew Lefton (Vignette #18), Mark Hopkins (Vignette #19), and Norm Whitley (Vignette #24). We will highlight this path in greater detail in Vignette #27 of this chapter.

2. The accelerated path

In some cases, a faculty member will be brought up for tenure one or two years in advance of the normal six years, as will be shown in Vignette #26 below. In very rare cases of so-called "superstars," tenure may be granted in as short a time as three or four years. Often, tenure in advance of the normal six years will be offered if a school wants to reward a professor for outstanding performance and/or wants to keep a professor who may have a tenure offer from another school.

3. The delayed entry path

The delayed entry path, sometimes referred to as the "bounce-around" path, is a situation in which a person enters the tenure track after a period, or periods, in other positions such as part-time, lecturer, or adjunct professor status at the same or another institutions [1, pp.129–130]. Ulrike Salzner (Vignette #6) and Martin Ramirez (Vignette #15) are examples of individuals who have followed or will follow this path.

4. The late practitioner path [1, pp.132–133]

The late practitioner path is one often taken by a person who has spent a considerable time in government or industry prior to choosing an academic career. Often, such individuals have had a part-time teaching affiliation with one or more colleges or universities over the intervening years, and now feel they are ready to make the shift to a full-time faculty position. Most of these people enter academia on the tenure track, although their tenure clock may be shortened by one or more years. Less frequently, an individual with an outstanding record in government or industry, coupled with good part-time teaching experiences, may start out as an associate or full professor with tenure.

Joseph Reichenberger (Vignette #7) and Mary Anne Carroll (Vignette #22) are examples of experienced persons entering academia with shortened tenure clocks.

5. The late career child-bearing path [1, pp.133–134]

This path was probably more common in the past than it is today. Some women may have postponed obtaining advanced degrees, or even if they did obtain them, delayed their entry into the job market for family and child-bearing reasons. Most women in this category, if they are able to enter academia, usually do so in a part-time or adjunct professor position rather than on the tenure track. Entry on the tenure track is made difficult by the fact that they often lack experiences in government or industry that would make them attractive as late practitioner entrants. Yet, some women in this situation have been able to teach part time while raising a family, and this experience has made entry onto the tenure track more likely. Eloise Hamann, whose story is described in Chapter 4, is one such example.

6. The from-one-school-to-another-school path

In this case, a faculty member leaves one school and goes to another school, sometimes with a stop in government or industry along the way. One approach is to leave a tenure position at one institution for a tenure position at another. This approach is the one Greg Petsko took (Vignette #20) when he left a tenured position at the Massachusetts Institute of Technology for a tenured position at Brandeis University.

An example of an individual who did not receive tenure at one school, left for a position in industry, and then followed up with a position at another institution that eventually led to tenure is given in Vignette #28 of this chapter.

7. The fail to get tenure—try again path

Some individuals who are not awarded tenure at their institution will appeal the decision, and on rare occasions, will have the decision reversed and be granted tenure.

8. The fail to get tenure—other career path

This path is the one taken by individuals who either recognize that they are not going to get tenure, or go through the full tenure evaluation process and are denied tenure. They then leave academia, at least for the time being, for another career. Examples of individuals in this category are presented in Vignettes #29 and #30 of this chapter.

9. The walk-away-from-tenure path

Some individuals choose to give up a tenured position for a career outside academia. Amir Buckhari, whose situation was described in Chapter 4, is one such person. I also know of a number of tenured faculty who have taken leaves of absence to start a business, or assume a position in government or industry, and who then decide to return to academia part time in a nontenured position.

10. The never-try-for-tenure path

Certain individuals, under certain circumstances, can have a very rewarding career in academia without becoming tenured professors. In Chapters 7 and 9, we discussed nonstandard faculty positions such as consulting professors, research professors, and adjunct professors. While these positions do not have the security of a tenured position, they also do not have the range of responsibilities and commitments of such positions, and for some people, this can be very attractive. Paul Losleben (Vignette #3) is an individual who, after a very successful career in government, came to Stanford University in a senior research scientist position. In such a role, he is able to direct research and supervise graduate students without the other service and formal teaching responsibilities associated with tenured faculty.

It is important to understand that there is no one correct path to tenure, and that many of you will follow paths that are combinations of two or three of those described above. Whichever path you do take, do so in a way that maximizes your chances of success, while also permitting you to live an enjoyable, rewarding, and balanced life. It is a mistake to sacrifice the pleasures of being a professor and of family and other relationships for six years on the assumption that you can "catch up" after the magical moment when you are awarded tenure. Just as your world did not change radically when you received your Ph.D., it will not do so once you get tenure. Decide now how you want to live your life, but do so with tenure as one of your long-term goals.

It might seem that the simplest way to go about obtaining tenure is to find out what your school and department require, and then establish a strategy for meeting those requirements. Unfortunately, it is rarely so easy. Tenure criteria are usually not described with a high degree of specificity. True, most schools have faculty handbooks setting forth the requirements for tenure and the documentation necessary to "make the tenure case." However, the wording in most of these publications is purposely vague with such phrases as "produces successful graduate students," "has positively reviewed publications," and "receives supportive letters from former students." Such wording exists because, as criteria become more specific, flexibility or "wiggle room" for the department decreases. As Lance Glasser notes in Vignette #29, one of the reasons he believes he did not receive tenure at the Massachusetts Institute of Technology (M.I.T.) was because he came across as too arrogant with his colleagues. Whether or not his tenure committee would agree, you can be sure "humility," or "lack of arrogance," are not listed as criteria in the M.I.T. tenure handbook.

Asking your colleagues questions about the tenure process is very important, yet, even here, answers can often be conflicting. As one faculty member at a large midwestern university recalls:

> We were told very different things by various members of the administration. The dean would say, "You need to be superb in teaching as well as research. If you are not great teachers, you cannot get tenure." But the department chairs would say the opposite, and the college executive chairs would agree with the department chairs. Because of such conflicting messages, you have to keep your eyes wide open. Talk to the people who will be deciding your tenure; they are the people who are going to be voting.

You also have to learn how to be proactive in managing your tenure case. You have to ask the right questions of the right people, seek necessary (and sometimes uncomfortable) feedback, know what adjustments to make along the way, and learn how to stay alert to cues that come in a variety of forms.

Much of the advice in the previous four chapters also applies to the tenure process. Yet, there are additional insights to be had, and these are reflected in the following vignettes and in the "In Addition" section that follows.

In our first vignette (#26), Eve Riskin, from the University of Washington (Research I institution), talks about how to leverage your teaching and research so as to support your drive toward tenure. She discusses the importance of standing firm for the things you need in order to be successful, and concludes with some specific suggestions on how to establish relationships with external reviewers who can have an important influence on your tenure success.

In Vignette #27, Keith Buffinton, of the Mechanical Engineering Department at Bucknell University (Liberal Arts I institution), points out that liberal arts schools such as his have a different set of priorities, and different measures of success, than do many of the institutions from which most of their beginning faculty have come. Understanding these differences is essential to getting tenure, as well as to being a satisfied and productive faculty member.

Vignette #28 features Ruthann Kibler who started her academic career at the University of Arizona in Tucson, AZ. After the normal six years on the tenure track, she was denied tenure. After three yeas as a research scientist in industry, she returned to academia at San Jose State University in San Jose, CA, where she is now a tenured associate professor of biology.

Our fourth vignette (#29) is with Lance Glasser, who discusses what happened to him when he did not receive tenure at the Massachusetts Institute of Technology. Although very disappointing at the time, Glasser made a decision to learn from the experience while moving on to something else in an immediate and forceful way. Both approaches have served him well in his new endeavors.

Our final vignette (#30) is with Sue Smith-Baish, who was denied tenure after six years in the Electrical Engineering Department at Bucknell University, the same school in which Keith Buffinton did receive tenure. The contrasts in their two situations offer important lessons for all beginning faculty who attempt to deal with pressures of professional and family life.

Vignette #26	Leveraging Wherever Possible

I believe almost anything you do for the first time can be frightening. As a graduate student, I was given opportunities to write my own research proposals, as well as to give a number of class lectures. Even though my first proposal wasn't funded, both of these experiences gave me a lot of confidence when it came to doing them for the first time as a professor.

This important point is made by Eve Riskin, an associate professor of electrical engineering at the University of Washington. in Seattle, WA. Riskin was recently awarded tenure a year ahead of schedule and, as she notes, "just in time to start my family." She has a number of suggestions on tenure for faculty who are just beginning their academic careers. The first has to do with initial conditions. Says Riskin:

Most faculty don't realize they can negotiate their start-up conditions in a way that will help them be more successful, as well as happy. My goal was to make my life easier, not richer. It wasn't the start-up salary, but the start-up funds I really cared about. I also wanted to arrange not to teach my first quarter. I was in a new city, a new home, and with a new employer, and having three months to get my bearings and write proposals made a big difference.

Riskin urges faculty to begin by making sure they do, in fact, have a tenure-track position. In her words, "Make sure, if in a few years you meet the tenure requirements, your position will become permanent. Promises to this effect are not good enough. Get it in writing." Riskin also suggests you avoid situations where you are competing with one, or perhaps two, other faculty for what will eventually be one tenured slot.

Since good research and good teaching are the two main tenure criteria at the University of Washington, anything you can do to get a head start in these areas is welcome. With respect to research, Riskin has some specific advice about leveraging:

> I began by writing a "baby grant" for $10,000 internal support for some ideas I had in medical image compression. The grant was rated the top in its category, and even though it was small, it felt good to be successful. I then used this success to apply for a $70,000, low-overhead National Science Foundation proposal which was also granted. In both cases, I got some very good feedback, which I made sure to put in my portfolio. I have since found I can use pieces of each proposal as the basis of the next proposal, even if I change directions slightly as I go along.

Riskin also urges faculty to work hard to get at least one Ph.D. student through before tenure decisions come up. Not only does this accomplishment help you with your own research, but it demonstrates to your colleagues and to potential students your effectiveness on this important "measure of success." Riskin believes it was a major factor in her early tenure award.

With respect to teaching, Riskin again emphasized a point made earlier in this book:

> As a graduate student, I volunteered to give some lectures which I didn't have to do. I kept all my notes, homework assignments, and where possible, the syllabi and other materials my advisor produced. As with research, this approach helped me get started on the right foot.

Riskin also has another important comment to make about teaching. "I don't think you have to prove you can teach a different course each semester," she says. "In my first five years, I only taught three *different* courses. I'm still not tired of them, and I can't begin to tell you the time savings coming from such an approach."

Regarding those all-important external reference letters, Riskin emphasizes the need to start publicizing your work early on. She has three specific suggestions in this regard:

1. Think of job interview trips as ways to publicize yourself. Even if you do not wind up going to that particular university, you can keep in touch with people there.
2. When you finish a paper, send a preprint to people in your area. This way, they will stay informed of your work, and can have more to say about you if they get asked to write a letter.

3. Take your journal paper reviewing very seriously. If you do a good job reviewing papers, it gets out that you know the field well. I also think there's a small "gratefulness" factor—if you help someone else out, maybe they will be grateful and help you out later on.

Riskin makes one other point not often discussed, particularly by women:

> You have to look out for yourself in the tenure process, and so my basic advice is, "don't always feel you have to be a good sport!" If you spend too much time trying to accommodate the wishes of your colleagues, you will get a lot of "ataboys," or "atagirls," and miss the big picture in terms of what it really takes be successful. I'm all for collegiality and getting along, but only so much. It's also important not to allow yourself to be taken advantage of. At first, I was pushed to do some things I didn't want to do, such as teach certain courses. I made a conscious decision to stand up for myself, and point out how what I wanted to do would contribute to the department's success as well as my own.

This excellent advice has clearly paid off well in Riskin's case.

Vignette #27 Understanding the Priorities

> It's very important for new faculty to understand our priorities and how much they differ from those of the schools from which most of them have come. At most research universities, the priorities are money, publishing, and teaching, in that order. But here the order is just the reverse. Solid teaching is absolutely number one, with quality scholarship right behind. Bringing in money is nice, as is departmental service, but both are way behind teaching and other forms of scholarship.

So advises Keith Buffinton, a tenured associate professor of mechanical engineering at Bucknell University, a liberal arts institution in Lewisburg, PA.

Because teaching is so important at Bucknell (see also Vignette #15), a faculty member's capability in this regard needs to be demonstrated fairly early. Faculty need to, as Buffinton puts it, "establish their ability to do something exciting in the classroom." Bucknell understands that this accomplishment may take a little time, but it needs to be every new professor's top priority.

With respect to scholarship, Bucknell is less concerned with quantity than with quality. "We do expect you to establish a national reputation as a scholar," explains Buffinton. "It is very important for you to show you have gone beyond what you did with your Ph.D. advisor. As soon as possible you need to establish a track record on your own."

There are many ways Bucknell will assist new faculty in their scholarship. For example, it will help out when it comes to the purchase of equipment and the retrofitting of laboratory space. The school will also support faculty efforts to collaborate and share resources with colleagues at larger research universities. It even has something called the Untenured Faculty Leave program that allows professors in either their third or fourth year to go off (at half salary) and work with individuals in other locations.

While there are no Ph.D. students at Bucknell, master's students, all of whom are required to do a thesis, are supported by the university and are available to faculty for various projects.

With respect to retention, tenure, and promotion, Bucknell has a 2–4–6 program. All faculty are reviewed in their second, fourth, and sixth years. The second year review provides an early opportunity for the university to give feedback, point out what is going right, and look at where the deficiencies are. Very few people fail to get past this hurdle.

Not so with the fourth year review. "Here we look for clear evidence of quality teaching, as well as evidence that you are on the right track with respect to scholarship," says Buffinton. "We want to see that you have separated from your advisor and have established your own reputation." Although most people survive their fourth-year review, a significant number receive such a gloomy evaluation of their progress toward tenure that they choose to leave the university on their own. The vast majority of the faculty who make it to the sixth year are awarded tenure.

When it comes to assessing teaching, Bucknell relies heavily on student evaluations, and in many cases input from alumni. "It is not uncommon in some departments for the review committees to write to almost all your previous students and ask for feedback," says Buffinton.

Like most schools, Bucknell seeks the opinion of external reviewers to assess scholarship quality. Faculty typically submit a list of 10–12 potential reviewers, from which the department, with the concurrence of the dean, will choose 3–5 for the actual review. "Of course, they immediately eliminate former advisors and friends," says Buffinton. But the committees also limit the number of reviewers from other small schools like Bucknell to one, and this limitation can be a problem. According to Buffinton:

> The departments want feedback from people at top research universities, and such reviewers tend to look at your work in terms of what goes on at their own schools. This approach can put us at a disadvantage because we are not likely to be as productive as they are. The key is to cultivate faculty contacts who understand the situation at liberal arts colleges. I selected people who I had talked with over the years at conferences and who understood the Bucknell situation.

Buffinton's final comments echo those of Eve Riskin and others about learning to say "no." "Our dean explicitly gives new faculty permission to say 'no' when it comes to excessive committee work and the like," says

Buffinton. "Yet, I have to admit this isn't always easy. You have to balance looking out for yourself with being a good department citizen, especially at a small school like ours."

Vignette #28 A Second Chance at Tenure

For me, one of the most striking differences between the two schools was their attention to the formalities of the tenure process. At one school, I really had no idea what I was supposed to do, the material I was supposed to prepare, and the procedures that would be followed. Oh sure, there were lots of rumors, and informal comments among the faculty, but the procedures were never explicitly spelled out, and no one ever sat down with me and said, "these are the kinds of things you need to do, this is the kind of material you need to prepare." The other school was almost the opposite extreme where the procedures were spelled out in detail and you knew where you stood at each step along the way. It was a lot more work, but I really appreciated knowing that what I was doing was going to matter one way or the other.

These are the comments of Ruthann Kibler, now a tenured associate professor of biology at San Jose State University (S.J.S.U.), a Master's I institution in San Jose, CA. Kibler has had the experience of being denied tenure at one university, spending time as a research scientist in industry, and then returning to academia at another institution where she subsequently received tenure. Her story has interesting lessons for all beginning professors.

Kibler received her bachelor of science degree in biology in 1964 from Marietta College in Marietta, OH. After a year as a research technician at the Institute of Hematology in Freiburg, Germany, she returned to the United States to earn a master of science degree in cell biology in 1967 from Purdue University in Lafayette, IN. She then spent a year as a research technician at the National Institutes of Health (NIH) in Bethesda, MD. Kibler's NIH experience was followed by a four-year period at the University of CA, at Berkeley, in Berkeley, California, where she received her Ph.D. in immunology in 1973. She then spent three additional years back in Germany, this time as a research associate at the University of Wurzburg Institute of Virology and Immunobiology in Wurzburg, Germany.

In 1976, Kibler accepted a two-and-one-half year nontenured research faculty appointment in the Department of Microbiology at the University of Arizona in Tucson, AZ. When her contract ended, she returned to Berkeley, CA, as a postdoctoral fellow with the American Academy of Microbiology, State Department of Health Services.

In the spring of 1980, Kibler had reached a point in her life where, as she puts it, " I needed to decide if I wanted to pursue an academic career or a research management position in the public health system." She decided to apply for both types of positions, and subsequently accepted a tenure-track assistant professorship in the Department of Microbiology and Immunology in the College of Medicine, at the University of Arizona in Tucson. (Note: her previous nontenured research faculty position was in the College of Liberal Arts at the University of Arizona.)

In her new position, Kibler had both teaching and research responsibilities. The teaching duties were all at the graduate level. They included an immunology class for Ph.D. students, a graduate seminar, and a course for medical students. Although, as Kibler puts it, "I had to start from scratch, having never taught before," she clearly enjoyed her teaching. In 1985, she received the Basic Science Educator of the Year Award from the College of Medicine. Two years later, she received the Outstanding Teacher in the Basic Sciences, Medical Student Graduating Class Award, also from the College of Medicine.

With respect to research, Kibler was expected to set up a research laboratory, obtain funds (she was given a $5000 start-up grant), and of course, supervise graduate students. As she put it, "I was so happy to have the job, that I didn't ask for much by way of support and that was a big mistake." (As noted in Chapter 9, "Getting the Results You Want," this kind of mistake is not uncommon.) Within a year, she had formed a relationship with the Southwestern Clinic and Research Institute in Tucson, which provided her with up to $50,000 per year in research support. Over the next few years, she also obtained additional research support as part of a National Institutes of Health SCOR grant.

In 1985, Kibler began putting material together for her tenure review. What was so striking about this exercise, as noted in Kibler's quote at the top of this vignette, was its lack of specificity on what to include in her portfolio. "Basically, I was told to 'say what you have done,' and little more," says Kibler.

This lack of specificity also extended to the feedback she received from various review committees. She was told informally that her department had unanimously recommended her for tenure. Her application then moved up to the Tri-College Committee consisting of representatives from the Colleges of Science, Medicine, and Agriculture. She heard, this time informally from her department head that "things had not gone well" at the college committee level, with no explanation of what this comment meant. Her papers were then brought to the University Tenure Committee, after which she received her first, and only, written response, a short letter thanking her for her service to the university, and informing her that she had not been granted tenure. No specific explanation was offered.

In spite of her extensive publications and her two teaching awards, Kibler suspected that her lack of research volume and her inability to obtain large research grants to sustain a research program were the main factors in her not obtaining tenure. Yet, the reasons were not spelled out in any specific way, so to this day, she is still not certain why she was denied tenure.

Kibler remained one more year at the University of Arizona to complete some of her research and finish up with her graduate students. She then decided to leave academia and try her hand at an industrial research career. She accepted a senior research scientist position at BioRad Laboratories in Hercules, CA, near San Francisco. After three years, she realized she missed academic life, particularly teaching. All of her prior teaching was at the graduate level, and as she comments:

> I wanted to find out if I would like teaching undergraduates. I needed to do so in a way that didn't require a long-term commitment if it turned out that I did not like working at this level. I decided to look for a position that was clearly temporary, say, one year at the most. Such a position would give me a way to find out what I liked, and at the same time provide a graceful way to exit if it turned out that I wanted to do something else.

Kibler applied for, and obtained, a one-year visiting assistant professorship at the University of Santa Clara in Santa Clara, CA, where she taught undergraduate courses in virology, immunology, and cellular biochemistry. She quickly realized that undergraduate teaching was something she really enjoyed. Encouraged by her Santa Clara experience, she applied for permanent faculty positions at four schools. Kibler was interviewed at three of them, and was offered positions at two, one of which was San Jose State University. She accepted the position at San Jose State University, where she began at the highest assistant professorship level, and was granted tenure and promoted to associate professor five years later.

At San Jose State University, Kibler teaches both undergraduate and graduate classes, supervises master's students, and maintains good research relationships with Becton-Dickinson Immunocytometry Systems, a local bioengineering company. She has spent the last five summers working for the company, as well as about one day a week during most semesters. Three of her students have also worked for the company on their master's degree projects.

The tenure evaluation system at San Jose State University is much more explicit than it is at the University of Arizona. Comments Kibler:

> At S.J.S.U., you know what's going on every step of the way. Each procedure is spelled out in detail. Faculty observe your classes, you are told what kinds of material to include in your portfolio, and you are given explicit feedback on how things are going. For example, I was observed by 13 faculty members over four years, and all but two of them came two or three times to either lecture or laboratory classes I taught. I also received written evaluations from each person every time they observed me. At the University of Arizona, I was never formally observed or evaluated by anyone.

> At S.J.S.U., I received written tenure reviews and a chance to reply to them at every step of the process. Such was not the case at the University of Arizona. While it is a much more laborious process at S.J.S.U., I welcome its openness and clarity.

For Kibler, the return to academia has worked out well. Her experiences at the University of Arizona, BioRad, and the University of Santa Clara taught her where the best fit would be—an institution where undergraduate and graduate teaching is valued, and where she can also maintain a research effort tied to industrial interests.

Vignette #29   Taking Another Direction

For me, the key was to move on to the next thing right away, and to do so in a forceful manner. In my case, it meant leaving Boston and going to Tokyo. I like to tell people I stayed on the East Coast, only it was on a different continent.

Lance Glasser, vice president for Advanced Programs at KLA Instruments in Santa Clara, CA and former director of the Electronics Technology Office at the Advanced Research Projects Agency (ARPA), remembers well what it was like not to receive tenure after six years at the Massachusetts Institute of Technology (M.I.T.). As Glasser comments:

At the time, it was a wrenching experience, and I don't want to minimize it. The department was like a club, populated in this case by people I really respected, and the basic message was, "We don't want you in the club."

Glasser finished his Ph.D. in electrical engineering/computer science at M.I.T. in 1979. After staying there for one year as a research associate, he applied for, and was offered, a tenure-track assistant professorship at the same institution. His work up to that time had been in microwaves and laser optics. However, M.I.T. wanted to hire him to work in the emerging field of VLSI (very large scale integration) circuit design. In fact, over the next two years, the department hired four assistant professors, all of whom had done research in other areas, to establish a home-grown expertise in VLSI.

While it is not impossible to switch fields early on, to do so and still make contributions sufficient for tenure in five–six years is very difficult. *None* of the four faculty referred to above received tenure. Looking back on the experience, Glasser likes to say, "I got four degrees from M.I.T.; Master's, Ph.D., and Assistant and Associate Professor."

At the time he was denied tenure, Glasser was offered a one-year terminal sabbatical (at half salary). At that point, he decided to do something quite different. With his wife and three children, he left for Japan to spend the next 13 months at the Hitachi Central Research Laboratories in Tokyo, studying superconducting transistors and Japanese research management.

His experience in Japan was profound. As Glasser puts it:

I came out of that period understanding more clearly the difference between a process orientation and a product orientation. In both the United States and Japan, people are judged by *where* and *how* they get to a certain point in life. However, in the United States, the first is more important than the second, and in Japan, the reverse is true. This difference means that in Japan, everyone can be part of the solution. In the United States, we lose the bottom two-thirds of our society because we say they didn't get where they were going.

The other thing Glasser came away with from Japan was a strong desire to enter public service, and as he puts it, "maximize my impact on the U.S. industrial base and to serve my country." Glasser laid the groundwork for this transition while in Tokyo. "About once a quarter, I would write a 10–15 page 'newsletter' and send it to people I knew back in the United States," he says. "I sent it to about 30 or so people, and I'm sure it got passed around quite a bit as well." The newsletter had a twofold purpose: (1) to document the interesting things Glasser was doing and finding out about, and (2) to keep him in front of people who might need his services when he came back to the United States.

Glasser did not have a specific job waiting for him on his return, so he did some consulting for a while. However, he kept up his contacts, and shortly thereafter, two quite different, but enticing opportunities materialized. The first was a research position at the IBM T.J. Watson Research Center in Yorktown Heights, NY, and the second a program manager position at the Advanced Research Projects Agency (ARPA) in Washington, DC. The choice between continuing on as an individual contributor (IBM) and going into management (ARPA) was a difficult one. Michael Dertouzus, a colleague at M.I.T., set the stage for Glasser when he remarked, "All the minor decisions in life are made by the head; all the major decisions are made by the heart." "This comment," says Glasser:

> freed me to make my decision with my heart, but it still wasn't clear what that decision should be. As I looked around in industry, I realized that for the most part, the highly respected people were managers, not individual contributors.

Glasser visited the ARPA headquarters in Washington, and as he puts it, "I resonated strongly with the place, the excitement in the air, the opportunity to make a difference." He accepted their offer, and since then had a meteoric rise in the organization. In less than two years, he moved from one of approximately 100 program managers to one of six office directors with a professional staff of 20 and an annual budget of $500 million.

Impressive as these accomplishments are, it is the changes from within made since his M.I.T. days of which Glasser is most proud:

> At ARPA, I became good at two things that I was terrible at during my time at M.I.T. I was better at politics and better at *not* being arrogant. My own analysis of my tenure situation at M.I.T. led me to the conclusion that some of it was

based on merit, but some of it was also based on my high level of arrogance. I was no longer able to do anything about the first, but I decided to work hard on the second.

Glasser has now moved on to industry where he is responsible for bringing a new mask inspection product to market at KLA Instruments. He is proof that it is possible to have careers in all three employment sectors: academia, government, and industry.

Vignette #30 Lessons Learned

I can see more clearly now why things didn't go the way I had hoped. Some events I had control over, others I didn't. I'm not really bitter because now I can concentrate on raising my children in a way that would not be possible if I had become a full-time tenured professor. Still, I miss teaching, and I want to return to it in some way in the future.

Interesting comments from Sue Smith-Baish, who spent seven years on the tenure track as an assistant professor of electrical engineering at Bucknell University, a Baccalaureate I institution in Lewisburg, PA. Smith-Baish did not receive tenure at Bucknell; however, she still lives in Lewisburg with her two children, Andrew and Erik, and her husband, Jim Baish. Her husband joined Bucknell a year after Smith-Baish came to the University, and received tenure in the Mechanical Engineering Department after six years on the faculty.

We looked at the Bucknell tenure situation from the point of view of Keith Buffinton, a successful candidate, in Vignette #27. Smith-Baish's story offers another perspective with additional lessons for those just beginning their academic careers.

Smith-Baish came to Bucknell in 1985 at a time when demand for electrical engineering professors was quite high. She had earned a bachelor of arts degree in physics from Franklin and Marshall College in Lancaster, PA, and a master of science degree in bioengineering from the University of Pennsylvania in Philadelphia, PA. She then continued on in the University of Pennsylvania's Ph.D. program in bioengineering. After completing her coursework and a good part of her research, her advisor suggested that she start looking for a faculty position on the assumption that it might take up to a year to find the right opportunity. To both her and her advisor's surprise, Smith-Baish received an offer from Bucknell a few weeks later. Bucknell's Electrical Engineering Department was interested in filling areas covered by two previous professors who advised projects in rehabilitation engineering and premed electrical engineering. Although Smith-Baish had no formal training in electrical engineering, her combined physics and bioengineering background was attractive to Bucknell.

Jim Baish, who was still working on his Ph.D. in mechanical engineering at the University of Pennsylvania, had attended Bucknell as an undergraduate, and liked the possibility that both he and his wife might end up teaching there. They decided to move to Lewisburg in September 1985. Smith-Baish would start as a full-time tenure-track professor, the first of two women ever hired in tenure-track positions in the Engineering College, and continue working part time on her dissertation. Her husband would commute back and forth to Philadelphia to finish his Ph.D. Baish did indeed complete his doctorate the following year, and in the fall of 1986, took a visiting, nontenure-track position at Bucknell.

Smith-Baish continued working part time on her dissertation. During the two-year period, 1987–1989, she and her husband worked out an arrangement in which each of them had a two-thirds appointment at Bucknell. Baish was put on the full-time tenure track in mechanical engineering in the fall of 1989, and chose to count his first-year visitor appointment as time on his tenure clock. As noted in Vignette #27, Bucknell has a 2–4–6 year promotion and tenure evaluation system. In 1987, Smith-Baish successfully passed her two-year review.

Smith-Baish applied for and was granted a 1989–1990 junior sabbatical, which is available to untenured faculty. However, she became pregnant in May 1989, and instead of taking the junior sabbatical, she applied for, and was granted, an unpaid leave for 1989–90. Smith-Baish's junior sabbatical was postponed until the 1990–1991 academic year. During her sabbatical, she worked on her dissertation, but the effort was hampered by having Erik at day-care where he frequently became sick. His health problems, which consisted mainly of viruses and recurring ear infections, were, of course, passed on to his mother, and this took its toll on her time and energy.

Smith-Baish completed her dissertation in January 1991, just in time to pass her fourth-year review, for which a completed doctorate was a minimum requirement. The following year, she was back to her full-time assistant professorship at Bucknell. In the fall of 1992–1993, she assembled her formal tenure dossier. While the review committee for the Electrical Engineering Department recommended her for tenure, the final decision is made by the University Review Committee (URC), a committee composed of the vice president for academic affairs, both deans, and four faculty members. In December 1992, the URC chose not to grant her tenure.

The primary reason for Smith-Baish's tenure denial was that she had not done enough research beyond her dissertation. Since she had believed, according to her interpretation of the faculty handbook, that she was brought up for tenure consideration one year too early, she appealed the tenure decision. (The full-year junior sabbatical should not have counted on her tenure clock.) She won this appeal, and was granted an additional year (the one during which she was appealing the decision) on her tenure clock. Her departmental review committee again recommended her for tenure. However, when the decision came up at the university level in December 1993, she was again not granted tenure.

Clearly she was disappointed, and by this time, had, in her words, "run out of steam." She decided not to go through another appeal. Reflecting on her experiences, Smith-Baish offers the following insights to faculty just starting out on the tenure track.

- Complete your dissertation before you start a full-time faculty position. If doing so means negotiating a later start-up date, a part-time start-up assignment, or a visiting appointment, then do so.
- Be cautious about going to an institution with research facilities significantly different from those that were required for your dissertation. If you do, be prepared to shift your emphasis, from more experimental to more theoretical undertakings, for example, or to find ways to collaborate with colleagues at other institutions that have the needed resources.
- Be careful about accepting a position requiring you to teach outside your field. Smith-Baish, whose background was in physics and bioengineering, had never taken an electrical engineering course. As she put it, " I certainly understood many electrical engineering principles, which derived from my physics background, yet I still had to scramble, and it took a great deal of my time."
- If you are a woman, seriously consider not having children until you have earned tenure. (See also Vignette #26.) If you do have children, then invest in regular help with such things as child care and household maintenance.
- Seek role models at your (or another) institution who can serve as confidants and mentors.
- Say "no" to most service assignments, particularly in your first few years. In her fist year, Smith-Baish was asked to serve on a very time-consuming search committee seeking a new dean in the College of Engineering.

Finally, Smith-Baish talks about the importance of looking for the best possible fit between you and a department. She notes that the more successful candidates for tenure had previous teaching, research, or industrial experiences beyond the dissertation, and a background that best fits the overall needs of the department.

IN ADDITION: THE TEN COMMANDMENTS OF TENURE SUCCESS, TENURE AS A POLITICAL PROCESS, AND GETTING HELP ALONG THE WAY

Let us now look further into the tenure process by examining insights from two books and from an unusual program underway at a university in British

Columbia. Specifically, we will look at: (1) the ten commandments of tenure success, (2) tenure as a political process, and (3) getting help along the way.

Ten Commandments of Tenure Success[2]

There are hundreds of published articles and many books on the subject of tenure. One of the best sources for beginning faculty is: *Getting Tenure*, by Marcia Whicker, Jennie Kronenfeld, and Ruth Strickland[1]. In it, the authors tell you how to manage your tenure case, and then step you carefully through the tenure process in a way that helps you meet your institution's research, teaching, and service criteria.

The book stresses the importance of keeping a record of everything related to tenure criteria. Such a record should not only include teaching evaluations, but also "letters from students about teaching, course syllabi, documents showing participation in service activities, copies of requests for reviews of articles and books, letters commending service activities, publications, and conference papers" [1, p.140].

The authors also talk about the process of identifying outside reviewers, whose comments will have a significant impact on your tenure case. They offer this advice [1, p.62, Table 4.4]:

- Pick people at equivalent peer or better institutions, but be sure they understand the norms of your kind of institution.
- Use no more than one referee from your doctoral institution.
- Use only one person from a group with which you have written articles.
- Referees should be at a higher rank than you.
- Include full professors along with associate professors.
- Contact people in advance before putting their name on your list.
- Do not include someone who indicates they are too busy or unfamiliar with your work.

Many of the key points in the book are summarized in the authors' Ten Commandments of Tenure Success. They are [1, p.138, Figure 9.1]:

1. Publish, publish, publish! (Pay attention to what does and does not count.)
2. View tenure as a political process. (It is more like a legislative process than a bureaucratic one.)

[2]Adapted from [1].

3. Find out the tenure norms. (Understand the difference between written standards and operational standards.)
4. Document everything. (As noted above.)
5. Rely on your record, not on promises of protection. (Remember that administrators come and go.)
6. Reinforce research with teaching and service. (Leverage each with the other for maximum effectiveness.)
7. Do not run your department or university until after tenure. (Skip the university policy meetings; go to the computer center, library, or advisory meeting with students instead.)
8. Be a good department citizen. (Determine where, when, and how to chip in and pull your weight.)
9. Manage your own professional image. (Image management is important, but not a substitute for productivity.)
10. Develop a marketable record. (Seek to develop a record that is tenurable anywhere.)

The Politics of Tenure

As noted above, you need to view promotion and tenure as a political process. By doing so, you are not ignoring substance; rather, you are recognizing that substance alone is not enough, that there is also a human side to the success equation. An excellent discussion of this element appears in Chapter 4, "Appreciating the Practical Politics of Getting Promoted," in *Mentor in a Manual—Climbing the Academic Ladder to Tenure*, by A. Clay Schoenfeld and Robert Magnan [2].

To the authors, politics is not a dirty word, but rather, "the total complex of relations between persons in society" [2, p.119]. While recognizing that political situations can differ widely from campus to campus, Schoenfeld and Magnan believe that you need to pay particular attention to "situations for which your principal challenge is to understand the going standards or requirements [of your department] and to adapt to them with insight and aplomb" [2, pp.120–121].

In learning and adapting to the dynamics of your tenure-granting department, the authors suggest that you try to determine the following [2, p.122]:

- What is expected of the department by the divisional CEO (usually the dean).
- What is expected of you by the department.
- The strengths and weaknesses of your compatriots.
- Other key people whose willing support is necessary to you and and your department.

Schoenfeld and Magnan then go on to list a dozen essential questions that you need to answer about your department. They are [2, p.122]:

1. What is the real power structure of the department?
2. Who are the informal leaders in the department? What is the source of their power as informal leaders? Are the informal leaders positive or negative forces in terms of my meeting my responsibilities? In other words, who are strategic sources of support? of sabotage?
3. Can I assume that my responsibilities, as outlined in my letter of appointment, are consonant with the understanding of the faculty members who will ultimately evaluate me?
4. How do my assigned responsibilities fit in with the responsibilities of the department? the college?
5. What specific functions am I personally responsible for on my own?
6. What standards must I meet in my first year? the second? beyond?
7. What policies and standard operating procedures exist to assist me?
8. What are the formal norms I am expected to comply with? Are these formal norms apt to be productive on my terms?
9. What informal norms am I expected to follow? Which are likely to be productive? counterproductive?
10. What are the strengths and weaknesses of department support staff personnel? of my graduate students, particularly any TAs or RAs? of my undergraduates, particularly advisees?
11. Who are the key people who support activities promoting the department mission? What are their positive and negative attributes?
12. What are the strengths and weaknesses of the department in terms of each required function—teaching, research, and service? [11]

Obviously, answering these questions will take time. Doing so will also test your astuteness and ability to identify and work with the right mentors and role models. The vignettes in this and previous chapters should give you assurance that others are indeed willing to help. Now, let us take a look at what one university is trying to do in this regard.

Getting Help Along the Way

Some schools actually go out of their way to help faculty understand their tenure and promotion process. One example is Simon Fraser University, a 13,000 Doctoral II institution in Burnaby, British Columbia, about an hour's drive from Vancouver. Simon Fraser has found that, if left to their own devices, most faculty will not do an adequate job when it comes to finding out about promotion and tenure. According to Susan Taylor, the executive director of the Faculty Association of Simon Fraser University:

> You'd be surprised how many young faculty just hide away in their offices. They believe all they have to do is work, work, work, and everything will be okay. But this is not the case. You have to talk to other faculty, ask questions, and not wait until the fifth or sixth year. You need to take a colleague to lunch, talk about your scholarship, learn from their experiences. You don't have time to reinvent the wheel.

Taylor, who has been at Simon Fraser since 1981, is responsible for a program that takes some of the mystery out of the renewal, promotion, and tenure process. Established in 1990, it is remarkable for its ability to attract participation from faculty, department chairs, and deans, as well as the university president.

The central element of the program is a day-long session at the beginning of the fall semester for new and tenure-track faculty. It features a series of speakers on such topics as the role of the Faculty Association, teaching, and obtaining research support. The highlight of the event is a panel presentation on tenure and promotion. "We give a lot of thought to who should be on the panel," notes Taylor. "Last year, we chose someone who had just achieved tenure, but who had not had a particularly easy time doing it. We also looked for a department chair, a member of the University Tenure Committee, and then a dean and the university president. Each had a particular perspective to offer" [3].

At Simon Fraser, a candidate's credentials are first reviewed by the Departmental Tenure Committee (DTC). The recommendations of the DTC, which do not have to be unanimous and which may be for or against tenure, are then passed on to the faculty dean. The dean reviews the candidate's file and passes his/her recommendation, which may be for or against tenure, on to the University Tenure Committee (UTC). The UTC evaluates the candidate and passes its recommendations, which again do not have to be unanimous and which may be for or against tenure, on to the university president. The president reviews the case, and then makes a recommendation to the Board of Governors. At any time along the way, a candidate may receive a request to appear before the DTC, the dean, the UTC, or the president. These appearances can be critical. At a recent information session put on by the Faculty Association, Paul Percival, professor of chemistry and former member of the UTC, recalled this situation [4]:

> It was a case where there was disagreement between the dean and the DTC. When the candidate came before us (the UTC), he gave what amounted to a short 15-minute talk on his research, and it absolutely thrilled the committee members. He turned out to be such a compelling teacher that he convinced us thoroughly, not only of the value of his research area, but of the fact that he was such a top-notch teacher. As a result, it went from the stage where the committee had been more or less evenly split to a unanimous decision in favor of the candidate after his talk.

Here are the key points with respect to tenure made by members of the Faculty Association's panel at a recent information session [4]:

- Keep in mind that the university did hire you in the first place. They did so at a time when they were able to be pretty selective, so they most likely want to keep you.
- I kept hearing the research–teaching–service mantra. Was it 40/40/20 or something else? Had I done enough on each every month? At some point, you have to just stop worrying, and say that this is all you can do. But, of course, you keep worrying.
- Try to add something to your CV every month. Doing so forces you to think about what you have accomplished, and to look at the kind of story you want to tell about your career.
- Pay attention to the presentation of your CV as well as the content. Remember, it is going to be seen by people outside your field, and it says something about how you are organized and how well you think.
- Explain to students at course evaluation time the significance of what they are about to do in your promotion and tenure process. You should not solicit support for your application for tenure, but it is important for students to realize that they are not writing private notes to you, that what they say can have a real impact on your future.
- You need to be seen as a good citizen of the department. Doing so means that you are going to feel a tension between speaking up or going along. Many times, you need to put your oar in the water with everyone else, but there are times when the courage of your convictions, well presented, can add an important dimension to your colleagues' understanding of your contribution to the department.
- Do double-duty wherever possible. Combine your work with graduate students and directed studies courses, with the kind of research you are doing. Go to conferences, and come out with names and research ideas. Bring speakers to campus; it gives both of you visibility, and it gives you a good future contact.
- Find a mentor. Stay away from current chairs; they are too busy anyway. But an ex-chair is ideal.
- Take your holidays, for Pete's sake! Nobody notices if you do not. But you will notice when everyone comes back rested and refreshed and you are ready to collapse!

All good advice, in no way unique to Simon Fraser. What is unique, or at least very unusual, is that Simon Fraser discusses these matters in such a public way, and in the process obtains the support of so many on the faculty and in the administration. As one professor put it after a recent session, "I've never been at such a helpful institution. I'm overwhelmed with the willingness to give advice and insights that I don't see how I could have obtained in any other way."

CONCLUSIONS

It is my hope that the 21 vignettes in Chapters 10–14 will provide you with useful insights on how to manage your time, and get underway with teaching, research, professional development, and tenure. If you are also able to peruse portions of at least a few of the books referred to in the "In Addition" sections of these chapters, you will have a good start on the road to a successful academic career.

However, books can only do so much. The main message of Chapters 10–14 is that you are not alone, and that help can come from many sources. These sources include members of your own department, others in your institution, and those at many other colleges and universities. Indeed, with near universal Internet access, advice and support are possible from faculty throughout the world, not just those in our North American neighborhood.

Remember Robert Boice's comments about "quick starters" in Chapter 10. These are the ones who took the time to ask for help, to socialize with colleagues, and who understood that they did not have to go it alone. They are the individuals, who in Steven Covey's words, "are moving from dependence to independence to interdependence." To help us all move further along on this path, I ask you to participate in the discussion groups and in the information exchange mechanisms available through my World Wide Web Home Page as noted at the beginning of this book. I look forward to hearing from many of you soon.

REFERENCES

[1] M. L. Whicker, J. J. Kronenfeld, and R. A. Strickland, *Getting Tenure*. Newbury Park, CA: Sage Publications, 1993. Copyright © 1993 by Sage Publications, Inc. Reprinted with permission of Sage Publications, Inc.
[2] A. C. Schoenfeld and R. Magnan, *Mentor in a Manual—Climbing the Academic Ladder to Tenure*. Madison, WI: Magna Publications, 1992.
[3] Telephone interview with author, Dec. 15, 1995.
[4] Correspondence with author, Jan. 29, 1996.

CHAPTER 15

Insights on Academia: Needed Changes

It is my hope that the material in Chapters 1–14 will help tomorrow's professors prepare for, find, and succeed at academic careers in science and engineering. Yet, for them to truly thrive, we as administrators and senior faculty must also do our part. We must serve as leaders, mentors, and partners in helping students, postdocs, and younger faculty colleagues move from dependence to independence to interdependence. In so doing, we will make an important contribution to the future of our academy.

Let us begin with a look at our role in helping students prepare for academic careers. This examination will be followed by a brief discussion of what we can do to assist graduate students and postdocs in finding the best possible faculty positions. We will conclude with a look at how to help beginning professors succeed in the current academic environment.

HELPING GRADUATE STUDENTS AND POSTDOCS PREPARE FOR ACADEMIC CAREERS

What advice and what level of encouragement should we provide students and postdocs regarding the pursuit of academic careers? Keep in mind that for many students, the possibility of such careers first emerges while they are college juniors or seniors. Their exposure to undergraduate teaching and the decision to go on to graduate school often combine to at least raise the possibility of following in their professors' footsteps. Remember, also, that the majority of Ph.D.

students did not go to research universities as undergraduates, and almost half of all new professors will not teach at such universities. This fact means that all of us, at all types of academic institutions, have a role to play in helping students make decisions about their future careers.

At present, most of our advice seems to fall into one of two categories. We either tell students to stay away from academia at all costs because of the poor job market, difficulty in obtaining research funding, realities about academic life, and/or our desire to have them go into industry, or we tell them an academic position—usually at a research university—is the only one worth having, industry is to be looked down on, our best students are here to replace us, and if they are really as good as they think they are, they will be able to find academic positions, just as we did.

Both of these positions are extreme. While we need realism, not fantasy, we do not need cynicism. We should discuss with our students what is actually going on in higher education, and the advantages and disadvantages of academic careers. In so doing, we need to support the Multiple-Option approach outlined in Chapter 4, which means not only looking at academic and nonacademic careers, but at academic careers at all kinds of institutions. Realistically, most of our students are going to examine these options anyway; they just will not tell us about it if they think we will disapprove. So why not get things out in the open and be part of the discussion? If we do not, many of them will come to their decision, for or against an academic career, based on incorrect or incomplete information about what it is actually like to be a professor in science or engineering.

Such discussions will inevitably lead to the question of whether we should limit enrollment of graduate students in certain fields because of the current excess of supply over demand. I believe that we should not do so. The market will take care of the situation provided we do two things: (1) provide the best information we can, including honest statements about our own motivations for wanting graduate students, and (2) provide the advice and help students need, i.e., be their partners in the career development process.

Once we have graduate students and postdocs who want to consider academic careers, what then do we do? One of the discoveries I made while writing this book was how little students and postdocs really know about academia despite their having spent most of their adult lives immersed in it. For this reason, we need to provide more Next-Stage experiences as outlined in Chapters 4– 6. In so doing, we must give up our reluctance to let students and postdocs in on what we are doing, and find ways for them to actively participate in "professor-like" activities. I understand the resistance in some quarters to giving such responsibilities to inexperienced individuals. The key, particularly in the case of teaching, is to view it as a complement to, not a substitute for, an experienced professor. It is possible to provide this complement without increasing the burden on existing faculty. The medical training and high school student-teaching models discussed in Chapter 6 are just two possible approaches.

During the past several years, I have been involved with a program at Stanford University called the Future Professors of Manufacturing. Among other things, it provides students with experiences similar to those they will have when they become professors. They are regularly called on to make presentations about their work and the program to potential donors, funding agencies, and university officials. They write many of their own research and financial aid proposals, and in many cases supervise master's-level RAs working in various areas of manufacturing research. And, of course, they are given teaching opportunities. These opportunities range from advanced TAships in manufacturing courses, to guest lectures in regular courses, to teaching segments or even full courses, individually or in teams, at Stanford University, the University of Santa Clara, San Jose State University, and even the University of California—San Diego, some 550 miles away.

We also need to provide students and postdocs with more Breadth-on-Top-of-Depth experiences as outlined in Chapters 4 and 5, particularly in view of the movement of some institutions toward the broader notion of scholarship (discovery, integration, application, and teaching) outlined by the Carnegie Foundation for the Advancement of Teaching. This movement raises an interesting question: Should we expand the concept of the Ph.D., particularly the dissertation, to include these other forms of scholarship?

The doctorate began as a way of qualifying individuals for research careers, but by the early part of the 20th century, it also became a requirement for teaching at the college or university level. This requirement was in place even for professors who wanted mainly to teach and who did little or no research. It did not make sense then, and it certainly does not make sense now. If we are going to promote all forms of scholarship, perhaps to a different degree at different institutions, then should we not do more to prepare future faculty to carry out such scholarship? Must we continue to have everyone do a standard "discovery of new knowledge" dissertation, and then have them try to adapt to other forms of scholarship later on? Or can we prepare some professors for these other forms of scholarship through a different kind of dissertation? Certainly, it is possible to make an original contribution through the integration, application, or teaching of knowledge.

How would we supervise such dissertations at traditional research universities? To some extent, we have a chicken-and-egg problem. Yet, as other forms of scholarship begin to take hold at some research universities, advances in this direction become possible. In other cases, research universities could look to form "dissertation alliances" with other schools, or even government and industry, where such forms of scholarship are more developed.

Would such a Ph.D. increase or decrease the recipient's career prospects? I believe that it will increase them if it maintains the rigor associated with the traditional Ph.D., and if it is better suited to the needs of particular institutions that are then prepared to reward such scholarship in the retention, promotion, and tenure process. (More on this topic in the section, "Helping Beginning Faculty Succeed."). Such Ph.D.s would also fit nicely with the Multiple-Option approach advocated in Chapters 4 and 5 by increasing the attractiveness of the recipient to government and industry. Let me be clear that I am not talking about eliminating the traditional Ph.D. or about substituting breadth for depth in a dissertation. But I do think that it is time to look at *additional* options to better prepare professors for the new academic reality.

HELPING GRADUATE STUDENTS AND POSTDOCS FIND ACADEMIC POSITIONS

What is our role in helping graduate students and postdocs find the best possible positions? We must begin by letting them know we want to be actively engaged in the process. This offer is not an invitation to take responsibility off their shoulders, but to play a role that complements their central activity. They can, and should, feel free to discuss all possibilities with us, including options in industry, in government, and at all kinds of academic institutions. "Best possible position" means what is best for them, not us. It means supporting their interest in looking at places other than Research I and II universities, if that is their choice. At times, it may mean encouraging them to apply to other types of institutions, even if they may not initially think to do so.

Another thing we can do is promote more programs that encourage a dialogue among students, faculty, administrators, and visitors about the academic application process. In particular, we can share with our students and postdocs the experiences at our institution with regard to current hiring efforts. We can include them in the process as observers, as participants on search committees, or as attendees at applicant academic job talks. Since not all of them will be able to participate this way, we can ask those who are to share their experiences with others at seminars or during informal gatherings.

We should also arrange for seminars and talks describing the procedures and processes at other institutions. An easy way to do so is to ask graduate students and postdocs from our institution to share their experiences in applying to other places.

We need to be proactive in introducing students and postdocs to opportunities we hear about through informal contacts. We should be sure to introduce them to colleagues who visit the department and to those we see at professional meetings. Even if these people do not have openings at their schools, they can often serve as secondary contacts to others who do have such positions.

We also need to go out of our way to obtain feedback about our students' and postdocs' written applications and performances during job interviews and campus visits. Often, we are the only one who can obtain this information, so we need to be proactive about getting it, and then sharing it in a way that is both honest and supportive.

Finally, we need to look beyond our own students and postdocs to the ones who are applying for faculty positions in our departments. We have a special responsibility here to provide honest and complete feedback. Giving such feedback can be difficult because the reasons for rejecting a finalist are often subjective and not easily explainable. Yet, if we begin by reviewing the candidates' strengths, and then move on to what did not make for a good match, or things in their presentation they could improve, it usually goes well, and the feedback is often welcome. I believe that it is the most important thing we do for such candidates.

HELPING BEGINNING FACULTY SUCCEED

The last topic I would like to address is what we as administrators and senior faculty can do to help new professors succeed in their chosen careers. As noted in Chapter 10, the central message from over 70 conversations with beginning faculty from schools across North America was their sense of being overwhelmed by the number of items on their plates, their belief that they were not doing any of their tasks well, and their frustration at having no time to think long term. Adding to their stress was their feeling of not having any clear understanding of what was expected of them, of not knowing what the really important things to concentrate on were, and where—indeed even whether—to seek help and guidance from others.

It would be a mistake for us to assume that all of this stress is just what we experienced 30, 20, or even 10 years ago when we started out as professors, and that like us, they will just have to work through it. Not only have the goalposts moved, the game itself has changed, and as we saw in Part I, this change is causing anxiety and uncertainty throughout all levels of higher education.

The first thing we can do is help set the proper context. We need to let our beginning colleagues know that while we expect them to act independently and accept responsibility, they are not alone, and they do not need to discover or accomplish everything by themselves. We need to assure them of our availability as mentors and partners, not only with research, but with other forms of scholarship, including teaching. Seeking feedback from more experienced faculty, learning from the mistakes of others, and asking for assistance when needed are not signs of weakness, but of wisdom. However, just letting beginning faculty know we are available is not enough. We need to seek them out on a periodic basis, inquire how they are doing, what difficulties and anxieties they are facing, and how we can be helpful with advice, advocacy, and specific decisions. Put another way, we need to insert ourselves into the vignettes of Chapters 10–14.

These comments may seem obvious, even a little trite. In my experience, however, we too often assume that someone else is taking care of these matters, only to find that they have fallen through the cracks and our beginning colleagues have been left to fend for themselves.

The second thing we can do is support beginning faculty in their efforts to set long-term goals, to do fewer things in greater depth, and to manage their time and their tasks more effectively. A trait shared by most successful faculty is their ability to set aside time in the present for longer term efforts such as establishing a program of scholarship, writing a successful proposal, developing a new course, and supervising Ph.D. dissertations. If I could suggest just one thing in this regard, it would be for deans, department chairs, and senior faculty to legitimize new faculty efforts to "establishing their absence," i.e., to set aside time away from the office to work on long-term important tasks. We should do so on an individual basis with junior faculty so that they know we are talking about them specifically, but also publicly so that everyone will know we are sincere.

Finally, we need to take a closer look at tenure. Questions about tenure are being raised not only by those outside academia (professionals in industry, state legislators, and the public at large), but also from whose within (some board of trustees members, administrators, and faculty). Some of these questions relate to the validity of the "institution of tenure" itself, while others deal more with the effectiveness of the tenure-granting process. The situation at the University of Minnesota reflects both of these concerns. Recently, the university's Board of Regents, supported by members of the state legislature, demanded a thorough review of the university's tenure code [1, p.A18]. Why? Because in an era of fiscal constraints, they felt that the university should be able to lay off tenured professors in financially troubled departments. To this demand, Nils Hasselmo, the university's president, responded: "[We need to] protect tenure, but be sure we make it an instrument that we can defend, with procedures that demonstrate we are not protecting incompetence, we are not protecting slovenly behavior; that's the only way I think we can build credibility [1, p.A18].

The view of many outside academia is that tenure is too easy to obtain, results in guaranteed lifetime employment, and demands little or no accountability from faculty once they receive it. As a vice president of a major high-technology company put it, "With all the downsizing and restructuring going on in industry, it's hard to support a system that, in a relatively short period of time, guarantees lifetime employment to people whose work may be out of date in just a few years."

As noted in Chapter 1, the main argument for tenure is that it provides faculty with freedom of expression, while at the same time giving them the needed security to develop a depth of expertise over time that is not possible in any other way.

However, there is more to tenure than academic freedom and long-term employment. The common perception that only faculty support tenure, and that if university administrators had their way, they would get rid of it in a minute, ignores the other benefits tenure brings to the institution. There are at least two additional ways the institution benefits from tenure: financial and managerial.

As we noted in Chapter 9, it is the total compensation package that is being negotiated at the time of employment. Tenure, or more accurately the promise of it, is part of this package. The security and freedom tenure offers comes at a price to the faculty member in terms of salary. If tenure did not exist, the institution would have to pay higher salaries to compensate for the lack of security tenure offers. If the institution does not do so, it increases the likelihood that faculty, particularly in the more high-demand fields of science and engineering, will go elsewhere, i.e., to government, industry, or other academic institutions.

While tenure does not guarantee that faculty will not spend some of their time on inappropriate outside activities, the lack of tenure increases the posibility that at least some faculty will engage in activities that could lead to a conflict of commitment, if not interest. Without tenure, some faculty will spend additional time "covering their bases" by engaging in external activities that take away from their academic duties. You have only to look at what happens to industrial employee activity, productivity, and morale when word gets out of a possible downsizing to understand how this might work in academia. Indeed, as academia rushes to embrace the corporate model of restructuring and outsourcing, it should think about what it is at risk of losing. A sense of community, a shared view of scholarship, a common history, and a commitment lasting over time do not always show up on the bottom line. However, they are at the heart of what the academy is all about, and are, I believe, closely tied to the security tenure brings.

The managerial benefit of having a system that rewards independence and encourages longevity is in the reduced overhead associated with not having formal annual reviews (as found in industry.) Of couse some form of faculty performance review is essential if we are to have accountability, it just does not need to be on an annual basis. An additional managerial benefit is in having senior talent available for administrative, governance, and mentoring reponsibilities.

The security tenure brings is attractive to most faculty for reasons that also go beyond academic freedom. As we noted in Part III, there is likely to be one (or at most two) school, of a given type in any geographic area. The existence of this constraint means that it is not easy for a professor to leave one institution and find another academic position without having to change locations. Today, the willingness of a faculty member to make such a move is mitigated by the likelihood that he or she has a spouse with a career tied to their current locality. Thus, tenure offers an added measure of employment security, not only for a faculty member, but also for his or her professional spouse.

What about overall accountability and productivity? Do faculty members reduce their efforts once they obtain tenure and lifetime security? The answer depends on the effectiveness of the tenure-granting process. Simply put, we get what we select. If a K–12 public school system grants tenure in two or three years to 90% of its beginning teachers, it should not be surprised to find a fairly normal distribution of productivity and innovation among its permanent faculty.

The more rigorous the tenure-granting process and the more it is explicitly linked to the mission of the department and the institution, the more likely it is to yield individuals who will continue with behaviors that were selected in the time prior to tenure. Indeed, it is rather hard to fake such behaviors for a period of up to a half dozen years.

Consider Colby College, a Baccalaureate I institution in Waterville, ME. It has a selective and periodically evaluated tenure-awarding process that results in a full-time faculty tenure percentage of 52. What does this mean in terms of productivity and creativity? According to William Cotter, Colby's president [2]:

> In virtually every case the granting of tenure has liberated that faculty member to become an even more productive and important contributor to the quality of academic and campus life, and her or his finest scholarly work is usually produced after the tenure decision, not before. Tenured faculty members are motivated by a pride in their profession, a sense of responsibility, and a recognition that they are the real "owners" of the college. In addition, the tenure selection process looks forward and tends to yield only those who are most likely to be stimulating teachers, productive scholars, and active participants over a thirty-year career.

Yet, how many departments and institutions are both willing and able to match their mission with their tenure criteria? To do so, there first needs to be an agreement on the mission at all tenure decision levels. The mission also needs to remain fairly constant over time. Both are difficult to do in the best of times, and are certainly not any easier in today's changing academic climate.

Recall the situation at San Jose State University (Master's I institution) in San Jose, CA. The Biology Department had always emphasized the scholarships of teaching and application. The Chemistry Department, on the other hand, the "UC wannabe," emphasized the scholarship of discovery. Both departments are in the College of Science, which raises the question of how a dean should respond to these different criteria when tenure recommendations come his/her way.

Presidents, provosts, and deans can articulate a vision and a mission, but keep in mind that tenure criteria originate at the department level. It is the faculty who must be clear about what they will support, and it is they who must make sure that their view is consistent with the rhetoric of the deans and the provost. It is essential to match the rhetoric with the reality up the entire tenure decision-making chain.

Matching the institution's mission with its tenure criteria would be an important step in the right direction. However, it is not enough. We also need to pay more attention to the fairness of the tenure-granting process. That the process is not fair in many instances can be seen in the number of faculty who thought they were doing the right things, and who in many cases received what seemed to them to be positive signals along the way, only to find after six years that they had been denied tenure. Fairness comes by being more explicit about what endeavors count toward tenure, what weight we give to each of them, and how their quality will be assessed. Help in these matters can come by returning to our notion of scholarship.

As we have discussed, four interlocking elements form the new scholarship paradigm developed by Ernest Boyer, late president of the Carnegie Foundation for the Advancement of Teaching. They are (1) the discovery of knowledge, (2) the integration of knowledge, (3) the application of knowledge, and (4) the scholarship of teaching. By recognizing and rewarding various forms of scholarship, colleges and universities can allow faculty to make contributions best suited to their interests, capabilities, resources, and goals. (See Figure 15-1.) Two recent tenure appointments in the School of Engineering at Stanford University reflect this possibility. In one case, a faculty member who did not have a Ph.D. was awarded tenure primarily for his "scholarship of application" involving significant contributions in the area of industrial design. The second case involved a professor whose primary scholarly contribution was the development of an innovative approach to undergraduate curriculum development involving Stanford and a number of other institutions.

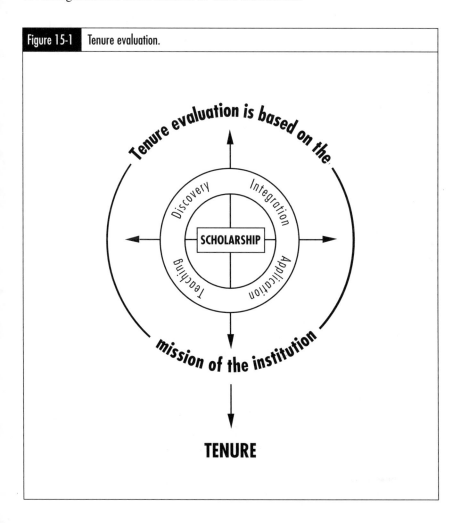

Figure 15-1 Tenure evaluation.

For the various forms of scholarship to be accepted in tenure decisions, there has to be an agreed-upon way of assessing their effectiveness. The Carnegie Foundation for the Advancement of Teaching has begun to study this issue. It has identified four principles related to faculty evaluation of scholarship: 1) qualities of a scholar, 2) standards of scholarly work, 3) documentation of scholarship, and 4) the credibility of the scholarship evaluation process [3, pp.23–24].

The Foundation suggests that we begin by looking at the quality of the person doing the work, rather than at procedures. In Boyer's words: "I suggest that scholarship relates, in the first instance, not to a catalog of accomplishments, but to qualities of character. Standards are, of course, critically important. But even more important are persistence, creativity, humility, and integrity, all of which are at the very heart of academic life" [3, pp.23–24].

In looking at the standards for scholarly work, the Foundation examined hundreds of documents and faculty handbooks, and surveyed dozens of editors and directors of scholarly presses. In so doing, it came up with the following six dimensions of good scholarship: knowledge, clear goals, appropriate methods and procedures, creative use of resources, effective communication, and significant results.

The documentation of scholarship must be rich and varied, and include such things as self-evaluation, peer review, student evaluation, and client evaluation. According to Boyer, "The documentation of scholarship should be a 'moving picture,' not a 'snapshot,' and that evidence should be gathered over time" [3, p.26].

Last, but not least, is the credibility of the scholarship evaluation process. It is important for such a process to be seen as fair, and for professors to have confidence in its procedures. Nevertheless, Boyer warns that the process must not become ridged and overly dependent upon quantitative measures. He puts it this way [3, p.26]:

> When all is said and done, what we must have, in faculty evaluation, are both good people and good procedures. We must clarify the standards and more effectively document the performance. A common language about scholarship will emerge, one that cuts across all the disciplines and helps create a true community of learning, not only nationally, but globally as well. For this to be accomplished, the evaluation of the professoriate requires rich experience, great wisdom, and even, perhaps, compassion.

An encouraging step in the above direction was taken recently by the 11-campus North Dakota University System. The state Board of Education tentatively approved a new tenure code that would allow professors to help chart their own paths to tenure, emphasizing, if they so choose, teaching and service over research [4, p.A16]. According to the new policy [4, p.A16]:

> When faculty members go on the tenure track, their contracts will describe their duties and goals and specify the weights to be given to research, teaching, and service in performance and evaluation. The mission of their institution—whether it is a research university or focuses more on teaching—will play a role.

Chapter 15 Insights on Academia: Needed Changes   361

The code also provides for a more explicit evaluation of the performance of faculty once they have tenure. They will be evaluated every three years, and institutions will be able to terminate professors who repeatedly receive poor reviews [4, p.A16].

This latter point leads to one more important question. How can we help faculty who want to develop professionally after they receive tenure? What do we do about those who might, for example, want to change their research field or scholarship emphasis? To some extent, tenure encourages such activity because it ties further promotions and salary increases to performance and not to scholarship that is out of date or no longer relevant. Indeed, as we saw in Chapter 3, funding agencies, both internal and external, will simply not support work that is no longer important and compelling, particularly in the current fiscal climate.

But changing fields is not easy, and for the most part, academia does not have a mechanism for supporting and rewarding such efforts. Most posttenure promotion and salary increases are based either on the same, "straight-ahead," in-depth criteria used in the granting of tenure, or on administrative service to the institution. As Michael Lightner, of the University of Colorado at Boulder, says, "We need a way of encouraging faculty professional development not unlike the way it is done in industry. It is ironic that academia is quite good at providing professional development courses for individuals from industry, but not nearly so good at doing the same thing for its own faculty" [5]. Lightner suggests that for interested faculty, we relax the annual publication/dissertation productivity measures, and think in terms of a three–five year performance contract in which the faculty member and the institution agree on a plan for moving into a new research field or scholarship area. Progress along the dimensions of the plan would be the basis for further promotion and salary increases. Such an idea is well worth considering.

CONCLUSIONS

By helping graduate students and postdocs prepare for, find, and succeed at academic careers in science and engineering, we not only fulfill our "intergenerational obligation," but we enhance the community of scholars of which we are all a part.

This community revealed itself in an unexpected way while I was doing research for this book. I examined dozens of college and university World Wide Web Home Pages, and found, interestingly enough, that none of them started out with the institutions' laboratories, multimedia center, or other high-technology facilities. They all began with the college or university insignia, followed by drawings or photographs of the oldest, most traditional buildings on campus, all intended, it seemed to me, to convey a sense of history, scholarship, and community.

Outside my window to my right are a number of buildings constructed over 100 years ago. Some are undergoing major reconstruction to completely change what they are like on the inside. However, care is being taken—at great expense —to preserve their exteriors so that they look as they did in the late 1800s. To my left is a brand new 150,000-square-foot computer science building that—thankfully—also looks from the outside like the 100-year-old sandstone buildings to my right. Inside the new building is the very latest in communications, teaching, and experiential technology, enabling, among other things, faculty and students to interact with their counterparts at institutions around the world. Yet, the major reason for constructing the building was to bring together under one roof students and faculty who had been dispersed among a dozen or so buildings across the campus. Let us all make the same effort at inclusion with those graduate students, postdocs, and faculty who would follow in our footsteps.

REFERENCES

[1] D. K. Magner, "A parlous time for tenure," *The Chronicle of Higher Education*, vol. XLII, no. 36, May 17, 1996.

[2] W. R. Cotter, "Why tenure works," *Academe*, vol. 82, no. 1, p.28, Jan.–Feb. 1996.

[3] E. Boyer, "Assessing scholarship," *ASEE Prism*, vol. 4, no. 7, Mar. 1995.

[4] D. K. Magner, "North Dakota revises tenure code to allow professors to emphasize teaching," *The Chronicle of Higher Education*, vol. XLII, no. 44, July 12, 1996.

[5] Discussion with the author, June 4, 1996.

APPENDIX A

Possible Items for Inclusion in a Teaching Portfolio[1]

Faculty members should recognize which of the items that might be included in a teaching dossier would most effectively give a favorable impression of teaching competence, and that might better be used for self-evaluation and improvement. The dossier should be compiled to make the best possible case for teaching effectiveness.

THE PRODUCTS OF GOOD TEACHING

1. Students' scores on teacher-made or standardized tests, possibly before and after a course has been taken as evidence of learning.
2. Student laboratory workbooks and other kinds of workbooks or logs.
3. Student essays, creative work, and project or fieldwork reports.
4. Publications by students on course-related work.
5. A record of students who select and succeed in advanced courses of study in the field.

[1] From R. Edgerton, P. Hutchings, and K. Quinlan, *The Teaching Portfolio: Capturing the Scholarship of Teaching*, a publication of the AAHE Teaching Initiative, American Association of Higher Education, One Dupont Circle, Washington, DC 20036, 1991, p. 8. Copyright © 1991 by the American Association for Higher Education. Reprinted with permission.

6. A record of students who elect another course with the same professor.
7. Evidence of effective supervision of Honors, Master's, or Ph.D. theses.
8. Setting up or running a successful internship program.
9. Documentary evidence of the effect of courses on student career choice.
10. Documentary evidence of help given by the professor to students in securing employment.
11. Evidence of help given to colleagues on teaching improvement.

MATERIALS FROM ONESELF

Descriptive material on current and recent teaching responsibilities and practices.

12. List of course titles and numbers, unit values or credits, enrollments with brief elaboration.
13. List of course materials prepared for students.
14. Information on professor's availability to students.
15. Report on identification of student difficulties and encouragement of student participation in courses or programs.
16. Description of how films, computers, or other nonprint materials were used in teaching.
17. Steps taken to emphasize the interrelatedness and relevance of different kinds of learning.

Description of steps taken to evaluate and improve one's teaching.

18. Maintaining a record of the changes resulting from self-evaluation.
19. Reading journals on improving teaching, and attempting to implement acquired ideas.
20. Reviewing new teaching materials for possible application.
21. Exchanging course materials with a colleague from another institution.
22. Conducting research on one's own teaching or course.
23. Becoming involved in an association or society concerned with the improvement of teaching and learning.
24. Attempting instructional innovations and evaluating their effectiveness.
25. Using general support services such as the Education Resource Information Center (ERIC) in improving one's teaching.
26. Participating in seminars, workshops, and professional meetings intended to improve teaching.
27. Participating in course or curriculum development.

Appendix A   Possible Items for Inclusion in a Teaching Portfolio   365

28. Pursuing a line of research that contributes directly to teaching.
29. Preparing a textbook or other instructional materials.
30. Editing or contributing to a professional journal on teaching one's subject.

INFORMATION FROM OTHERS

Students

31. Student course and teaching evaluation data that suggest improvements or produce an overall rating of effectiveness or satisfaction.
32. Written comments from a student committee to evaluate courses and provide feedback.
33. Unstructured (and possibly solicited) written evaluations by students, including written comments on exams and letters received after a course has been completed.
34. Documented reports of satisfaction with out-of-class contacts.
35. Interview data collected from students after completion of a course.
36. Honors received from students, such as being elected "teacher of the year."

Colleagues

37. Statements from colleagues who have observed teaching either as members of a teaching team or as independent observers of a particular course, or who teach other sections of the same course.
38. Written comments from those who teach courses for which a particular course is a prerequisite.
39. Evaluation of contributions to course development and improvement.
40. Statements from colleagues from other institutions on such matters as how well students have been prepared for graduate studies.
41. Honors or recognition such as a distinguished teacher award or election to a committee on teaching.
42. Requests for advice or acknowledgment of advice received by a committee on teaching or similar body.

Other Sources

43. Statements about teaching achievements from administrators at one's own institution or from other institutions.
44. Alumni ratings or other graduate feedback.

45. Comments from parents of students.
46. Reports from employers of students (e.g., in a work–study or "cooperative" program).
47. Invitations to teach for outside agencies.
48. Invitations to contribute to the teaching literature.
49. Other kinds of invitations based on one's reputation as a teacher (for example, a media interview on a successful teaching innovation).

Appendix B
Statement of Personal Philosophy Regarding Teaching and Learning[1]

Mary Anne Carroll
Atmospheric, Ocean and Space Sciences Department
University of Michigan

My move from a research laboratory to an academic setting was motivated by a desire to teach. Therefore, being a member of a university where attention to teaching and learning has a high priority is important to me. My interest in teaching comes from my own positive experience as an undergraduate and from a love of learning. It also stems from a belief that environmental issues are intricately connected with technology, and a sense of responsibility to educate a citizenry that is "literate" in environment sciences.

In considering how one goes about sharing one's love of learning, it is important to consider that learning strategies differ widely, and that teaching strategies are not always easily matched with students' needs. In addition, students bring widely varying knowledge bases to the table in each course, so each course is different according to the background and learning preferences of that particular class. A further complication is that students also bring different levels of maturity, interest, and motivation. The challenge is to make course materials accessible to all students, and to be responsive to individuals who are having difficulty integrating new material without boring others. Is this possible?!

[1] Reprinted with permission of Mary Anne Carroll, 1996.

I believe that learning can and should be fun, and that students who are active participants learn much more than those whose participation is largely passive. Teaching and learning involve an inherent contract. Students must agree to take responsibility for their learning in order to engage, and teachers must be willing to be engaged, as well. When students are so engaged, their learning is not solely dependent upon the rate of the delivery of lectures, so a mix and match of pace and teaching strategies designed to meet the needs of a range of learning skills need not be debilitating to the progress of any students. I welcome a group of students who are actively involved, thinking and questioning the material presented to them, whether presented by me or by another student.

Part of the contract involves the completion of homework assignments so that classroom periods can be used for group work and other activities that involve students and encourage their learning from each other. Although I initially felt the need to lecture in detail on all topics covered, my perspective has changed as my level of familiarity with the course material has increased. I believe that a teacher is not a giver of knowledge, but rather a facilitator or a guide for the student. As a guide, it is my responsibility to find or create alternate presentations of the material that I feel help clarify key points, and to design in-class activities that provide students with the opportunity to integrate material and to practice using new concepts. I have used take-home exams (even with an enrollment of 57) because I believe that students learn more and are not disadvantaged if they find timed exams difficult.

The concept of teacher as guide and the goal to encourage students to achieve carry over into my research group as well. I have high expectations of my research group's performance, and I would like my students to have high expectations of themselves as well. While I believe that assigning dissertation topics is disadvantageous for the student, I am actively involved as a resource during the selection process, as well as someone who provides encouragement, constructive criticism, and praise—all important components of nurturing the training process.

APPENDIX C

Professional Associations for Academic Job Seekers in Science and Engineering

The following professional associations provide members with support and information about academic job listings. Listings may be found in "employment" sections of the association's journals or newsletters, or in separate employment bulletins mailed periodically throughout the year.

SCIENCE

American Astronomical Society
2000 Florida Ave. N.W., Suite 300
Washington, DC 20009
(202) 328-2010
http://www.aas.org/

American Chemical Society
1155 16th St. N.W.
Washington, DC 20036
(202) 872-4600
http://www.acs.org/

American Institute of Biological Sciences
1401 Wilson Blvd.
Arlington, VA 22209
(703) 415-1751
http://www.reston.com/aibs/aibs.html

American Mathematical Society
Box 6248
Providence, RI 02940
(401) 455-4000
http://www.ams.org/

American Physical Society
335 East 45th St.
New York, NY 10017
(212) 682-7341
http://www.aps.org/

American Societies for Cell Biology/American Society for Biochemical Molecular Biology/ Federation of American Societies of Experimental Biology
9650 Rockville Pike
Bethedsa, MD 20814
(301) 530-7153
http://www.FASEB.org/ascb

American Society for Information Science
1010 16th St. N.W.
Washington, DC 20006
(301) 495-0900
http://www.uiowa.edu/~libsci/asis/asis.html

Botanical Society of America
School of Biological Sciences
University of Kentucky
Lexington, KY 40506
(606) 257-9000
http://www.uky.edu

Ecological Society of America
Center for Environmental Studies
Arizona State University

Tempe, AZ 85187-1201
(602) 965-3000
http://www.sdsc.edu/

Geological Society of America
3300 Penrose Place, Box 9140
Boulder, CO 80301
(303) 447-2020
http://www.geosociety.org/

Mathematical Association of America
1529 18th St. N.W.
Washington, DC 20036
(202) 387-5200
http://www.maa.org/

National Association of Environmental Professionals
5165 MacArthur Blvd. N.W.
Washington, DC 20016
(202) 966-1500
http://enfo.com/NAEP/

ENGINEERING

American Institute of Chemical Engineers
345 East 47th St.
New York, NY 10017
(212) 705-7338
http://www.che.ufl.edu

American Institute of Mining, Metallurgical &
Petroleum Engineers
345 East 47th St.
New York, NY 10017
(212) 705-7695
aime_ny@aol.com

American Society of Engineering Education
11 Dupont Circle N.W., Suite 200
Washington, DC 20036
(202) 331-3500
http://www.asee.org/

American Society of Mechanical Engineers
345 East 47th St.
New York, NY 10017
(212) 705-7722
http://www.eng.usf.edu/studorgs/asme/

Association for Computing Machinery
11 West 42nd St.
New York, NY 10036
(212) 869-7440
http://csclub.waterloo.ca/

Association of Environmental Engineering Professors
Department of Civil Engineering
Villanova University
Villanova, PA 19085
(906) 487-3495
http://www.bigmac.civil.mtu.edu/aeep.html

Institute of Electrical and Electronics Engineers (IEEE)
345 East 47th St.
New York, NY 10017
(212) 705-7900
http://www.ieee.org/

Institute of Transporation Engineers
525 School St. S.W., Suite 410
Washington, DC 20024
(202) 554-8050
http://www.10.com/~itehq/

Society of Industrial and Applied Mathematics (SIAM)
1405 Architects Building
117 South 17th St.
Philadelphia, PA 19103
(215) 382-9800
http://www.siam.org/

Society of Women Engineers
345 East 47th St., Room 305
New York, NY 10017
(212) 705-7855
http://www.swe.org/

APPENDIX D

Questions to Ask Before Accepting a Faculty Position[1]

DEMOGRAPHICS

- Is the institution public, semi-public, or private?
- How does the control of the institution (public or private) impact the college and department?
- What are the university/college/department's mission, vision, and goals? How well are they accepted by the administration and faculty?
- What is the undergraduate, master's student, and doctoral student enrollment of the university/college/department? What are the retention and graduation rates?
- Is the institution on a quarter or semester system?
- Are there satellite campuses staffed with a separate administration and faculty from the main campus? Will you be asked to teach at the satellite campuses?

---

[1] From "Is this an offer you can't refuse? Criteria for selecting an engineering faculty position," by Timothy Greene, Marilyn S. Jones, John G. Casali, and Nancy E. Van Kuren, *ASEE Prism*, pp. 30–33, Sept. 1994. Copyright © 1994 by the American Society of Engineering Education. Reprinted with permission.

- Does the institution offer a night program at the graduate or undergraduate level?
- What engineering disciplines are studied in the department?
- Are there other engineering institutions in the state that are competing for state, federal, and private dollars?
- What percentage of the institution's income is from tuition, endowment, research, or extension revenues?
- Is there an institutional foundation? How is it operated? If so, are there restrictions on receiving and expending funds?
- What is the financial status (current or future) of the university/college/department?

FACULTY

- What is the number of full professors, associate professors, assistant professors, instructors?
- How many faculty members are currently tenured at each professorial level?
- Is there a quota on the position levels or on tenure?
- What percentage of untenured faculty members who have reached their mandatory year have not achieved tenure in the past five years?
- What is the expected faculty growth in the university/college/department?
- Is the department administered by a head or chair? If so, what is the term of service for the position?
- What is the background of the department head or chair (education, teaching interests, and research interests)?
- What is the department head's philosophy regarding teaching, research, and publication? What is his or her vision for the future?
- Who on the faculty is doing research and teaching in your area of expertise?
- Are faculty awards given by the university/college/department?
- What are the sabbatical policies? Are sabbaticals encouraged?
- How are faculty members evaluated on a regular basis? Is an annual report required?

FACULTY DUTIES

- What constitutes a full-time teaching load?
- What is a typical teaching–research–service–extension load/ratio?

Appendix D Questions to Ask Before Accepting a Faculty Position   375

- What is the ratio between teaching graduate and undergraduate courses?
- How are teaching assignments made? (Who makes them? What are the criteria on which they are based—areas of expertise, research load, etc.?)
- How are summer assignments made?
- What is the pay rate associated with a summer appointment?
- On what committees can you expect to serve (e.g., graduate program, undergraduate curriculum, graduate admittance, college)?
- Are junior faculty included in committee assignments during the first year?
- Are professional activities encouraged and financially supported?
- Are TV courses taught? If so, how are they assigned to each faculty member? How does a TV course relate to a regular teaching load?

FACILITIES

- What laboratories are available in your area of expertise?
- To what extent are these laboratories currently committed to research projects and/or course support?
- Are there technical personnel to support the laboratories?
- Are funds available to alter or upgrade the laboratories?
- What are the priorities on purchasing new equipment?
- Are the facilities conducive to teaching classes and/or conducting research?
- Who is responsible for the laboratories' activities and expenditures?
- What computer facilities are available for faculty members, undergraduate students, and graduate students?
- Does the library carry the appropriate engineering journals?
- Does the library possess a computer-automated literature search capability?
- Is there a separate engineering library?
- Is there a departmental, college, or university computer network? If so, how is it supported? What software is available?
- Does the institution provide an e-mail system that is connected to either Internet or Bitnet?
- Is clerical and/or technical support available?
- Is support provided for patent searches and securements?
- Is support provided for writing textbooks?
- How is clerical and/or technical support assigned to a faculty member?

UNDERGRADUATE PROGRAM

- Is the degree program accredited? Since when? What were the critiques for the last visit?
- What is a nominal class size?
- What is the retention rate?
- Are graduate student teaching assistants available?
- Who is responsible for undergraduate academic advising? What is the faculty role in academic advising?
- Which courses can you expect to teach? (Be aware that this question is usually asked of the interviewee.)
- Is there a cooperative education program with industry?
- Are special study options offered?
- Can new undergraduate courses be initiated? If so, what is the procedure?
- Is documentation available concerning degree requirements, course descriptions, and teaching schedules?
- Is an honors program available?
- When do students enter the program (as freshmen, sophomores, or juniors)?
- Are professional and honor societies available for students? If so, how active are these societies?

GRADUATE PROGRAM

- Is there a separate graduate college?
- If there is a graduate faculty, must you join it before guiding graduate students?
- What graduate degrees are offered (master of engineering, master of science, doctor of philosophy, doctor of engineering)?
- Are special options (concentrations) of study offered to graduate students?
- What are the requirements for entrance into the master's program for B.S. engineers?
- What are the requirements for entrance into the master's program for B.S. nonengineers?
- What is the ratio between U.S. and foreign graduate students?
- Are there restrictions on when a new faculty member can begin to direct Ph.D. candidates?

- What is the graduate student stipend? Are there tuition waivers?
- Is there a graduate student cooperative education or internship program?
- What fellowships are available to support graduate students?
- What is the nominal size of a graduate class?
- How many master's and doctoral students does each faculty member typically direct?
- Can new graduate courses be initiated? If so, what is the procedure?

RESEARCH

- Is there a research division? If so, what is its function?
- Are "seed" funds available for research program initiation?
- Is state or university core research funding available? If so, what is the priority for funds distribution?
- What companies have established research contacts with the college or department in your area of expertise?
- Is assistance available to write research proposals and to create related budgets?
- Does the institution obtain research opportunity announcements and formal requests for proposals (RFPs), screen them, and then direct them to the appropriate faculty?
- To what level is the university, college, or department willing to cost-share on research proposals?
- Is the cost-sharing in "hard dollars," release time, or another category?
- What are the fringe benefits and overhead rates of researchers?
- What percentage of overhead is returned to the university/college/department?
- Is there an institutional foundation to maintain scholarships, equipment donations, and so on?
- Is release time (from teaching, committee appointments, and administration assignments) available to conduct funded research?
- Is interdisciplinary research encouraged? Are there opportunities and/or expectations for collaborating with other departments within or outside the institution?
- Are travel monies available to visit potential sponsors and/or technical conferences?
- Are these monies controlled at the department level?

PROMOTION AND TENURE

- What are the general requirements for promotion and tenure in regard to:
  Number of refereed publications?
  Number of nonrefereed publications?
  Number of published textbooks and/or chapters?
  Number of national or international presentations?
  Teaching quality and quantity?
  Extension activities?
  Research dollars?
  Number of master's students directed?
  Number of doctoral students directed?
  University service?
  Community service?
  Professional societal service?
- What is the review process for promotion and tenure (university, college, department)?
- What is the time schedule for promotion?
- Does tenure automatically come with promotion?
- Are faculty tenured without being promoted?
- Are student evaluations of faculty members' teaching used in the promotion process?
- Are peer evaluations used in the promotion process?
- How is consulting viewed in the promotion and tenure decision?

PROFESSIONAL DEVELOPMENT

- Are senior chapters of appropriate technical societies located nearby?
- Are funds available to send faculty to continuing education courses or conferences?
- Can a faculty member take courses at the university? If so, how many? What is the cost?
- Are leaves of absence available to work in industry, study abroad, and so on?

BENEFITS AND CONTRACTUAL ISSUES

(This section does not cover questions associated with medical, dental, or life insurance, issues common to any hire. Be sure to investigate these as well.)

- Is the contractual appointment for 9, 10, 11, or 12 months?

Appendix D   Questions to Ask Before Accepting a Faculty Position

- If the appointment is for less than a year, is pay spread over 12 months?
- Is the retirement program tied to a national program such as TIAA/CREF or VALIC?
- Is consulting permissible? How many days per week/month?
- Can any of your dependents enroll in the institution at reduced rates?
- How are intellectual properties handled?
- What is the institution's policy on patents and faculty-developed inventions?
- Can provisions be made to compensate for moving expenses?
- Is state income tax waived?
- What options are available for dual-career couples?
- Will the institution provide any assistance in helping faculty spouses find employment at the institution or in the general area?
- Does the institution have child care or elementary education options available through its educational programs?
- Is transitional housing available through the institution's housing department?

APPENDIX E

Sample Offer Letters

Note: Certain identifying features of each letter have been changed to protect confidentiality.

SAMPLE 1: OFFER FROM A LIBERAL ARTS COLLEGE

June 18, 1996

                    Memorandum of Agreement

(Offering) University is pleased to offer Mr. John Smith a tenure-track appointment to the Faculty of (offering) University as an Assistant Professor of Biology. The initial appointment is for three years, from August 1996 through June 1999.

If degree requirements are completed before the beginning of the fall 1996 semester, your salary will be $XX,XXX. If degree requirements have not been completed by that time, your first year salary will be ($XX,XXX − $1000). Upon completion of degree requirements during the first year, your salary will be increased and prorated on the basis of $XX,XXX for the remainder of the academic year 1996–1997.

No renewal of the initial contract will be considered if the Ph.D. has not been completed by December 15, 1997. If degree requirements are completed during the academic year 1996–1997, you may choose whether or not to count that year toward tenure. If degree requirements have not been completed before June 1, 1997, the academic year 1996–1997 will not count as credit toward tenure.

In addition, you will be named a recipient of the endowed (name) Fellowship. As such, you will receive an annual summer stipend equivalent to 1/9th of your salary, to begin in the summer of 1997, provided you accept no other employment during one summer month. You will also receive an annual research stipend of $6000 to use for travel, student assistance, and material and equipment in support of your research and teaching. The Department's (name) Fellowship holder, currently (name), will serve as a mentor for you, and assist you in the development of plans for the use of these Fellowship funds. Renewal of the Fellowship for the final year is dependent upon a successful third-year review in the fall of 1998. The three years of Fellowship will terminate at the end of the summer of the academic year 1999–2000. Upon presentation of appropriate receipts, your moving expenses will be reimbursed up to 6% of your starting salary.

A tenure-track appointment is a probationary period, one that does not necessarily include the right to permanent employment. The minimum requirements and conditions for retention and the award of tenure by the Board of Trustees upon completion of the probationary period are set forth in each department's stated criteria in the University policies on academic freedom, responsibility, retention, and tenure found in the enclosed Faculty Handbook.

Your acceptance of this offer will be indicated by your signing both copies of this letter and returning one original to this office by June 10, 1996. We look forward with pleasure to your acceptance.

_____   _____
Vice President for Academic Affairs   Dean, College of Science

I accept the above appointment.

_____   _____
Signature   Date

SAMPLE 2: OFFER FROM A MASTER'S UNIVERSITY

May 31, 1996

Ms. (name of candidate)
Address

Dear Ms. (name of candidate):

On the recommendation of the Department of Materials Engineering, I am most pleased to offer you a position as faculty, Step 4, with academic year employment. The present salary is $XX,XXX per year. This is a probationary position, which could lead to tenure.

As an academic year employee, you will be entitled to regular health and dental benefits, you will accrue sick leave in accordance with the policies of the (name) State University System, and you will become a member of the state of (name) Public employees Retirement System. Please contact the Personnel Officer at (XXX) XXX-XXXX if you have any questions regarding these or other benefits. In extending this offer to you, I wish to stress the importance of faculty involvement in teaching, student advising, laboratory development, research, and service activities. Faculty members are expected to: (a) contribute to the instructional program through effective teaching, development of courses, curricula, and laboratories at both graduate and undergraduate levels; (b) develop professionally by initiating successful proposals and conducting sponsored research leading to presentations at professional society meetings and publications in recognized journals through the peer review process; and (c) take part in service activities by serving on Department, School, and University committees and to participate in extracurricular activities sponsored by the Department, School, and University.

It is anticipated your efforts will primarily be concentrated in the area of Electronic Materials.

This appointment offer reflects my belief that you will be an active and effective contributor to the Department, the School of Engineering, and to the engineering profession. The Materials Engineering Faculty and I look forward to your joining our University and helping in the continued development of a high-quality engineering education program. I would appreciate receiving a letter of acceptance as soon as possible.

Sincerely,

(name), Dean

SAMPLE 3: OFFER FROM A RESEARCH UNIVERSITY

June 2, 1996

Dr. (candidate name)
Address

Dear (candidate name):

The Department of Physics at the University of (name) is very pleased to offer you a tenure-track assistant professor position with a start date of September 1, 1996. We are very impressed with you as a young scientist and with your potential as a teacher. The initial appointment period is for three years at a starting academic-year salary of $XX,XXX paid in biweekly increments over the full 52-week year. The University allows you to supplement your academic-year salary with up to an additional 2/9 of this amount derived from grants, etc. The University offers the usual benefits ranging from health insurance to a State of (name) retirement system. This letter outlines the commitments we are able to make to you in the context of this offer. To formally accept our offer, please sign this letter at the bottom and return the appropriate signed copy to me.

As I have noted, this offer is for an initial period of three years. This is a typical "initial offer" of faculty status in the Department. For a typical such appointment, during the second year, an internal review of your progress is carried out, and a one-year reappointment is then considered. Then, during the next year, the third appointment year, a more extensive review of progress is carried out. This "mini-tenure review" usually includes seeking external reference letters to help determine research progress and taking a close look at teaching effectiveness. The purpose of this review is both to determine whether we wish to renew the candidate's appointment and to provide the candidate with significant feedback so that he/she can try to make any needed "midcourse corrections" on the way to the tenure decision. A successful review results in a further three-year appointment (i.e., through the seventh year from the initial appointment). Tenure and promotion to associate professor are normally considered in the sixth year of appointment. If granted, tenure begins with the start of the seventh year; if denied, this seventh year would normally be the last, with further appointment highly unlikely.

As we have discussed, I am aware that you are considering the possibility of postponing your start date beyond September 1 (i.e., at least to January 1). We would still begin your appointment on September 1, but immediately put you on "leave without pay" status until your actual arrival to take up the position. In this event, all of the above timelines (e.g., time to tenure decision) would slip accordingly. You should also be aware that there can be a waiting period of (approximately) 60 days for, e.g., health insurance coverage to take effect once you are on payroll; once you have a definite start date, I can let you know what you must do to ensure continuous coverage.

The start-up funding package we can offer is, I believe, consistent with what we have discussed informally. The School of Humanities and Sciences (SH&S) will provide up to $50,000 in your first year for (computing) equipment, one full-year research assistantship (approximately $15,000 to the student) each of your first two years, and $6000 per year in your first two years for "other personnel." This last item is intended to provide support, most likely in the form of student help, in setting up and maintaining the computer system(s) you intend to obtain with the equipment money; "expert" help is available from SH&S staff as well as from the central campus information technology groups, but the "routine" operational issues should be addressed by your own personnel. In addition to these commitments, our Dean has agreed to provide you with your first year's summer salary (in the usual amount of two-ninths of your academic year salary) and with travel funds up to $8000 per year for your first two years. The commitments to summer salary and travel are "backstop" commitments in that he would be relieved of these commitments should you receive external funding to cover them. However, in the event that you do receive external funding, he has agreed to offer half the amount of any unused backstop commitments for use as "matching funds" to help leverage additional external funding. The timeline for these commitments should be viewed as definite. While I am fairly certain that the Dean would respond favorably to a well-reasoned request to delay a given expenditure, he most definitely does not want any of these commitments to be viewed as "open ended"; without prior agreement, these commitments will expire as described.

In addition to the School commitments, the Department will provide you with up to a total of $15,000 over your first three years for "miscellaneous" expenses. Based on my growing experience with start-up funding issues, I believe you will find it useful to have modest funds to use for unexpected opportunities or unexpected expenses.

Finally, the University is able to pay "reasonable" moving expenses provided you follow the appropriate procedures and use approved carriers. Please contact me before you begin planning to move.

As we have discussed, I am aware that you are considering different possible approaches to research funding. Should you wish to begin the process of proposal preparation before your arrival, I am prepared to assist you with the relevant University forms and policies. I see no difficulty in processing proposals here in advance of your arrival should that be your desire. Similarly, there are a number of possible fellowship and Young Investigator programs that we might wish to explore in advance of your arrival on campus. If you wish, we can explore these further over the summer. Finally, along these lines, I am prepared to consider a reduced teaching assignment during your first or second semester here if that would facilitate preparation of research proposals.

The Department of Physics is very excited about the possibility of your joining us. I hope that you were able to see that the Department is committed to growth and improvement. We have been hiring excellent, enthusiastic young faculty members, and will continue to do so. Your ideas should lead to successful grant proposals and an outstanding academic career. We sincerely hope you will be able to accept our offer. If we do not hear from you by July 10, I will call to see where you stand on the offer. In the meantime, if there are any questions, please feel free to call any of us.

Sincerely,

Professor and Chair of Physics

I accept the terms and conditions of the above offer.

_____   _____

Name of Candidate   Date

APPENDIX F

Elements Found in Most Successful Proposals

The following list (collected from various sources by and published with permission from Assistant Chair, Rebecca Claycamp, Department of Chemistry, University of Pittsburgh) may help Principal Investigators anticipate areas where their proposals could be strengthened.

GENERAL CONSIDERATIONS

1. Relates to the purposes and goals of the applicant agency.
2. Strictly adheres to the content and format guidelines of the applicant agency.
3. Is directed toward the appropriate audience—that is, those who will review the proposal.
4. Obviously addresses the review criteria of the funding source.
5. Is interesting to read.
6. Uses a clear, concise, coherent writing style, free of jargon, superfluous information, and undefined acronyms—that is, it is easy to read.
7. Is organized in a logical manner that is easy to follow.

8. Calls attention to the most significant points in the proposal through the use of underlining, differences in type, spacing, titles, and appropriate summaries.
9. Is paginated from beginning to end, including appendix when directly appended to the proposal.
10. Makes appropriate use of figures, graphs, charts, and other visual materials.
11. Is meticulously proofread so that it has few (if any) grammatical errors, misspellings, or typos.

THE PROPOSAL

12. Has title that is appropriate, descriptive, and (perhaps) imaginative.
13. Unless it is brief, has a table of contents that is straightforward and accurate.
14. Has a clear, concise, informative abstract that can stand alone.
15. Has clearly stated goals and objectives that are not buried in a morass of narrative.
16. Follows naturally from previous/current programs or research.
17. Documents the need to be met or problems to be solved by the proposed project.
18. Indicates that the project's hypotheses rest on sufficient evidence and are conceptually sound.
19. Clearly describes who will do the work (who), the methods that will be employed (how), which facilities or location will be used (where), and a timetable of events (when).
20. Justifies the significance and/or contribution of the project on current scientific knowledge or a given population of people or a body of writing/art.
21. Includes appropriate and sufficient citations to prior work, ongoing studies, and related literature.
22. Establishes the competence and scholarship of the individual(s) involved.
23. Does not assume that reviewers "know what you mean."
24. Makes no unsupported assumptions.
25. Discusses potential pitfalls and alternative approaches.
26. Presents a plan for evaluating data or the success of project.
27. Is of reasonable dimensions—not trying to answer all the questions at once.
28. Proposes work that can be accomplished in the time allotted.

29. Demonstrates that the individual(s) and the organization are qualified to perform the proposed project; does not assume that the applicant agency "knows all about you."
30. Includes vitae that demonstrate the credentials required (e.g., do not use promotion and tenure vitae replete with institutional committee assignments for a research proposal).
31. Documents facilities necessary for the success of the project.
32. Includes necessary letters of support and other supporting documentation.
33. Includes a bibliography of cited references.

THE BUDGET

34. Has a budget that corresponds to the narrative: all major elements detailed in the budget are described in the narrative and vice versa.
35. Has a budget sufficient to perform the tasks described in the narrative.
36. Has a budget that corresponds to the applicant agency's guidelines with respect to content and detail.

APPENDIX G

Common Shortcomings of Grant Proposals[1]

Table G.1

Proposal Section	Problem
Budget	Excessive funds requested; capital equipment request unjustified; funds requested are insufficient to complete described project
Biographical sketch and backgrounds of investigators	Investigator inexperienced; insufficient number of investigators
Research plan	Poorly organized; too long; too narrow; improperly focused; poorly written; sloppy preparation; inadequate detail
Aims	Project scientifically premature—requires more pilot work; validity questioned; vague or unsound scientifically; too ambitious; hypotheses poor
Significance and background	Problem of little significance or repeats previous work; assumptions questionable; rationale poor; literature background poor or inadequate
Pilot studies	Pilot work ill-conceived; data inappropriately analyzed; experiments lack imagination

[1] From R. V. Smith, *Graduate Research: A Guide for Students in the Sciences*. New York: Plenum, 1990, p. 243, Table 12-2. Copyright © 1990 by Robert V. Smith. Reprinted with permission. Smith's table was constructed in part from E.M. Allen, "Why are research grant applications disapproved?," *Science*, vol. 132, pp. 1532–1534, 1960. Copyright © 1960 by the American Association for the Advancement of Science. Reprinted with permission.

Table G.1 (*Continued*)

Methods	Methods unsuited to stated objective; unethical or hazardous procedure proposed; controls poorly conceived or inadequately described; some problems not realized or dealt with properly; results will be confusing, difficult to interpret, or meaningless; emphasis on data collection rather than interpretaton
Collaboration	Cooperative agreement inadequate, vague, or poorly conceived; no letters of support
Facilities	Equipment lacking, too old, or insufficiently robust for project
All sections	Poor editing; poor reproduction

Index

NOTE: Bold page numbers indicate figures and illustrations.

A

Abler, Ronald F., 290
Abstract, 304–305, *See also* Publication
"Academic Author's Checklist", 127
Academic duty, and professional responsibility, 320–322
Academic employment, *See also* Job offer
  acceptance of, checklist of questions for, 373–379
  advantages/disadvantages of, 99–100
  application materials preparation, 192–206
    cover letter, 192–**195**, 208
    curriculum vitae, 190–191, 196–204, 208
    letters of recommendation, 190–191, 204–206, 334
  application procedures, 182–218
    academic job talk, 191, 209, 211–214, 215
    campus visit, 207–211
    conferences, 206–207, *See also* Conferences
    departmental expectations of new-hires, 186–188, 212, 223
    discovery of job openings, 188–189
    establishment of new positions, 185–186
    networking, 190, *See also* Networking
    rifle, shotgun, splatter approach, 184–185
    time frame for academic positions, 190–192
  assistance to graduate student/postdoc, 354–355, 364
  choices for in job search, 170–173, *See also* Personal/professional considerations
    institution choices, 170–171
    personal/professional objectives, 172–173, 221–240
    types of appointments, 171–172
  compared to government/industry employment, 3–4, 83, 86–90, 93, 98–103, 144, 214–217, 246, 266–268, 356–357
  compensation in. *See* Compensation
  as corollary to industry employment, 98, 100, 111
  job announcement (sample), **194**
  leave of absence, 331
  negotiating for professional/personal success, 172–173, 221–240
    dealing with rejection, 233–237
    dual-career couples, 232–233
    negotiating approach, 222–223
    response to job offer, 224–232
  offer letters
    liberal arts college, **381–382**
    master's university, **382–383**
    research I university, **383–385**
  part time, 6, 98, 171–172, 178, 329–330
  pre-employment preparation, 176–180, *See also* Preparation strategy
  pre-employment research, 168, 173–174
    background reading, 173–174

393

pre-employment research (*Continued*)
  campus visits, 175–176
  networking academic contacts, 174–175
professoriate challenges, 59–78, *See also* Department; Faculty
  changes in teaching and research, computational tools use, 63–65
  communications tools use, 61–63
  cooperation/competition balance, 72–74
  government funding elimination, 67–69
  high-risk/low-risk behaviors balance, 74–76
  implications for faculty scholarship, basic/applied research balance, 74
  industry role in academia, 69–71
  interdisciplinary programs, 65–67
  research cost increases, 68–69
  rewards of, 99–100
  tenure-track vs. non-tenure, 171–172, 236–237, *See also* Tenure
  transition to from industry, 177–180
  workload requirements, 180, 227–228
Academic institutions, 3–29, *See also* Colleges and universities; specific institutions
  changes in, 351–362
  assistance to graduate students/postdocs, 351–354
  governance and decision making, 13–18
  multiuniversity, 15
  organizational structure, 13–15, 53–54
  power and money, 16–18, 37
  tenure, 15–16
  historical perspective, 4–6, 21
  institutional issues, 18–20
  key characteristics, 6–13
    Carnegie Classification, 6–9, **10–11**
    teaching and research emphasis, 9–13
  politics, 13–15, *See also* Politics
  pre-employment investigation of, 175–176, 373–379
  sample schools, 22–29, **23**
    Bucknell University, 23–24
    Memorial University of Newfoundland, 24
    Rochester Institute of Technology, 25
    San Jose State University, 25–26
    Stanford University, 26
    University of Michigan, Ann Arbor, 24–25

  University of New Orleans, 26–27
  scholarship issues, 20–22
  in undergraduate education, 27–29
Academic job talk, 132, 191, *See also* Communication skills; Conferences and presentations
  "five minute drill", 209
  preparation for, 211–214, 215
The Academic Job Search Handbook, 170
Accountability, 4, 18, 261, 320, *See also* Productivity
  required for funding, 291
Adams, Tom, 154
Administrator, *See also* Department; Governance and decision making
  compensation for, 17
  relation to new faculty, 300
  reputation of, compared to departmental reputation, 110
  statements from in teaching portfolio, 365
Advanced Research Projects Agency (ARPA), 70, 340–342
Advisor, *See also* Mentor; Supervision
  considerations for choice of, 96, 110, 118–123, 129
  recommendation by, 184, 204
  relation with graduate students, 94, 138
  responsibilities of, 247
  role of in employment process, 188, 191–192, 196, 232, 235, 334, 336
Aeronautical engineering, 25, 49–51, **50**, 271, *See also* Science and engineering
Agricultural and mining (A&M) colleges, 5
Agriculture, 37, 110, 174, 271
Aldrich, Jeff, 62
Allen, Emily, 192
Altbach, Philip, 16, 37
Ambrose, Susan, 112, 114, 132, 275, 276–279, 294, 303
American Association for the Advancement of Science, 65, 321, 324
American Association of University Professors (AAUP), 329
American Astronomical Society, 217
  URL address, 369
American Chemical Society, 129, 245
  URL address, 369
American Institute of Biological Sciences, 129

URL address, 370
American Institute of Chemical Engineers,
    URL address, 371
American Institute of Mining, Metallurgical &
    Petroleum Engineers, URL address,
    371
American Institute of Physics (AIP), 174
American Mathematical Society, URL address,
    370
American Physical Society (APS), 129, 174
    URL address, 370
American Societies for Cell Biology, URL
    address, 370
American Society of Engineering Education,
    URL address, 371
American Society of Engineering Education
    (ASEE), 174
American Society for Information Science,
    URL address, 370
American Society of Mechanical Engineers,
    129
    URL address, 371
Anagnos, Thalia, 232, 246
Anthropology, 39, *See also* Social sciences
Application
    for academic employment, 168, 183–218,
        See also Academic employment
    in scholarship model, 20–22, 27–29, 54
Applied Materials Corp., 273
Arizona, 5
Arizona State University, 67
ASEE Prism, 127, 309
Associate of Arts colleges (AA), defintion and
    statistics, 6–9, **7–8**
Association of American Colleges and
    Universities, Preparing Future Faculty
    program, 147
Association for Computing Machinery, URL
    address, 371
Association of Environmental Engineering
    Professors, URL address, 371
Astronomical Society of the Pacific, 84, 154
Astronomy, 47, 84, 109, 113, 150, 216–217
AT&T Bell Laboratories, 172
Atmospheric science, 110
Auburn University, 67
Automation, 113
Aylesworth, Kevin, 94

B

Baccalaureate colleges I&II (BA I&II), *See
    also* Colleges and universities
    defintion and statistics, 6–9, **7–8**
    faculty statistics, 9
    organizational structure, 14
    teaching experience gained at, 153
    teaching and research emphasis, **11**–12, 22
Background research, *See also* Preparation
    strategy
    for pre-employment preparation, 173–176
Baerwald, Thomas J., 290
Bailar, John C. III, 324
Baish, Jim, 342–344
Balogh, Deborah Ware, 322
Becton–Dickinson Immunocytometry Systems,
    339
Benefits package, *See also* Compensation
    pre-employment investigation of, 378–379
Benowitz, Steve, 53
Bernstein, Bianca L., 147
Beyer, J., 40
Billets, *See also* Academic employment
    establishment of, 185–186
Biochemistry, 39–40, 66
Bioengineering, 66
Biology, 100, 129, 150, 174, 188, 266,
    303–304
    discipline development in, 39–40
    interdisciplinary opportunities in, 53, 66
    undergraduate requirements, **47**–48
Biomedical engineering, 49–51, **50**, *See also*
    Science and engineering
BioRad Laboratories, 339
Biotechnology, 113, 116
Blaylock, Guy, 136
Boering, Kriste, 143–144
Boice, Robert, 253, 258–259, 275
"Bolstering", 231–232
Boston College, 37
Botanical Society of America, URL address, 370
Botany, 39
Boyce, Peter B., 84
Boyer, Ernest L., 3, 6, 12, 20–22, 27–29, 270, 358
Brandeis University, Rosenthiel Center, 73,
    110, 294, 298–300, 330
Breadth. *See* Preparation strategy

Brei, Diann, 25
Bridger, Alison, 115, 170, 246–248, 329
Brinkman, Ralf, 110
Broad, William J., 68
Brown, John Seely, 65
Bube, Richard, 99–100
Bucknell University, 23–24, 152, 266,
    268–270, 332–333, 335–337, 342–344
    URL address, 174
Buffington, Keith, 332–333, 335–337, 342
Bukhari, S. Amir, 89, 331
Burn-out, 245, 258, *See also* Time management
Burroughs, Robert, 125
Business department, 40, 53, 54, *See also*
    Industry

C

California, 5, 93
California Department of Fish and Game, 154
California Institute of Technology, 48, 92
California Polytechnic State University, San
    Luis Obispo, 266, 272–275
California State Polytechnic University, 67
California State University, 25
Cambridge University (Gr. Britain), 4
Campus visit, 175–176, 207–211, *See also*
    Academic employment
Canada
    academic opportunities, 19–20
    colleges and universities, 4–6
    contemporary education statistics, 6, 9
    foreign graduate student types, 41
    government funding decreases, 68
Cardinal Technologies, Inc., 89
Carnegie Foundation for Advancement of
    Teaching
    Carnegie Classification of U.S. institutions,
        6–9, **7–8, 10–11**, 360
    *Scholarship Reconsidered: Priorities of the
        Professoriate*, 20
Carnegie-Mellon University, 65, 112, 123, 276
Carroll, Mary Anne, 88, 312, 313–315, 330,
    367–368
Casali, John G., 210
Casper, Gerhard, 13
Certificate programs, 25–26, 154

Chaos theory, 66–67
Charles Frankel Prize in the Humanities, 28
Chemical engineering, 22, 25, 40, 49–51, **50**,
    188, *See also* Science and engineering
Chemistry, 100, 129, 282
    discipline development in, 39–40
    undergraduate requirements, **47**–48
Chronicle of Higher Education (CHE), 67,
    148, 174, 189, 191
Chrysler Corp., 267
Citizenship, *See also* Service; Society
    relation to graduate degrees, **41**
Civil engineering, 40, 178, *See also* Science
    and engineering
    course preparation in, 276–279
    departmental emphasis on, 48–51, **50**
Class notes, interdisciplinary sharing of, 67
Class. *See* Course; Lecture; Teaching
Claycamp, Rebecca, 293
Clayton, Geoffrey C., 216–217
Colby College, 358
Cole, Steven, 39, 45
Collaboration. *See* Interdisciplinary cooperation
College catalogs, 173–174, 207–208
Colleges and universities, *See also* Academic
    institutions; Department; Faculty
    baccalaureate colleges I&II,
        defintion and statistics, 6–9, **7–8**
        faculty statistics, 9
        organizational structure, 14
        teaching and research emphasis, **11**–12, 22
    Carnegie Classification of U.S. institutions,
        6–9, **7–8**
    doctoral universities I&II,
        defintion and statistics, 6–9, **7–8**
        faculty statistics, 9
        organizational structure, 13
        teaching and research emphasis, 21–22
    faculty relation with, 38–40
    as faculty source and destination, 9
    for graduate/postdoc work, 108–111
    historical perspective, 4–6
        agricultural and mining (A&M), 5
        land grant colleges, 5
        post WWII expansion, 5–6
        private research–oriented, 5
    multiuniversity, 15
    organizational structure, 13–15, **14**

private
  relations with alumni, 16
  research grant funding, 17
public
  downsizing of, 18
  income sources for, 12, 16–17, 55, 68
  research grant funding, 12
public/private with accreditation, 6
relations with industry, 18–19, 25, 52, 54,
  111, 130, 171–172, 177–180, 272–275,
  315–318, 323–324
  changing role of, 69–71, 72–74, 296–298
research universities I&II
  defintion and statistics, 6–9, **7–8**
  faculty statistics, 9
  organizational structure, 13
  teaching and research emphasis,**11**–12, 21–22
Colorado State University, 67
Commerce Business Daily, 291
Committee on Science, Engineering and Public
  Policy, 68, 93–94, 100
Communication skills, 101, 143–144, 178–179,
  212–213, 226–227
Communications tools, 61–63, 64, 75–78, 93,
  112–113, 116, 265, 364, *See also*
  e-mail; Internet
Community college, 281
  employment possibilities at, 86, 89, 153
  part-time faculty use, 6
Compensation
  for administration, 17
  benefits package, 378–379
  from consulting, 316–317
  for graduate student/postdoc teaching, 144,
    154, 161
  industrial/academic, compared, 52
  negotiation for in new-hire, 223, 225–230,
    333, 375
  in public/private institutions, 17, 55
  relation to discipline, 37, **44**–45
Competition, *See also* Interdisciplinary cooperation
  academic
    compared to cooperation, 72–74
    gaining advantage in, 147–148
  affect on of communications tools, 62
  consideration of in preparation strategy,
    96–98, 207

for funding, 38, 292–293, 374
industrial, 70
inter-disciplinary/inter-departmental, 38
international, 18, 70
in job search, 168, 177, 207
in research, 304–305, 310–311
for tenure, 334
Computational engineering, 48, *See also*
  Computer science
"Computational Prototyping for the 21st
  Century Semiconductor Structures", 65
Computational tools, affect on teaching and
  research, 63–65, 68, 281–282, 364
Computer-Aided Instruction (CAI), 281–282
Computer engineering, 25
Computers, 5, 6, 63–64, 223, 227, 265, 279,
  281, 318, 364, 375, *See also*
  Communications tools; Computational
  tools
Computer science, 100, 150
  departmental emphasis on, 49–51, **50**
  graduate student statistics, 41, 43
Conferences and presentations, *See also*
  Publication
  academic job talk, 132, 191, 209, 211–214, 215
  checklist for, 130–131
  departmental expectations for, 26
  documentation of in teaching portfolio,159, 364
  in pre-employment process, 206–207
  preparation strategy for, 97, 112, 114, 123,
    129–133, 147, 191, 197, 353
  research strategy for, 295–296, 314
  undergraduate, 28–29
Confidentiality, *See also* Ethics
  in consultation and publishing, 317, 322
  in response to job offer, 226–227
Conflict of interest, 309, 320–324, 357,
  See also Professional responsibility
Consulting, *See also* Industry employment
  as alternative to tenure-track employment,
    171–172, 331, 341
  ethical considerations for, 312, 315–318
  obtaining contractual rights for, 225, 229, 379
Contract, 291, 317, *See also* Funding
  teaching contract appointment, 379
Coping With Faculty Stress, 253, 254–255
Cornell University, 5, 60, 67
  Ph.D. enrollment limits, 94

Costs
for conferences, 129
institutional evaluation of, 18–19
for research, 68–69, 72–73, 77, 117, 251
Cotter, William, 358
Council of Graduate Schools, Preparing Future Faculty program, 147
Couples, See also Personal/professional considerations; Women
dual-career couple considerations, 232–233, 236, 237–239, 342–344, 379
job search considerations for, 170, 172, 208
professional concerns for, 24, 87, 329, 333
Course, See also Lecture; Teaching
preparation considerations for, 247–250, 268, 276–279
teaching of for teaching experience, 151–152
Course catalogs, evaluation of, 48, 49
Course syllabi, 319, 345
inclusion of in teaching portfolio, 158, 364
interdisciplinary sharing of, 67
preparation of, 154, 334
Cover letter, 184, 192–**195**, 208
preparation of, 192–**195**
Covey, Steven, 253
Cross, Patricia, 262
Curriculum
general education requirements, 47
interdisciplinary sharing of, 67, 364
modularization of, 62
multicultural emphasis for, 18
relation to discipline development, 40
Curriculum materials, interdisciplinary sharing of, 67
Curriculum vitae (CV), See also Cover letter
considerations for submission of, 175, 189, 208
preparation of, 184, 190–191, 196–204, **198–203**, 208, 349
Czujiko, Roman, 93

D

Dando, William, 291
Darley, John, 122, 209
Database, See also Networking
for job search information, 174–175
Davidson, Cliff, 112, 114, 132, 275, 276–279,
294, 303
Demographics, pre-employment investigation of, 373–374
Department, See also Colleges and universities; Faculty
affect on of communications tools, 62
consideration of in job search, 171
downsizing and/or elimination of, 6
engineering department, 48–51, **50**
faculty relations within, 38–40, 256
new hire
assistance to, 252–253, 336, 355–361
expectations for, 186–188, 212, 223
position establishment for, 185–186
organizational structure affecting, 13–15, **14**, 37, 53–54
politics and influence within, 16–18, 37, 40
relation to, academic institution, 16–18, 171
reputation of, compared to school reputation, 108–110
science department, 47–48
service to, 312, 313–315
Department of Defense (DoD), 291
Dertouzus, Michael, 341
DeSieno, Robert, 280, 283
Dickert, Sanford, 114
Discipline, See also Department; Faculty; Interdisciplinary cooperation
degree of development affecting, 39–40
faculty commitment to, 38
relation to
compensation, **44**–45
employment in alternate department, 172
graduate student types, 40–43, **41–42**
organizational structure, 37, 53–54
postdoc appointments, **43**
publication rates, 45
scholarship goals, 53–54, 344
Dissertation, 131, 210, 302, 353, See also Research topic
choosing advisor for, 118–123
completion of, 232, 343–344
Distance learning, 6, 148, 375, See also Communications tools; Video instruction
interdisciplinary, 67
teaching considerations for, 64, 157, 267
Diversity. See Gender issues; Minorities; Women

Doctoral degree, *See also* Postdocs;
  Preparation strategy
  annual issuance, 9
  choices for and against, 84–90, 94
  by citizenship, **41**
  completion of before hire, 232, 343–344
  definition and process for, 85–86
  doctorate-granting institutions, 30–32
  enthusiasm required for attainment of, 84, 86, 94, 108, 112–114, 121
  for industry/government employment, 83, 85–90, 98–103
  postdoc appointments and, **43**
  by race/ethnicity and discipline, **42**
  requirements for in employment, 9, 83, 85, 98–102, 177–180, 353
  by sex and discipline, **42**
  supply and demand issues, 19–20, 84, 90–95, 352
  enrollment limits, 93–94
Doctoral universities I&II (Doc. I&II), *See also*
  Colleges and universities
  defintion and statistics, 6–9, **7–8**
  faculty statistics, 9
  organizational structure, 13
  teaching and research emphasis, 21–22
Downsizing
  in colleges and universities, 6, 18
  in industry, 93
Drotleff, Elizabeth, 134–135
Duke University, 51, 65

E

Ecological Society of America, URL address, 370–371
Economics, 40, 67
Edgerton, Russell, 158, 275, 283
Educational Resource Information Center (ERIC), 364
Education department, 53, 54
Education (discipline), 37
Efficiency considerations, 255–256, *See also*
  Productivity; Time management
Electrical engineering, 22, 25, 40, 48–51, **50**, 100, 125, 188, 342, *See also* Science and engineering
Electronics, 5

The Elements of Style, 305
e–mail, 62, 268, 278, 292, 295, 375, *See also*
  Communications tools; Internet
  considerations for use of, 175, 196, 281
Emotions
  in job offer consideration, 231–232, 341
  in tenure process, 327
Employee. *See* Faculty
"Employment Opportunities and Job Resources on the Internet", 189
Employment. *See* Academic employment
Endowment, 16–17, *See also* Funding
Engineering, 174, 282, *See also* Science and engineering
  discipline development in, 39–40
  interdisciplinary opportunities in, 53, 54
  R&D expenditures for, **46**
Engineering department, disciplines and employment potentials, 48–51, **50**
Engineering Graduate Studies and Research Directory, 174
Engineering-Science, Inc., 178–179
Enthusiasm, 295
  consideration of, in choice of advisor, 120–122
  in employment interview, 212, 213
  required for doctoral degree award, 84, 86, 94, 108, 112–114, 121
Environmental Protection Agency (EPA), grant funding process, 291
Ercolano, Vincent, 309, 323–324
Ethics, 27, 211, 309, 320, *See also* Fraud;
  Professional responsibility
  "ethically problematic behaviors", 310
  student/faculty ethical code, 318–319
The Ethics of Teaching–A Casebook, 322
Examinations
  inclusion of in teaching portfolio, 158, 363, 365
  preparation of, 151, 154, 156–157, 368
  use of previously-prepared, 227
Extension course, 153–154

F

Faculty, *See also* Colleges and universities;
  Department
  compensation, *See also* Compensation
  in public/private institutions, 17

compensation (*Continued*)
  relation to discipline, **44**–45
  evaluation of in job search, 171, 176, 187, 208–209
  networking among. See Networking
  new position establishment for, 185–186
  numbers of
    by discipline, 43–44
    post-WW II, 5–6
  organizational structure affecting, 13–15, **14**, 53–54
  part-time, 6, 98, 171–172, 178, 329–330
  post-WWII increase in, 5–6
  pre-employment investigation of, 374–375
  professional spouse concerns, 24, 87
  relations with
    discipline, department, and university, 38–40
    graduate students, 321, 322–323, 364
    industry, 70
  retirement/new-hire concerns, 5–6, 93, 94, 147, 185–186
  scholarship concerns, *See also* Scholarship
    basic/applied research balance, 74
    cooperation/competition balance, 72–74
    high-risk/low-risk behaviors balance, 74–76
  search committee functions, 186–187, 222–223
  as source of organizational power, 13–15
  sources and destinations for, 9
  supply and demand issues, 90–95
  teaching portfolios for, 284–285
  women/minority, mentoring for, 42–43
Family. *See* Couples; Women
Farn, Michael, 125–126
Federal funding. *See* Funding; Government funding
Feedback, *See also* Peer review; Student evaluation
  inclusion in teaching portfolio, 365
  for interviews, 213–214
  obtaining at conferences, 131–133, 295
  for publications, 305
  on teaching, 156, 268, 270, 272, 282, 336, 338, 354–355
Feibelman, Peter J., 86, 116, 126, 133, 294, 302–303
Felder, Richard, 162, 267, 270
Fellowship, 291–292, *See also* Funding

Ferris, Timothy, 129
Fiber optics, 116
Finn, Robert, 189
Fire protection engineering, 48, *See also* Science and engineering
First Things First, 253, 256–258
Fitzpatrick, Susan M., 293
Fleet, James C., 86
Florida, 4
Florida State University, 157
Ford, Martin, 224–232
Ford Motor Company, 64
Four-year institutions 4, 9, 47, *See also* Baccalaureate colleges I&II
Fraknoi, Andrew, 154
Franklin and Marshall College, 342
Fraud, 18, *See also* Ethics; Professional responsibility in publication and research, 304, 310–311, 321
Freedom, *See also* Tenure
  relation to
    ethics, 320
    tenure, 357–358
  in university employment, 3, 37
Freedom of Information Act, 324
Freeman, Jo Anne, 266, 272–275
Fruchter, Renate, 118
Funding, *See also* Government funding; Grant; Proposal
  competition for, 38, 292–293, 374
  as conditon of employment, 4, 186–187, 200, 210, 288, 299–300, 353
  ethical considerations for, 310, 321–322
  pre-employment investigation of, 377
  reductions in
    affecting compensation, 55
    affecting scholarship, 67–68, 72
  sources for, 290–291
    Illinois Research Information System (IRIS), 292
  for university-based research centers, 51

G

Garment industry, 274
Gender issues, 171, 208, 213, 216–217, *See also* Women

Index 401

General Motors Corp., 273–274
Genetic Engineering News, 188
Genetics, 66
Geological Society of America, URL address, 371
Geology, 39–40, 47, 100
Geophysics, 47
George Mason University, 224
Georgia Institute of Technology, 67, 110, 274
Getting Tenure, 327, 345
Gibbons, John H., 72, 74
GI Bill, 5
Gift, 290–291, 317, See also Funding
Gittler, Mark, 154
Glasser, Lance, 332, 333, 340–342
Gmelch, Walter, 253, 254–255
Goodenough, Ursula, 75
Goodman, Joseph, 125
Goodstein, David, 85, 92
Governance and decision making, 13–18, 37, 320, 373
  affected by discipline, 40
  organizational structure, 13–15, 14, 53–54
  tenure. See Tenure
Government employment, See also, Academic employment; Industry employment
  compared to academic employment, 3, 45, 88–90, 100, 214–217, 235–236, 246
  as corollary to academic/industry employment, 98, 100, 316, 330–331
Government funding, 51–52, 200, See also Funding; Grant; Proposal
  Carnegie classification for, 7
  "cooperative agreements", 291
  decrease in, 16–18, 67–68, 72–74
  for direct/indirect costs, 17
  post-WWII, 5–6
  for research and development, 45–46
Graduate programs
  guides to, 174
  pre-employment investigation of, 376–377
Graduate Research: A Guide for Students in the Sciences , 113, 290–291
Graduate school, 108–111, See also Academic institutions; Colleges and universities
Graduate Student Packet , 174
Graduate students, See also Postdocs; Research
  assistance to, 351–355, 364

  by citizenship, **41**, 93
  costs of maintaining, 68–69
  funding for, 291–292
  presentations by, 133
  relations with
    advisor, 94
    faculty, 321, 322–323, 334, 338
    request for by new professor, 228, 229
  school choice for, 108–111
  teaching as component of, 144–158
  transition of to professor, 86–90, 117, 187, 248, 250
  types of, 40–43
Grant, 290–291, 334, 338, See also Funding; Proposal
  research grant
    as conditon of employment, 4, 186–187, 200, 210, 288, 299–300, 353
    role in public/private institutions, 12, 17
  "sugar grants", 75
Green, Timothy J., 210
Guthrie, Denise H., 214–215

H

Haddara, Mahmoud, 24
Hahnemann University, 4
Hall, Russ, 175
Hamann, Eloise, 87, 330
Hate crimes, 18
Hawaii, 4
Health, 113, 274
Heiberger, Mary M., 170, 209
Hemmenway, Cynthia, 237
Henning, Albert, 230
High-energy physics, 45, 69, 113, 187, 234–235, See also Physics
Higher education. See Academic employment
Hitachi Central Research Laboratories, 340–341
Holden, Constance, 214–215
Homework assignments, 368
  preparation of, 151, 154, 156, 334
  use of previously-prepared, 227
Honors awards, inclusion in teaching portfolio, 365
Hopcroft, John, 60

Hopkins, Mark, 25, 173, 294, 296–298, 329
Hopkins, Rick, 115
Horowitz, Mark, 122
Human Genome project, 65
Humanities, 37–39, *See also* Social sciences
Humke, Paul, 246, 250–251, 329
Hurd, Paul DeHart, 66
Hutchings, Patricia, 158, 275, 283
Huyser, Karen, 122

I

IBM Corp., 88, 341
Illinois, 5
Illinois Research Information System (IRIS), *See also* Funding
 URL address, 292
Independence, 3, 37, 176, *See also* Tenure
Industrial engineering, 40, 49–51, **50**, 64–65, 110, 148, 248–250, 266, 271–275, *See also* Science and engineering
Industrial technology, 113
Industry
 affect on university research, 18–19, 52–54, 65, 69–71, 72–74, 315–318, 323–324
 interdisciplinary applications in, 49–52, 66, 69, 171, 215
 preferences of for research topic, 69–71, 100–102, 111, 112–117, 125–126, 129
 private teaching companies, 59
 relations with colleges and universities, 25, 111, 130, 171–172, 177–180, 272–275, 377
 postdoc internships, 97
Industry employment, *See also* Academic Employment; Government employment
 compared to academic employment, 3–4, 45, 83, 86–90, 93, 98–103, 144, 214–217, 235–236, 246, 356–357
 co-op programs, 171, 297
 as corollary to academic emplyment, 98, 100, 111, 172, 177–180, 246–248, 266–268, 296–298, 330–331, 337–340
 internships for, 97, 101
Information, *See also* Knowledge; Learning; Publication

freedom of, 324–325
 perception of by student, 262–266, **263**
Information technology, 112–113, *See also* Communications tools; Computational tools; Distance learning
Institute of Electrical and Electronics Engineers (IEEE), URL address, 371
Institute for the Study of Advanced Development, 262
Institute of Transportation Engineers, URL address, 371
Institution. *See* Academic institutions; Colleges and universities
Instructional television. *See* Communications tools; Distance learning
Integration
 interdisciplinary, 67
 in scholarship model, 20–22, 27–29, 53
Intel Corp., 65, 88, 102–103
Interdisciplinary cooperation, *See also* Competition
 among students, 265, 268–270, 273–274, 368
 departmental, 38, 271–272, 294, 298–300
 departmental/institutional, 294–296, 300–302, 314, 364
 for funding, 293, 317
 industrial, 66, 69
 industrial/academic, 49–52, 171, 215
 for new professor, 243, 258–259, 296, 299–300
 in publications, 45, 304–305
 in research, 51–52, 62, 65–67
 in research topic, 97, 110–111, 113–115, 117–118, 122
 scholarship in, 20–22, 53–56, 62
 in team teaching, 148, 151–152, 161, 227, 249–250, 268, 271–272, 365
 in tenure and promotion, 48, 67, 73, 131
Internet, 61–63, *See also* Communications tools
 as research tool, 76–78, 295, 324–325, 375
 use in funding, 292
 use in job search, 48, 49, 112, 174, 187, 189, 191, 206, 208–209
Internship, *See also* Industry employment; Interdisciplinary cooperation
 academic, 119
 documentation of in teaching portfolio, 364

Index 403

with industry, 97, 101
Interview, *See also* Preparation strategy
   at conferences, 129, 206
   for course evaluation, 365, 366
   for employment, 168, 191, 207–208, 210, 213–215, 334
Iowa State University, 266, 271–272

J

Jaco, William, 100
Japan, 340–341
Jenison, Rollie, 266, 271–272, 329
Job announcement, sample, **194**
Job offer, *See also* Academic employment; Preparation strategy
   acceptance of, checklist of questions for, 373–379
   advertised positions for, 188–189
   dual-career couples consideration, 232–233, 237–239
   rejection of, 233–237
      decision to try again, 234–235
      examination of options, 235–236
   responding to, 224–232
      communication of needs and wants, 226–227
      confidentiality concerns, 226–227
      determining certainty of job offer, 224–225
      format for request presentation, 228–229
      "in bounds"/"out of bounds" negotiation, 229
      institutional prestige considerations, 230
      logic and emotion in decision making, 231–232
      maintaining professional options, 231
      personal and professional needs consideration, 225
      productivity/quality rationales in negotiation, 227–228
      salary negotiation, 230
      tenure considerations, 229–230
   search committee considerations, 186–187, 222–223
Job talk. *See* Academic job talk
Johansen-Trottier, Kelly, 190, 193, 196, 208
Johns Hopkins University, 5
Jones, Marilyn S., 210
Journals. *See* Publication; Writing skills

K

Kazmer, David, 228
Keith-Spiegel, Patricia, 322
Kelly, David, 22
Kelly, Nancy, 153
Kennedy, Donald, 313, 320–322
Kerr, Clark, 3, 37
Kibler, Ruthann, 333, 337–340
Kitt Peak National Observatory, 73, 150
KLA Instruments, 340–342
Knowledge, *See also* Information; Scholarship
   costs of creating, 69
   developing of, 39–40, 96
Komives, Elizabeth, 135
Koseff, Jeff, 232
Kraut, Alan G., 67
Kraut, Joe, 124
Krishna, Nety, 136
Kronenfeld, Jennie, 327, 345

L

Land grant colleges, 5, 21, 25
Landis, Ray, 99–100
Laudan, Lawrence, 39
Leadership potential, 186–187
Learning, *See also* Teaching
   characteristics of successful teachers, 275–276
   course planning for, 276–279
   developing engaged and responsive learners, 268–270
   elements of effective teaching, 266–268
   interdisciplinary team teaching, 268, 271–272, 365
   introduction, 261–266
   professional responsibilities in, 318–320, 367–368
   relation to teaching portfolio, 283–285
   teaching styles and learning styles, 262–266, **263**

Learning (*Continued*)
  with technology, 279–283
  "upside-down curriculum", 272–275
Leave of absence, 331, *See also* Academic employment
Lectures, 62, 132, 368, *See also* Course; Curriculum; Teaching
  guest lecturing, 148, 150, 155
  preparation considerations for, 156, 258, 274, 276, 334
  use of previously-prepared, 227, 247–248
Lederman, Leon M., 69, 72, 321
Lee, Hau, 312, 315–318
Lefton, Lew, 232, 293, 294–296, 329
Lennon, Tava, 152
Letters of recommendation, 190–191, 204–206, 334, 345, *See also* Academic employment; Preparation strategy
Leveraging
  in consulting arangements, 315–318
  in industry-university collaboration, 296–298
  in job search, 175
  in teaching assistants, 149
  for tenure, 332, 333–335, 346, 349
Library 375, *See also* Internet
  digital, 61, 63–64, 295
  as source of employment information, 173–174
Life sciences, 51, 66
Lifestyle. *See* Personal/professional considerations
Ligare, Martin, 117
Lightner, Michael, 261, 361
Linde, Andrei, 129
Lindquester, Terri, 245
Lodahl, T., 40
Lord, Susan, 192
Losleben, Paul, 65, 76, 331
Louisiana Tech University, 87, 183
Love, Brian and Nancy, 115, 233, 237–239, 329
Loyola Marymont University, 177–180, 269

M

McDonnell, James S. Foundation, 293
McGill University, 324
McGinn, Robert, 54, 310

MacLachlan, Ann, 206–207
Magnan, Robert, 346
Magnanti, Thomas, 61
Management, *See also* Time management
  in research, 4, 111, 135–137
Management engineering, 40
Marincovich, Michele, 213
Martin, Joanne, 120
Masnari, Nino, 22, 294, 300–302
Massachusetts Institute of Technology, 5, 61, 63, 67, 153, 330, 332, 333, 340–342
Master's degree, *See also* Doctoral degree
  employment opportunities with, 86, 111
  relation of to doctoral degree, 89
Master's universities and colleges I&II (MA I&II),
  defintion and statistics, 6–9, **7–8**
  faculty statistics, 9
  organizational structure, 14
  teaching experience gained at, 153
  teaching and research emphasis, **11**–12, 22
Materials sciences and engineering, 49–**50**, 271
Mathematical Association of America, URL address, 371
Mathematics, 174, 250–251, 282, 319
  graduate student statistics, 41, 43, 93
  non-academic employment in, 100
  research time-management in, 294–296
  undergraduate requirements, **47**–48
Mechanical engineering, 22, 40, 48–51, **50**, 115, 118, 129, 150, 265, 266–268, 271–272, 318, *See also* Science and engineering
Media, 130
Medical technology, 113
Memorial University of Newfoundland, 160–162
  Ocean Engineering Research Centre (OERC), 24
  URL address, 174
Mentor, *See also* Advisor; Supervision
  interdisciplinary, for faculty, 38, 272, 284–285
  letters of recommendation by, 204–206
  for minorities/women, 42–43, 102–103, 344
  role and responsibilities of, 146–147, 152, 236, 252, 276, 349
  stress and efficiency issues for, 256

Mentor in a Manual, 346
Mercury, The Journal of the Astronomical Society of the Pacific, 84, 216
Merola, Joseph S., 94
Merrill, Rebecca, 253, 256–258
Merrill, Roger, 253, 256–258
Meteorology, 47, 49, 110, 115, 150, 170, 246–248
Metropolitan university, *See also* San Jose State University
 definition and statistics, 6–9, **7–8**
 professional couple employment at, 233
 science curriculum at, 54–56
 teaching and research opportunities, 22, 111, 252–253
Michigan Technological University, 173, 186, 232
Microbiology, 303
Microelectronics Center of North Carolina, 301
Minorities
 doctoral degrees received by, **42–43**
 employment considerations for, 171, 216–217
 in graduate programs, 302
 in undergraduate admission, 18
Minsky, Marvin, 63
Misconduct. *See* Fraud; Professional responsibility
Moffat, Anne Simon, 124
Montgomery, Susan, 22, 221
Morrill Act, 5
Multiculturalism, 18
Multidisciplinary issues. *See* Interdisciplinary cooperation
Multiuniversity, 15, 300–302, *See also* Interdisciplinary cooperation

N

Narayanamurti, Venkatesh, 66
National Aeronautics and Space Administration (NASA), 247, 292
National Association of Environmental Professionals, URL address, 371
National Coalition for Manufacturing Leadership, 67
National Institutes of Health (NIH), 70, 75, 292
National Overseas Schools Advisory Council, 28
National Research Council, Committee on Science, Engineering, and Public Policy (COSEPUP), 68, 93–94, 100
National Science Foundation (NSF), 27, 70, 75, 301
 grant funding process, 290–291, 292
 Synthesis Coalition, 62, 67
National Technical Institute for the Deaf, 25
National Technological University (NTU), 27, 281
Natural sciences, *See also* Science; Social science
 doctoral degrees awarded in, **41–42**
 faculty statistics, 43–44
 interdisciplinary opportunities in, 54
 publication opportunities in, 45
 R&D expenditures for, **46**
 science department requirements in, **47–48**
 tenure-track opportunities in, 43
Needy, Kim, 246, 248–250, 256, 329
Negotiation. *See* Academic employment
Networking
 at conferences, 129–131, 295–296
 during graduate and postdoctoral research, 137
 for funding, 293
 for job search, 174–175, 177, 188, 190–191, 207, 341–342
 by new faculty, 258–259, 314, 334–336
 for teaching opportunities, 155
Neural systems engineering, 48, *See also* Science and engineering
The New Faculty Member, 253, 258–259, 275
The New Professor's Handbook, 275, 276–279, 294, 302–303
New York Times, 68
North Carolina Agricultural and Technical State University, 65, 301
North Carolina State University, 22, 237, 262
 Advanced Electronic Materials Processing (AEMP), 65, 294, 300–302
 "Preparing the Professoriate" program, 146–147
North Dakota University, 360
Northwestern University, 67
Nuclear engineering, 25, 49–**50**, *See also* Science and engineering

O

Office of Science and Technology Policy, 72
Ohio State University, 157
Operations research, 40
Oreovicz, Frank S., 253, 255–256, 275, 280–281
Organizational skills, 144, 187, 215
Organizational structure, of colleges and universities, 13–15, **14**, 53–54
Outsourcing, 18
Overheads, 212, 279, 282
Oxford University, 4

P

Papers. *See* Publication
Part-time employment, 6, 98, 171–172, 178, 329–330
Peer review, *See also* Feedback; Student evaluation
  affect on tenure, 53, 54, 272, 332, 339–340, 345, 378
  ethical considerations for, 310–311, 321–322
  inclusion of in teaching portfolio, 159, 365
  practice in as preparation strategy, 97
  of research paper, 303, 315, 321–322, 334–335
Pennsylvania State University, 67, 290
Perkins, David V., 322
Personal/professional considerations, *See also* Couples; Women
  efficiency considerations, 255–256
  in job search, 134, 168, 170–173, 208, 210, 221–240
  learning to say "no", 296, 313–315, 332, 335, 336–337, 344
  stress considerations, 254–259, 349
  in teaching career, 246, 251, 332, 378
Peterson's graduate program guides, 174
Petroleum engineering, **50**–51
Petsko, Greg, 53, 73, 294, 298–300, 301, 330
Pew Charitable Trusts, 147, 266
Pfeffer, Jeffrey, 17, 40
A Ph.D. is Not Enough, 86, 116, 294, 302
Ph.D. *See* Doctoral degree
Physical sciences, 46, 51, 53, 174
Physics, 49, 282, *See also* High-energy physics
  discipline development in, 39–40
  publication opportunities in, 45
  undergraduate requirements, **47**–48
Physics Letters, 45
Plagiarism, 310, 322–323, *See also* Professional responsibility
Plummer, James, 112
Political science, 39
Politics, 40, 67
  of academic institutions, 13–15
  affect on
    foreign students, 93
    interdisciplinary issues, 67
  departmental
    determination of in on-campus visit, 176
    relation to discipline development, 40
  "politically neutral" disciplines, 41
  of tenure, 341, 345, 346–347
Pomona College, 152, 269
"Postdecisional regret", 231–232
Postdocs, *See also* Doctoral degree; Graduate students; Research, graduate
  assistance for, 351–355
  costs of maintaining, 68–69
  postdoc appointments and doctoral degrees, **43**, 85, 89–90
  school choice for, 108–111
  teaching as, 144–158, 160–162
  transition of to professor, 87, 117, 187
Prentice-Hall, 175
Preparation strategy, *See also* Academic employment; Teaching
  decision for academic career, 84, 85–90
  choices for and against, 86–90
  identification of possibilities, 167–180
  exploration and search procedures, 168–169
  job search preparation, 176–180
  needs and wants determination, 170–173
  researching the job market, 173–176
  Ph.D. supply and demand, 84, 90–95, 352
  3–prong strategy, 84, 95–102, **109**, 126–127, 135–140, 143, 190
    Breadth-on-Top-of-Depth approach, 96, 100, 107, 112, 116, 119, 124, 353
    Multiple-Option approach, 111, 116, 119,

123, 153, 177, 214, 235–236, 352
Next-Stage approach, 96–98, 111, 119,
    123, 127, 152, 352
Preparing Future Faculty program, 147
'Preparing the Professoriate" program,
    146–147
Presentations. *See* Conferences and presentations
Prestige, 37, 230, 311
    departmental, compared to institutional, 108–110
Princeton University, 153
Prinz, Fritz, 115
Private teaching companies, 59
Productivity, *See also* Accountability;
    Research; Teaching
as basis of compensation negotiation, 224,
    227–228
communications tools affecting, 64, 77
computational tools affecting, 63
increase in requirements for in university,
    4, 19, 56
industrial emphasis on, 70
relation to
    faculty retirement/new hire, 5–6, 93
    interdisciplinary programs, 65–67
Professional development. *See* Personal/professional considerations
Professional responsibility, 309–325, *See also*
    Ethics
and academic duty, 320–322
appropriation of ideas, 322–323
conflict of interest, 309, 320–324, 357
consulting and industry relationships,
    315–318
departmental service, 313–315
freedom of information, 324–325
introduction, 309–313
teaching and learning standards, 318–320
Professional schools (Prof.)
    definition and statistics, 6–9, **7–8**
    employment possibilities at, 86, 89
Professional societies
    in engineering, 371–372
    in science, 369–371
    service in, 364
        compared to departmental service, 313–315
    as source of employment information, 174, 188
Professoriate challenges. *See* Academic
    employment

Promotion. *See* Tenure
Proposal, *See also* Funding; Grant; Research
    topic
interdisciplinary preparation of, 258
preparation of, 72, 310–311, 333–334
research proposal preparation, 123–125,
    138–140, 236, 276, 290, 353
resubmission of, 290
shortcomings of, 391–392
successful proposal elements, 387–389
writing of as preparation strategy, 97, 111, 119
Publication, *See also* Conferences
abstracts, 304–305
competition and cooperation in, 72–73
departmental expectations for, 26
discussion of in lectures, 268–270
electronic, 77–78
ethical considerations for, 310–311
by foreign Ph.D.s, 93
inclusion of in teaching portfolio, 363–365
interdisciplinary, 45, 72, 127, 317, 320
in natural sciences and social sciences, compared, 45
in preparation strategy, 97, 111, 112,
    126–128, 193, 197, 236–237
    "Academic Author's Checklist", 127–128
required for tenure, 19, 21–22, 26, 121, 171,
    297, 334–335, 345, 378
research papers, 302–305, 315
Public school system, 357
Pulley, Shon, 87, 108, 137–140, 223, 329
Purdue University, 67, 111

Q

"Quick starters", 258–259, *See also* Teaching;
    Time management
    characteristics of, 275–276
Quinlan, Kathleen, 158, 275, 283

R

R&D. *See* Research and development
Race, *See also* Minorities
    effect on undergraduate admission, 18

Radar, 5
Raji, Prasad, 116
Ramirez, Martin, 25, 143, 266, 268–270, 277, 330
Randall, Doug, 66
Razavi, Bezhad, 172
Reed, Michael, 123
Reichenberger, Joseph, 177–180, 330
Reitt, B.B., 127
Rejection. *See* Job offer
Religion, 171, 208
Reputation. *See* Prestige
Research, *See also* Productivity; Teaching
  basic vs. applied, 74, 99–100, 113
  commercialization of, 19
  describing in employment interview, 210, 215
  ethical considerations for, 310–311, *See also* Ethics
  forces for change in
    communications tools use, 61–63
    computational tools use, 63–65
    government funding decrease, 67–69, 72
    industry role in research 69–71, *See also* Industry
    research costs increase, 68–69, 72–73, 77, 117, 251
  graduate and postdoctoral
    choosing an advisor, 118–123
    choosing research topic, 97, 110, 111–118, *See also* Research topic
    choosing a school, 108–111
    conferences and presentations, 129–133, *See also* Conferences
    networking, 137, *See also* Networking
    publishing, 126–128, *See also* Publication
    research continuum, 137–140
    research procedures example, 125–126
    research project management, 111, 135–137
    research proposal preparation, 119, 123–125, 131
    supervising other researchers, 111, 119, 133–135, 138
  high-risk vs. low-risk, 74–76
  increase in emphasis on, 9–13, **11**, 19
  industry support of. *See* Industry
  in-house research offices, 293
  interdisciplinary. *See* Interdisciplinary cooperation
  Internet use for, 76–78, 295, 324–325
  pre-employment investigation of, 377
  relation to teaching, 9, 19–22, 53, 55, 146, 256, 289–306, 375
  relation to tenure, 19, 21–22, 121, 171, 297, 334
  "teleresearch", 62–63
Research and development (R&D), *See also* Government funding; Industry
  funding expenditures for, 51–52
  relation to discipline, 45–47, **46**
Research-Doctorate Programs in the United States, 110
Research paper. *See* Publication
Research project management, 135–137., *See also* Time management
Research topic, *See also* Dissertation; Proposal
  choosing, 97, 110, 111–118, 138–140, 236, 290, 368
  choosing advisor for, 119–123
  industry preferences for, 69–71, 100–102, 111, 112–117, 125–126, 129
Research Triangle Institute, 301
Research universities I&II (Res. I&II), *See also* Colleges and universities
  defintion and statistics, 6–9, **7–8**
  faculty statistics, 9
  organizational structure, 13
  teaching and research emphasis, **11**–12, 21–22
Reshaping the Graduate Education of Scientists and Engineers, 70
Resources. *See* Funding
Responsibility. *See* Professional responsibility
Resume, *See also* Curriculum vitae
  compared to curriculm vitae, 196
Retention and promotion. *See* Tenure
Retirement
  California early retirement, 253
  relation to new hires, 93, 147, 185, 379
Rhodes College, 245
Riskin, Eve, 332, 333–335, 336–337
Robotics, 114
Rochester Institute of Technology, 25, 173, 294, 296–298
URL address, 174
Rogers, Mark C., 51

Rooren, Kenneth J., 12
Roos, Leonard L., 4
Rubbia, Carlo, 45
Rutgers University, 157

S

St. Olaf College, 246, 250–251
Salancik, Gerald, 16
Salary. *See* Compensation
Salzner, Ulrike, 160–162, 330
Sandage, Allan, 129
Sandia National Laboratories, 126
San Francisco State University, 154
San Jose State University, 83, 87, 170, 183, 232, 333, 358
  certificate programs, 25–26
  faculty statistics, 9
  meteorology department, 246–248
  new faculty mentoring, 252–253
  research and tenure procedures, 337–340
  science department, 54–56
  URL address, 174
Savory, Paul, 212
Schoenfeld, A. Clay, 346
Scholarship, *See also* Knowledge; Research; Teaching
  compared to service, 28
  of discovery, integration, application, teaching, 20–22, **21**, 27–29, 61, 76, 229–230, 243, 353–354
  excellence and breadth concerns, 12–13, 359–360
  faculty concerns for, 71–78, 335–336
    basic/applied research balance, 74
    cooperation/competition balance, 72–74
    high-risk/low-risk behaviors balance, 74–76
    multiuniversity collaboration program, 76–78
  inter-/intra-disciplinary, 38, 53–54, 63
  relation of to publication, 127
Scholarship Reconsidered: Priorities of the Professoriate, 20
Scholarships, 291–292, *See also* Funding
Schultz, Noel and Kirk, 173, 186, 232, 233
Science: The End of the Frontier?, 75

Science, 188, 214, 324
  URL address, 325
Science, Technology, and Society, 310
Science and engineering in higher education, 6, 37–57, 54, *See also* Natural science; Social science
  cross-institutional comparisons, 38–47
  degree of discipline development, 39–40
  ease of publication, 45
  financial compensation, **44**–45
  number of faculty, 43–44
  number of postdocs, **43**
  politics and influence, 40
  R&D expenditures, 45–47, **46**
  types of graduate students, 40–43, **41**–**42**
  engineering department, 48–51, **50**, 174
  interdisciplinary collaboration. *See* interdisciplinary cooperation
  science department, **47**–48, 54–56, 174
The Scientist, 53, 188
Search committee procedures, 183, 192, 234–235, *See also* Academic employment; Job offer
  new-hire expectations of, 186–187
  relation to applicant, 222–223
Secondary school, teaching and employment at, 86, 89, 146
"Secrets from the Other Side", 214
Self-confidence, 123–124
Selter, Gerald, 55
Semiconductor processing, 113, 116
Semiconductor Research Corp. (SRC), 70
Seminars. *See* Conferences
Service
  compared to scholarship, 28
  departmental, 312, 313–315, 320, 349
  expectations for, 22, 197, 314, 349, 375, 378
  public service, 341–342
  time-management for, 249, 252–253, 276
Sharp, James, 68
Shavers, Cheryl, 88, 102–103
Shelton, Robert, 73
Sheppard, Sheri, 265, 266–268, 329
Shore, B.M., 158
Shultz, George P., 3, 28
Siemens Corp., 110
Silverman, Linda, 262, 267, 270
Simon Fraser University, 347–350

Smith, Robert, 113, 118, 120, 123, 290–291, 293
Smith, Roger V., 290–291
Smith-Baish, Sue, 333, 342–344
Social life. *See* Personal/professional considerations
Social sciences, 37–39, 45, 53
  doctoral degrees awarded in, **41–43**
  R&D expenditures for, **46**
Society, *See also* Citizenship; Service
  academic institution obligations to, 15, 367–368
  affect on of reduced research funding, 68
  professional responsibilities to, 309–310
  relation to industrial research, 70
Society of Industrial and Applied Mathematics (SIAM), URL address, 371
Society of Women Engineers, URL address, 371
Sociology
  discipline development in, 39–40
  "eight–meter rule", 63, 78
  publication in, 45
Southwestern Clinic and Research Institute, 338
Specialized institutions (Spec.), defintion and statistics, 6–9, **7–8**
Springer, George, 26, 111
Stacks, Pam, 83, 183
Stanford University, 5, 9, 17, 22, 26, 54, 99, 111, 151, 225, 232, 236, 274, 312, 313, 320–322, 331, 359
  Center for Integrated Systems (CIS), 125, 133
  Center for Teaching and Learning, 157, 213
  consultation and industry relationships, 315–318
  Future Professors of Manufacturing, 353
  Integrated Manufacturing Association (SIMA), 52, 118
  interdisciplinary programs, 65–66
  mechanical engineering department, 266–268
  undergraduate math and science requirements, **47**
  University Teacher Assessment Project, 284–285
Statement of Principles on Academic Freedom and Tenure, 15

Steinbart, Enid, 232
Stiger, Stephen, 62
Stress considerations, 254–259, 349, *See also* Time management
Strickland, Ruth, 327, 345
Strunk, William Jr., 305
Student evaluation, *See also* Feedback; Peer review
  inclusion in teaching portfolio, 365
  of new professor, 156–157, 268, 270–272
  for tenure, 22, 53, 336, 345, 349, 378
Students, 19, 38, 76, 373, *See also* Graduate students; Learning; Postdocs
  evaluation of in on–campus visit, 176, 208–209
  supervision of as preparation strategy, 97, 111, 119, 133–135
Success. *See* Personal/professional considerations
Summer school teaching, 152–153, 375
Supervision, *See also* Advisor; Mentor
  documentation of in teaching portfolio, 364
  of research, 289–290, 338
  supervisory practice, 111, 119, 133–140, 353
Supervisor. *See also* Administrator; Advisor; Mentor
Sweeney, James, 314
Syllabi. *See* Course syllabi
Syntex Corp., 154
Syracuse University, 284

T

Taylor, Susan, 348
Teaching, 4–6, 9–13, **11**, *See also* Learning
  case-study use in, 267–268
  "deep teaching", 122
  documentation of, 157–159, 191, 193, 210, 283–285
  effective teaching, 262, 266–268
  forces for change in, 60–70, **71**
    communications tools use, 61–63
    computational tools use, 63–65
    government funding decrease, 67–69
    industry role in research, 69–71
    interdisciplinary programs use, 65–67
    research costs increase, 68–69

and learning. *See* Learning
new teachers
　advice for, **263**–264, 275–276, 277
　assistance to, 252–253
　expectations for, 186–188, 212, 223
　five-point advice for, 267–268
　orientation for, 156–157, 243
　requirements for, 171, 186–188, 223, 227–228, 333–334
　part-time, 6, 98, 171–172, 178, 329–330
　practice teaching, 98, 111, 119
　by private companies, 59
　relation to research, 9, 19–22, 53, 146, 256
　scholarship model for, 20–22, 55
　styles and techniques, **263**–264, 277
　successful teachers characteristics, 275–276
　teaching assignment preparation, 156–159, 236
　teaching portfolio, 157–159, 191
　teaching practice, 98, 111, 119
　team teaching, 148, 151–152, 161, 227, 249–250, 268, 271–272, 365
　tenure-track vs. non-tenure, 171–172, 236–237
　types of, graduate student/postdoc, 147–148
　types of
　　cautionary advice, 155, 236
　　class segment/module, 151
　　extension course, 153–154
　　graduate student/postdoc teaching, 145–147
　　guest lecturing, 150
　　as postdoc, 160–162
　　regular course, 151–152
　　summer school, 152–153
　　teaching assistantships, 148–149
　　teaching at other institutions, 153
　　team teaching, 151–152, 161, 227, 249–250, 268, 271–272, 365
Teaching assistants (TAs), 144, 156, 228, *See also* Graduate students; Postdocs; Teaching
　experience provided by, 148–149
　labor strike affecting, 18
　orientation for, 156
Teaching Engineering, 253, 255–256, 275
Teaching philosophy, 206–207, 367–368
Teaching portfolio, 191, 193, 210

initial preparation of, 157–159
institutional use of, 283–285
items for inclusion in, 363–366
The Teaching Portfolio: Capturing the Scholarship of Teaching, 158, 275, 283
Technical review. *See* Peer review
Technology, *See also* Communications tools
　relation to teaching and learning, 279–283, 364
　computer-aided instruction (CAI), 281–282
　interactive laserdisk (ILV), 282–283
　video instruction, 281
Telecommunications, 116, *See also* Communications tools
Tenure
　assistance with, 347–350
　criteria for, 332
　issues affecting
　　academic appointment, 171–172, 185, 187, 224, 229–230, 236–237
　　academic governance, 15–16, 356–361
　　competition, 334
　　discipline, 43, 111
　　funding cuts, 55
　　interdisciplinary cooperation, 48, 67, 73, 131, 300, 301–302
　　peer review, 53, 54, 296
　　politics of, 341, 345, 346–347
　　research and publication, 4, 19, 21–22, 26, 121, 171, 297, 334–335, 345
　　scholarship, 21–22
　　service, 22, 197, 314, 349
　　stress, 254–255
　　student evaluation, 22, 53, 272
　　teaching portfolio, 157–159
　leveraging for, 332, 333–335, 346, 349
　pre-employment investigation of, 378
　priorities for, 335–337
　relation to, technology use, 279–280
　requirements for, 21–22
　10 commandments for, 345–346
　tenure evaluation, **359**
　tenure paths, 328–333
　　accelerated path, 329
　　delayed-entry path, 329–330
　　fail-other career path, 331
　　fail-try again path, 331, 337–340

tenure paths (*Continued*)
  from-one-school-to-another-school path, 330
  late career child-bearing path, 330
  late-practioner path, 330
  never-try-for-tenure path, 331
  traditional path, 329
  walk-away-from-tenure path, 331
Texas, agricultural and mining (A&M) colleges, 5
Texas Instruments, 125–126
Textbooks
  authoring, 303, 365, 366, 375, 378
  digital, 63–64, 282
  selection of, 268, 278
  use in multiple-section course, 152
Thelin, John, 5
Thermodynamics, **50**–51, 62
Time management, 135, 152, 155, 180, 245–259, 356
  achievement of balance, 258–259
  in consultation, 316
  departmental assistance with, 252–253
  with distance learning, 281
  efficiency considerations, 255–256
  introduction, 245–253
  off-campus time, 246–250, 356
  for research, 290, 293, 294–296
  stress considerations, 254–255, 258
  Time Management Matrix, 256–**257**
  urgency addiction, 256–258
"Town and Gown", 4
Travel, 255, 295, 377
  as component of compensation package, 223, 379
  in job search, 175–176, 211
Tschirgi, Robert, 52
Tufts University, 86
"Two-body problem". *See* Couples

U

Undergraduate education
  pre-employment investigation of, 376
  scholarship definition for, 27–29
Undergraduates. *See* Students
United States
  academic opportunities, 19–20
  colleges and universities
    historical perspective, 4–6
    public/private with accreditation, 6
  compared to Japan, 340–341
  contemporary education statistics, 6–9, **7–8**
  foreign graduate students, **41**
  R&D funding, 45–**46**
University of Alabama, Birmingham, 12
University of Arizona, 13–**14**, 333, 337–340
University of Auckland (N.Z.), 152
University of California
  Berkeley, 9, 65, 150, 187, 206, 262, 337
  compared to California State University, 25
  Davis, 69–70
  Irvine, 12
  Los Angeles, 172
  San Diego, 67, 135, 353
  Santa Cruz, 134, 152, 154, 269
University of Chicago, 5, 233
University of Colorado, 261, 361
University of Maryland, 27, 48
University of Massachusetts, 136, 228
University of Michigan, 22, 67, 152, 187, 221, 312, 313–315, 367–368
  Ann Arbor, 24–25, 88
  URL address, 174
University of Minnesota, 356
University of Missouri-Columbia, 87, 108, 137–140
University of Nebraska, 233, 238
University of New Mexico, 67
University of New Orleans, 86, 172–173, 232, 293, 294–296, 313, 318–320
  URL address, 174
University of North Carolina, 65
University of Pennsylvania, 342–343
University of Pittsburgh, 246, 248–250, 293
University of Santa Clara, 339, 353
University of Southern California, 274
University of Virginia, 59
University of Washington, 332, 333–335
Urgency addiction, 256–258, *See also* Time management
U.S. Department of Agriculture (USDA), grant funding process, 291
Usenet newsgroups, 189, *See also* Internet
Using the Internet in Your Research, 174

U.S. Navy, 273
U.S. News and World Report, 28

V

van de Meer, Simon, 45
Van Kuren, Nancy E., 210
Varahramyan, Kody, 87, 183
Veregee, Sally, 26
Verhelst, Roger, 88
Vick, Julia M., 170, 209
Video instruction, 61, 280–281, 282, 375, See also Communications tools; Distance learning
Videotape
  including in teaching portfolio, 158
  observation of for teaching experience, 156
  for practice employment interview, 191
Vinci, Rick and Michelle, 236
Virginia Polytechnic Institute and State University (VA Tech), 150, 237–239, 274
Voice Processing Corp., 153

W

Wankat, Philip C., 253, 255–256, 275, 280–281
Washington, A&M colleges, 5
Wayne State University, 64, 67
Whicker, Marcia, 327, 345
Whistle blowing, 311, See also Ethics; Professional responsibility
White, E.B., 305
Whitley, Bernard E. Jr., 322
Whitley, Norm, 26–27, 86, 172, 313, 318–320, 329
William and Flora Hewlett Foundation, 266
Witten, Edward, 129
Wittig, Arno F., 322
Wolf, Kenneth, 284
Women, See also Couples
  doctoral degrees received by, 41–43, **42**
  employment considerations for, 171, 216–217, 330, 344
  expectations for, 249
  as percentage of faculty, 43–44

Worcester Polytechnic Institute, 48
  URL address, 189
Work-study, 366, See also Government employment; Industry employment
World War II, university expansion following, 5–6
World Wide Web. See Internet
Writing skills, 178–179, 196, 269, See also Cover letter; Proposal
  for cover letters, 193–194
  for research proposals, 123–125
Wulf, William A., 59, 63–64, 75

X

Xerox Corp., 25, 65, 173, 296–298

Y

Yano, Candice, 187
Young Scientists Network, 94

Z

Zanna, Mark, 122, 209
Zolla-Pazner, Susan, 70, 74
Zoology, 39
Zusman, Ami, 15, 52

About the Author

Richard M. Reis has had a long-standing interest in helping individuals prepare for, find, and succeed at academic careers in science and engineering. He is currently the executive director of the Stanford Integrated Manufacturing Association and associate director of Global Learning Partnerships for the Stanford University Learning Laboratory. While much of his work involves helping graduate students prepare for professional careers outside academia, as Dr. Reis notes: "Industry understands the need to preserve its seed corn, and one way to do this is to help develop professors who are well prepared, highly motivated, and strongly supported."

Dr. Reis is also a consulting professor in both the Stanford University Electrical Engineering and Mechanical Engineering Departments. Among his many responsibilities is the teaching of a year-round seminar on preparing graduate students for academic careers in science, engineering, and business. The seminar is part of the Stanford University Future Professors of Manufacturing program. He is also a part-time professor of astronomy at the College of San Mateo, a community college in San Mateo, CA, and a curriculum consultant at Menlo College, a liberal arts institution in Atherton, CA.

From 1987 to 1989, Dr. Reis served part time as the associate dean for professional development in the Stanford School of Engineering, and full time from 1978 to 1982 as the director of science and engineering at the Stanford Career Planning and Placement Center.

Prior to coming to Stanford, he was the executive officer and editor of the astronomy magazine, *Mercury*, for the Astronomical Society of the Pacific in San Francisco, CA, a tenured professor of science education at Memorial University of Newfoundland in Newfoundland, Canada, an instructor in astronomy and physics at California State University at Los Angeles in Los Angeles, CA, and a high school physics teacher at University High School in Los Angeles, CA.

Dr. Reis holds bachelor's degrees in physical geography (1964) and physics (1965), both with honors, and a master's degree in science education (1968) from

California State University at Los Angeles in Los Angeles, CA. He also holds a master's degree in physical science (1969) and a Ph.D. in science education (1971) from Stanford University in Stanford, CA.